AF705404

Topological Phases of Matter and Quantum Computation

CONTEMPORARY MATHEMATICS

747

Topological Phases of Matter and Quantum Computation

AMS Special Session
Topological Phases of Matter and Quantum Computation
September 24–25, 2016
Bowdoin College, Brunswick, Maine

Paul Bruillard
Carlos Ortiz Marrero
Julia Plavnik
Editors

AMERICAN MATHEMATICAL SOCIETY

EDITORIAL COMMITTEE

Dennis DeTurck, Managing Editor

Michael Loss Kailash Misra Catherine Yan

2010 *Mathematics Subject Classification.* Primary 81R50, 16D90, 81T05, 20G42, 18D10, 19D23.

Library of Congress Cataloging-in-Publication Data

Names: AMS Special Session on Topological Phases of Matter and Quantum Computation (2016: Brunswick, ME), author. | Bruillard, Paul, 1984– editor. | Ortiz Marrero, Carlos, 1989– editor. | Plavnik, Julia, 1985– editor.

Title: Topological phases of matter and quantum computation: AMS Special Session on Topological Phases of Matter and Quantum Computation, September 24–25, 2016, Brunswick, Maine / Paul Bruillard, Carlos Ortiz Marrero, Julia Plavnik, editors.

Description: Providence, Rhode Island: American Mathematical Society, [2020] | Series: Contemporary mathematics, 0271-4132; volume 747 | Includes bibliographical references.

Identifiers: LCCN 2019040079 | ISBN 9781470440749 (paperback) | ISBN 9781470454579 (ebook)

Subjects: LCSH: Quantum computing–Congresses. | Topological groups–Congresses. | Quantum groups–Congresses. | Categories (Mathematics)–Congresses. | AMS: Quantum theory – Groups and algebras in quantum theory – Quantum groups and related algebraic methods. | Associative rings and algebras – Modules, bimodules and ideals – Module categories; module theory in a category-theoretic context; Morit | Quantum theory – Quantum field theory; related classical field theories – Axiomatic quantum field theory; operator algebras. | Group theory and generalizations – Linear algebraic groups and related topics | Category theory; homological algebra | K-theory – Higher algebraic K-theory – Symmetric monoidal categories.

Classification: LCC QA76.889 .A467 2020 | DDC 006.3/843–dc23

LC record available at https://lccn.loc.gov/2019040079

DOI: https://doi.org/10.1090/conm/747

Color graphic policy. Any graphics created in color will be rendered in grayscale for the printed version unless color printing is authorized by the Publisher. In general, color graphics will appear in color in the online version.

Copying and reprinting. Individual readers of this publication, and nonprofit libraries acting for them, are permitted to make fair use of the material, such as to copy select pages for use in teaching or research. Permission is granted to quote brief passages from this publication in reviews, provided the customary acknowledgment of the source is given.

Republication, systematic copying, or multiple reproduction of any material in this publication is permitted only under license from the American Mathematical Society. Requests for permission to reuse portions of AMS publication content are handled by the Copyright Clearance Center. For more information, please visit `www.ams.org/publications/pubpermissions`.

Send requests for translation rights and licensed reprints to `reprint-permission@ams.org`.

© 2020 by the American Mathematical Society. All rights reserved.
The American Mathematical Society retains all rights
except those granted to the United States Government.
Printed in the United States of America.

♾ The paper used in this book is acid-free and falls within the guidelines
established to ensure permanence and durability.
Visit the AMS home page at `https://www.ams.org/`

10 9 8 7 6 5 4 3 2 1 25 24 23 22 21 20

To our dear friend and colleague, John I. Haas

Contents

Preface

These days it is impossible to turn on the news, open a book, or browse the internet without seeing some mention of quantum mechanics and quantum computation (see Figure 1). Even when delving into the dusty recesses of math and physics academic journals, quantum computing is lurking. Aside from our cultural fascination with the strangeness of quantum mechanics, quantum computing seems to be so pervasive because of its futuristic claims and its near-term realization. Indeed, not only have researchers constructed small devices, but these devices will eventually enable breaking conventional encryption techniques through the rapid factorization of large integers and provide a means to combat global warming through the development of novel catalysts for fertilizer and carbon fixation [**1**]. Even in the academic realm it is no longer a niche subject, and in many ways quantum computing has become a crosscutting discipline. We would venture to guess that there is no other subject studied by physicists, mathematicians, computer scientists, chemists, and engineers.

Quantum computing comes in a variety of flavors, from the qubit model to adiabatic constructions to the topological paradigm. The qubit model is arguably the most well known and operates through a seemingly simple extension of our classical binary computer systems. This model was arguably the first model for quantum computing and has been widely studied. A host of algorithms exist for qubit based devices and small scale qubit systems have been constructed in the laboratory. However, for all of their promise, they have yet to live up to expectations. This is due to a phenomenon known as *decoherence* wherein a quantum system couples with its environment and information "leaks out."

In the qubit realm decoherence has been combated in two ways; the first is through engineering. By operating these devices in highly isolated cryogenic environments, researchers are able to stave off the effects of decoherence for a short time. This volume will have little to say about these engineering solutions. However, articles in this text will touch on a software based solution, *error correcting codes*. In an error correcting code, one uses a collection of physical qubits to simulate a logical qubit. By incorporating redundancy into the physical qubits, one can correct errors in the logical qubits. Intuitively, the simplest mechanism for accomplishing this would simply be to repeat each qubit several times, operate on each qubit in the collection in an identical manner, periodically vote on the correct value, and alter the values in the minority. Such a system can be easily implemented in a classical computer, but there are several obstructions in the quantum realm. For instance, the no-cloning theory forbids the cloning of a qubit while the measurement problem essentially prevents voting in the obvious way. Nonetheless, these

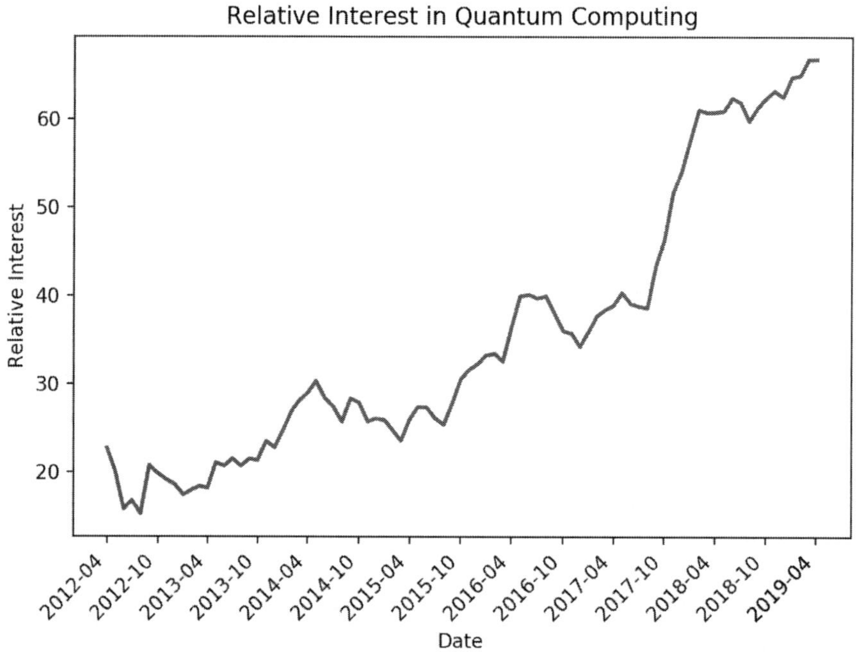

FIGURE 1. Interest in quantum computing as gauged by Google Trends for "What is quantum computing" [**7**]. A value of 100 indicates peak popularity while lower values indicate the popularity as a fraction of the peak popularity. For instance, a value of 50 indicates that the term is half as popular as it was at its peak. The data in this plot has been exponentially windowed ($\beta = 0.9$) to illustrate the trend.

ideas can be modified and refined to produce exotic new coding schemes unique to quantum computers [**2**].

An alternative to solving the decoherence problem in the qubit system is to reimagine quantum computing altogether. This was done in the early 2000s by Alexei Kitaev [**3**]. Rather than trying to devise a clever new coding scheme, Kitaev asked for a physical system that would resist environmental disruption at the hardware level and enable the simulation of a qubit system. This led to the birth of *topological quantum computing.* Under this paradigm one produces quasiparticles known as anyons in (2+1)-dimensions. As these anyons are moved around one another, their worldlines trace out a braid (see Figure 2). For simple topological reasons, it is possible to produce nontrivial braids and thus particle exchange in these systems is governed by the braid group rather than the symmetric group. The amazing fact is that if the anyons have the correct particle statistics, then a quantum computation can be encoded in such a braid and evaluated by fusing the particles [**4**]. The major advantage of producing a calculation in this manner is that it depends only on the topology of the braid and thus is resistant to thermal effects. Indeed, thermal fluctuations will tend to have one of two effects, the first being that the anyons will jitter, but this will not affect the topology of the braid if the jitter is small enough. Second, thermal effects may cause the pair production of anyons which will wander through the system. This is problematic only if the errant particles braid nontrivially with the computational anyons and then recombine; the likelihood of this happening can be made quite small with careful engineering. While the major hurdle to overcome in the qubit model is one of engineering, the obstruction to a topological quantum computer appears to be one of existence. It has long been known that anyons of the correct statistics exist in theory, but we have yet to find them in nature.

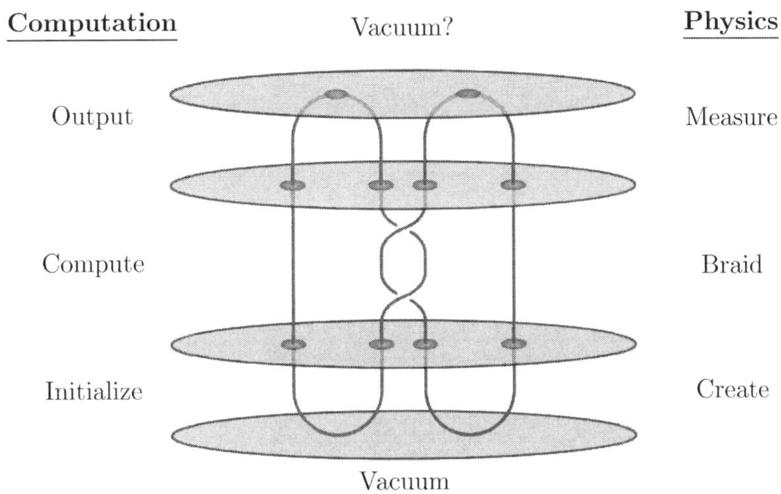

FIGURE 2. In topological quantum computing, particles are created in a (2+1)-dimensional space, exchanged so that their world lines form a braid, and then pairwise fused. Given the correct particle statistics, a computation can be embedded in the topology of these braids. In fact, not only can the qubit model be simulated, but such a process is inherently resistant to decoherence.

The study of topological quantum computation is related to two new branches of mathematics, *modular categories* and *subfactors*. These theories are really two sides of the same coin, and in fact one can often translate between them [**5**]. Their major distinction is that the study of modular categories has its roots in representation theory, algebra, and topology, while subfactors owe their lineage to analysis. A large portion of the articles in this text are devoted to these two subjects. While these mathematical theories are interesting in their own right, the main focus here will be on their use as mathematical models for topological quantum computing. These models allow researchers to understand properties of topological quantum computers, such as universality, as well as provide a taxonomy of possible systems [**6**]. Furthermore, these theories can lead to insights in modern mathematics and computer science, such as the computational complexity of computing link invariants and new representations of the braid group.

Quantum computing occupies the triple point of mathematics, physics, and computer science. We have made an effort to include articles coming from each of these disciplines. Furthermore, in the mathematics realm we have included articles on both modular categories and subfactors. As such we believe there is something that almost everyone will find familiar, in addition to new topics closely related to their discipline from which researchers can draw inspiration. We sincerely hope that you enjoy.

Bibliography

[1] Troyer, Matthias, "What Can We Do with a Quantum Computer?" Presentation available at https://www.microsoft.com/en-us/research/wp-content/uploads/2015/03/Matthias-Troyer_Quantum-applications.pdf (2017).

[2] Michael A. Nielsen and Isaac L. Chuang, *Quantum computation and quantum information*, Cambridge University Press, Cambridge, 2000. MR1796805

[3] A. Yu. Kitaev, *Fault-tolerant quantum computation by anyons*, Ann. Physics **303** (2003), no. 1, 2–30, DOI 10.1016/S0003-4916(02)00018-0. MR1951039

[4] Zhenghan Wang, *Topological quantum computation*, CBMS Regional Conference Series in Mathematics, vol. 112, Published for the Conference Board of the Mathematical Sciences, Washington, DC; by the American Mathematical Society, Providence, RI, 2010. MR2640343

[5] S. Popa, *Classification of subfactors: the reduction to commuting squares*, Invent. Math. **101** (1990), no. 1, 19–43, DOI 10.1007/BF01231494. MR1055708

[6] Paul Bruillard, Siu-Hung Ng, Eric C. Rowell, and Zhenghan Wang, *Rank-finiteness for modular categories*, J. Amer. Math. Soc. **29** (2016), no. 3, 857–881, DOI 10.1090/jams/842. MR3486174

[7] Google Trends, "What is quantum computing" Available at https://trends.google.com/trends/explore?date=2012-01-01%202019-04-14&geo=US&q=what%20is%20quantum%20computing (2012).

Contemporary Mathematics
Volume **747**, 2020
https://doi.org/10.1090/conm/747/15036

Lie theory for fusion categories: A research primer

Andrew Schopieray

A diverse collection of fusion categories, in the language of [**22**], may be realized by the representation theory of quantum groups. There is substantial literature where one will find detailed constructions of quantum groups, and proofs of the representation-theoretic properties these algebras possess. Here we will forego technical intricacy as a growing number of researchers study fusion categories disjoint from Lie theory, representation theory, and a laundry list of other obstacles to understanding the mostly combinatorial, geometric, and numerical descriptions of the examples of fusion categories arising from quantum groups. Our expository piece aims to create a self-contained guide for researchers to study from a computational standpoint with only the prerequisite knowledge of fusion categories. A multitude of figures and worked examples are included to elucidate the material, and additional references are abundant for those readers looking to delve deeper. Note that in general our chosen references are intended to provide useable resources for the reader and do not always indicate provenance. Lastly we have included several open and approachable questions of general interest throughout the final sections.

The organization of this paper is as follows: Sections 1 and 2 summarize the classical representation theory of semisimple Lie algebras in the spirit of [**37**] to introduce the chosen language and notation used extensively in what follows. Those unfamiliar with Lie algebras are encouraged to work through the provided examples themselves, while readers who possesses this prerequisite knowledge can safely begin reading in Section 3 referring back to earlier sections as needed. Terminology most relevant to future explanation is italicized for this purpose. Section 3 explains computationally relevant subtleties of the modern generalization of the representation theory of quantum groups including quantum dimensions and the affine Weyl group, followed by Section 4 which defines our primary objects of study: the fusion categories $\mathcal{C}(\mathfrak{g},\ell,q)$ where $\mathfrak{g}$ is a finite-dimensional simple complex Lie algebra and q is a root of unity such that q^2 has order $\ell\in\mathbb{Z}_{\geq 1}$. Section 5 discusses the fusion rules of $\mathcal{C}(\mathfrak{g},\ell,q)$, the classification of fusion subcategories, and simple factorizations. Modular data and the Galois symmetry thereof is covered in Section 6, while tensor autoequivalences, module categories, and commutative algebras are contained in Section 7.

2010 *Mathematics Subject Classification.* Primary 18-02; Secondary 18D10, 20G05.

©2020 American Mathematical Society

1. Lie algebras

1.1. Basics and examples. A *Lie algebra* (complex and finite-dimensional for our purposes) is a complex vector space $\mathfrak{g}$ equipped with an anti-symmetric bilinear map $[\cdot,\cdot] : \mathfrak{g} \times \mathfrak{g} \to \mathfrak{g}$ (bracket operation) which satisfies the Jacobi identity

$$[x,[y,z]] + [y,[z,x]] + [z,[x,y]] = 0.$$

Any vector space, henceforth assumed to be complex, can be given a trivial Lie algebra structure by declaring the bracket operation to be the zero function. The following example illustrates why such a Lie algebra is referred to as abelian.

EXAMPLE 1 ($\mathfrak{gl}_n$). For a vector space V, let $\mathfrak{gl}(V)$ be the Lie algebra of endomorphisms of V equipped with the commutator bracket $[f,g] = fg - gf$ for all $f, g \in \mathfrak{gl}(V)$. If $\dim(V) = n$, $\mathfrak{gl}(V)$ can be considered as the space of $n \times n$ complex matrices which we denote by $\mathfrak{gl}_n$. This example is universal in what follows as all finite-dimensional Lie algebras can be seen as subalgebras of $\mathfrak{gl}_n$ for some $n \in \mathbb{Z}_{\geq 1}$. Refer to [**36**] for a brief but general proof of this fact, originally due to I.D. Ado.

An ideal in a Lie algebra $\mathfrak{g}$ is a vector subspace $\mathfrak{h} \subset \mathfrak{g}$ such that $[x,y] \in \mathfrak{h}$ for all $x \in \mathfrak{g}$ and $y \in \mathfrak{h}$ while there is a weaker notion of a Lie subalgebra: a vector subspace closed under the bracket operation. Those nonabelian Lie algebras without proper nontrivial ideals are known as *simple*. Other standard constructions such as direct sums, homomorphisms, centers, etc. of Lie algebras can be formulated by the reader or referenced as needed [**37**, Section 2].

EXAMPLE 2 ($\mathfrak{sl}_n$). One decomposes $\mathfrak{gl}_n \cong \mathbb{C} \oplus \mathfrak{g}$ into a direct sum of ideals, where the trivial ideal consists of scalar multiples of the identity matrix I_n and $\mathfrak{g}$ is the simple (n^2-1)-dimensional Lie algebra $\mathfrak{sl}_n$ ($x \in \mathfrak{gl}_n$ with $\mathrm{Tr}(x) = 0$). Denote the $n \times n$ matrix units e_{ij} for $1 \leq i,j \leq n$. A distinguished basis for $\mathfrak{sl}_n$ [**37**, Section 25] consists of e_{ij} for $i < j$, $f_{ij} := e_{ij}$ for $i > j$, and $h_i := e_{ii} - e_{i+1,i+1}$ for $1 \leq i \leq n-1$.

1.2. Semisimple Lie algebras. Each Lie algebra $\mathfrak{g}$ naturally acts on itself yielding the *adjoint representation*, a Lie algebra homomorphism $\mathrm{ad}_{\mathfrak{g}} : \mathfrak{g} \to \mathfrak{gl}(\mathfrak{g})$ via $x \mapsto \{y \mapsto [x,y]\}$. This construction is paramount as we will study Lie algebras which act on themselves nondegenerately. To this end the Killing form of $\mathfrak{g}$ is the invariant complex symmetric bilinear form $\kappa(x,y) := \mathrm{Tr}(\mathrm{ad}_{\mathfrak{g}}(x)\mathrm{ad}_{\mathfrak{g}}(y))$ for all $x, y \in \mathfrak{g}$. A Lie algebra is *semisimple* if its Killing form is nondegenerate.

Stipulating that simple Lie algebras be nonabelian prevents, for instance, $\mathbb{C}$ from being simple while also not semisimple. Semisimple Lie algebras are a tractable class of Lie algebras as each decomposes in an essentially unique way into a direct sum of simple ideals upon which the Killing form is again nondegenerate [**37**, Theorem 5.2].

EXAMPLE 3. Recall the basis for $\mathfrak{gl}_2$ in Example 2 ordered I_2, e_{12}, h_1, f_{21}. The kernel of $\mathrm{ad}_{\mathfrak{gl}_2}$ is the span of I_2 and we compute the following Gram matrix for the Killing form of $\mathfrak{gl}_2$, partitioned to illustrate the simple summand $\mathfrak{sl}_2$. In general $\mathfrak{gl}_n$ is a (non-semisimple) central extension of (semisimple) $\mathfrak{sl}_n$.

$$\left[\begin{array}{c|ccc} 0 & 0 & 0 & 0 \\ \hline 0 & 0 & 0 & 4 \\ 0 & 0 & 8 & 0 \\ 0 & 4 & 0 & 0 \end{array}\right]$$

Despite the language, being a semisimple Lie algebra $\mathfrak{g}$ does not imply that all $x \in \mathfrak{g}$ are semisimple, which is to say $\mathrm{ad}_{\mathfrak{g}}(x)$ is a diagonalizable endomorphism of $\mathfrak{g}$. Any subalgebra of $\mathfrak{g} \subset \mathfrak{gl}_n$ generated by semisimple elements, known as a toral subalgebra, is necessarily abelian and moreover simultaneously diagonalizable.

1.3. Root space decomposition. The adjoint action of a maximal toral subalgebra $\mathfrak{t} \subset \mathfrak{g}$ provides a uniform way to decompose a semisimple Lie algebra (as a vector space) into generalized eigenspaces or *root spaces* [**37**, Section 8]:

$$\mathfrak{g} \cong \mathfrak{t} \oplus \bigoplus_{\alpha \in \mathfrak{t}^* \setminus \{0\}} \mathfrak{g}_\alpha \tag{1}$$

where for each $\alpha \in \mathfrak{t}^* = \mathrm{Hom}(\mathfrak{t}, \mathbb{C})$, $\mathfrak{g}_\alpha := \{x \in \mathfrak{g} : [t, x] = \alpha(t)x \text{ for all } t \in \mathfrak{t}\}$. It is nontrivial that $\mathfrak{t} = \mathfrak{g}_0$ in (1) and that the Killing form is nondegenerate when restricted to $\mathfrak{t}$ (note $\mathfrak{t}$ is not an ideal of $\mathfrak{g}$). We define Φ, the *root system* of $\mathfrak{g}$ to be the collection of all nonzero functionals $\alpha \in \mathfrak{t}^*$ such that $\mathfrak{g}_\alpha \neq 0$. Through the nondegenerate form on $\mathfrak{t}$, $\mathfrak{t}^*$ becomes a real Euclidean space, allowing geometric tools and reasoning to be applied to semisimple Lie algebras.

If $\mathfrak{g} \subset \mathfrak{gl}_n$ with $h_1, \ldots, h_r$ a basis for $\mathfrak{t}$, there is an obvious (thinking of elements of $\mathfrak{t}$ as diagonal matrices) spanning set for $\mathfrak{t}^*$: the functionals $\varepsilon_1, \ldots, \varepsilon_n$ where $\varepsilon_i(h_j)$ is the ith diagonal entry of h_j. But knowing $\mathfrak{g}$ is nonabelian, then $r < n$ and there are relations amongst the ε_i. The root system Φ is irreducible [**37**, Section 10.4] for all simple Lie algebras $\mathfrak{g}$, and there exists a basis of $\mathfrak{t}^*$ called *simple roots* such that every root is either a sum of simple roots with nonnegative coefficients (positive roots), or a sum of simple roots with nonpositive coefficients (negative roots). We refer to the set of simple roots as a base of Φ, denoted by Δ.

There is a partial *dominance ordering* on the *root lattice* $P := \mathbb{Z}\Phi$ for which $\alpha \prec \beta$ if and only if $\beta - \alpha$ is a sum of positive roots. For example positive roots are those $\alpha \in \Phi$ such that $\alpha \succ 0$ and negative roots are those $\alpha \in \Phi$ such that $\alpha \prec 0$.

EXAMPLE 4 ($\mathfrak{sl}_3$). The elements h_1, h_2 generate a maximal toral subalgebra $\mathfrak{t} \subset \mathfrak{sl}_3$. The functionals $\varepsilon_1, \varepsilon_2, \varepsilon_3$ span $\mathfrak{t}^*$ but the zero trace condition of $\mathfrak{g} := \mathfrak{sl}_3$ guarantees $\varepsilon_1 + \varepsilon_2 + \varepsilon_3 = 0$ (remember this is an equality of functionals on $\mathfrak{t}$). It is a straightforward computation to check that all non-empty root spaces are one-dimensional and that with simple roots $\alpha_1 := \varepsilon_1 - \varepsilon_2$ and $\alpha_2 := \varepsilon_2 - \varepsilon_3$,

$$\mathfrak{g} \cong \mathfrak{g}_{\alpha_1} \oplus \mathfrak{g}_{\alpha_2} \oplus \mathfrak{g}_{\alpha_1+\alpha_2} \oplus \mathfrak{t} \oplus \mathfrak{g}_{-\alpha_1-\alpha_2} \oplus \mathfrak{g}_{-\alpha_2} \oplus \mathfrak{g}_{-\alpha_1}. \tag{2}$$

For instance $\mathfrak{g}_{\alpha_1}$ is the span of e_{12}, $\mathfrak{g}_{-\alpha_2}$ is the span of f_{32}, etc. We compute $[e_{12}f_{21}] = h_1$ and $[e_{23}f_{32}] = h_2$ which implies $\kappa^*(\alpha_1, \alpha_2) = \mathrm{Tr}(\mathrm{ad}(h_1)\mathrm{ad}(h_2)) = -6$. But $\|\alpha_1\| = \sqrt{\mathrm{Tr}(\mathrm{ad}(h_1)\mathrm{ad}(h_1))} = 2\sqrt{3}$ and thus the angle formed between α_1 and α_2 is $\arccos(-1/2) = 120$ degrees. This produces the following geometric realization of (the irreducible root system corresponding to) $\mathfrak{sl}_3$.

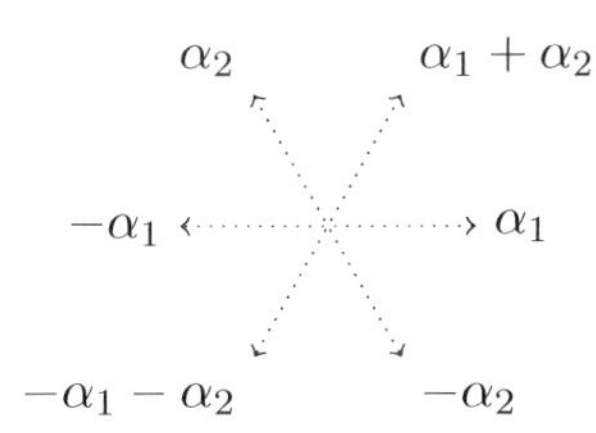

1.4. The classification theorem. Schur's Lemma implies if $\mathfrak{g}$ is a simple Lie algebra, then the (dual) Killing form is the unique nondegenerate symmetric bilinear form on $\mathfrak{t}^*$ up to a scalar. Once and for all we define $\langle \cdot, \cdot \rangle$ to be the form normalized so the shortest root has squared length 2. We can then define two combinatorial devices: a *Cartan matrix* and *Dynkin diagram* for each simple Lie algebra and corresponding irreducible root system. If $\Delta = \{\alpha_1, \ldots, \alpha_r\}$ is a base of irreducible Φ, the Cartan matrix has entries $c_{ij} := \langle \alpha_i, \alpha_j^\vee \rangle$ where $\alpha_j^\vee := 2\|\alpha_j\|^{-2}\alpha_j$ is the coroot of α_j. The corresponding Dynkin diagram is a connected graph with r vertices, $c_{ij}c_{ji}$ edges between vertex i and j, and an arrow placed on each multiple edge pointing toward the shorter of the two simple roots should they differ in length (refer to [**38**, Chapter 2] for an intrinsically motivated definition of these diagrams). In the case all roots are the same length the Dynkin diagram, irreducible root system, and corresponding simple Lie algebra are referred to as *simply-laced.*

There is a one-to-one correspondence between isomorphism classes of simple complex finite-dimensional Lie algebras and finite Dynkin diagrams.

The classification of (finite) Dynkin diagrams consists of four infinite *classical* families indexed by *rank* $r := |\Delta|$ and five *exceptional* cases. [**39**, Pg. 53]

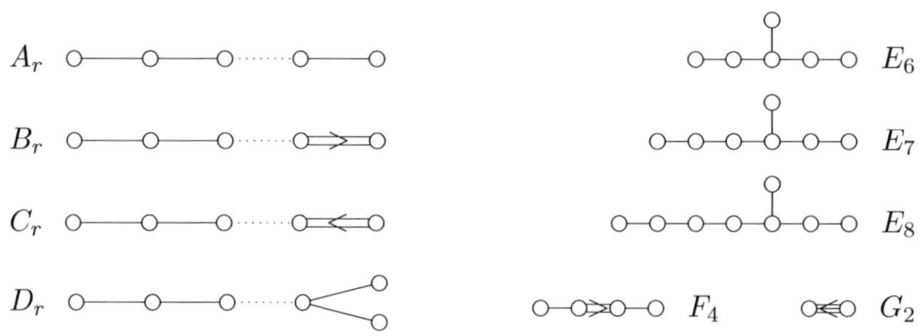

There is a cornucopia of ways authors refer to the simple Lie algebras and the fusion categories we will associate with them (Section 3), the most common being the Dynkin diagram type, the simple Lie algebra as a subalgebra of $\mathfrak{gl}_n$ [**37**, Section 1], or its associated simply-connected compact Lie group (Example 15).

2. Representation theory of semisimple Lie algebras

With most algebraic objects the study of representation theory is equivalent to the study of *modules* over a particular unital associative algebra. In the case of representations of a group G one speaks of $\mathbb{C}G$-modules and in the case of representations of a Lie algebra $\mathfrak{g}$ one speaks of $\mathcal{U}(\mathfrak{g})$-modules where $\mathcal{U}(\mathfrak{g})$ is the universal enveloping algebra of $\mathfrak{g}$. An explicit basis for $\mathcal{U}(\mathfrak{g})$ can be written in terms of an ordered basis for $\mathfrak{g}$ by the Poincaré-Birkhoff-Witt Theorem [**37**, Section 17.3]. For $x \in \mathcal{U}(\mathfrak{g})$ we denote the module action by $x.v$ for any v in a $\mathcal{U}(\mathfrak{g})$-module V.

2.1. Weight spaces. The maximal toral subalgebra $\mathfrak{t} \subset \mathfrak{g}$ (Section 1.2) acts on any $\mathcal{U}(\mathfrak{g})$-module V, decomposing (as a vector space) into *weight spaces*

$$V \cong \bigoplus_{\lambda \in \mathfrak{t}^*} V_\lambda \tag{3}$$

where for $\lambda \in \mathfrak{t}^*$ the weight space $V_\lambda := \{x \in V : h.x = \lambda(h)x \text{ for all } h \in \mathfrak{t}\}$. In this language the nonzero weights in the adjoint representation of $\mathfrak{g}$ are precisely the roots Φ. If $x \in \mathfrak{g}_\alpha$ for some root $\alpha \in \Phi$ and $v \in V_\lambda$, one should verify $x.v \in V_{\lambda+\alpha}$.

EXAMPLE 5. ($\mathcal{U}(\mathfrak{sl}_2)$-modules) The Lie algebra $\mathfrak{sl}_2$ has a maximal toral subalgebra spanned by $h := h_1$ (Example 2) and as convention we choose the root α corresponding to $e := e_{12}$ as a base, hence $f := f_{21}$ spans the $-\alpha$ root space. Let V be any n-dimensional irreducible $\mathcal{U}(\mathfrak{sl}_2)$-module. Weights of this module correspond to real numbers since $\mathfrak{t}^*$ is one-dimensional. Acting on elements of V by e increases their weight space by 2 (since $[he] = 2e$), so there exists a maximal nonzero $v \in V$ such that $e.v = 0$ because $\dim(V) < \infty$. The vectors $f^i.v$ for $0 \leq i \leq n-1$ then form a basis for V. If $v \in V_\lambda$, then one computes that up to a nonzero scalar, $0 = e.(f^n.v) = (\lambda - n + 1)(f^{n-1}.v)$ and thus $\lambda = n - 1 \in \mathbb{Z}_{\geq 0}$. Symmetric powers of the two-dimensional natural representation (with maximal weight 1) give a construction of a module of each dimension, and these constitute all finite-dimensional complex $\mathcal{U}(\mathfrak{sl}_2)$-modules up to isomorphism.

The representation theory for general $\mathfrak{g}$ follows from this example. To see this let $\eta^+ := \bigoplus_{\alpha \succ 0} \mathfrak{g}_\alpha$. If V is an irreducible finite-dimensional $\mathcal{U}(\mathfrak{g})$-module, there is a non-zero *highest-weight* vector $v \in V$ such that $\eta^+.v = 0$. The highest weights appearing in this way are very limited [**37**, Section 21.1]. In particular $\langle \lambda, \alpha^\vee \rangle \in \mathbb{Z}_{\geq 0}$ for all $\alpha \in \Delta$ inspiring the name *dominant integral weights*, the collection of which is denoted Λ^+. The collection of roots dual (via $\langle \cdot, \cdot \rangle$) to the coroots $\alpha_i^\vee$ for $\alpha_i \in \Delta$ generate all dominant integral weights via nonnegative integer linear combinations and thus are referred to as *fundamental weights* λ_i, while the $\mathbb{Z}$-linear span of the fundamental weights, the *weight lattice*, will be denoted Q.

EXAMPLE 6 ($\mathfrak{sl}_n$, continued from Example 2). The elements $h_1, \ldots, h_{n-1}$ generate a maximal toral subalgebra of $\mathfrak{sl}_n$ while the roots $\alpha_k := \varepsilon_k - \varepsilon_{k+1}$ for $1 \leq k \leq n-1$ are a base for Φ. We then compute the fundamental weights

$$\lambda_k = \sum_{j=1}^{k} \varepsilon_j - \frac{k}{n} \sum_{j=1}^{n} \varepsilon_j .$$

2.2. The classification theorem. Sufficient notation has been established to succinctly describe the representation theory of semisimple Lie algebras.

There is a one-to-one correspondence between dominant integral weights $\lambda \in \Lambda^+$ and isomorphism classes of finite-dimensional irreducible $\mathcal{U}(\mathfrak{g})$-modules $V(\lambda)$.

NOTE. A rigorous and generalizable approach to this classification is through the study of *Verma modules* [**35**, Section 9.2].

Lastly we compute exactly which weight spaces appear in $V(\lambda)$, the finite-dimensional $\mathcal{U}(\mathfrak{g})$-module of highest weight $\lambda \in \Lambda^+$, and the dimensions, or *multiplicities* $m_\lambda(\mu)$ of its weight spaces for any $\mu \in Q$. For this task we rely on the *Weyl group* $\mathcal{W}$ of the simple Lie algebra $\mathfrak{g}$ (tabulated in [**37**, Section 12.2]). If $\alpha_i \in \Delta$, let σ_i be the reflection of $\mathfrak{t}^*$ perpendicular to α_i and let $\mathcal{W}$ be the group generated by all σ_i. The dimensions of weight spaces of any finite-dimensional irreducible $\mathcal{U}(\mathfrak{g})$-module are fixed by $\mathcal{W}$, i.e. for all $\mu \in Q$ and $\sigma \in \mathcal{W}$, $m_\lambda(\sigma\mu) = m_\lambda(\mu)$. Thus to compute the weights which appear in a finite-dimensional irreducible $\mathcal{U}(\mathfrak{g})$-module of highest weight λ one computes all dominant integral weights of the form $\lambda - \sum_{\alpha \succ 0} n_\alpha \alpha$ for nonnegative integers n_α, then acts upon this set with the Weyl group to generate the remainder of the weights. The multiplicities of all nonzero weight spaces can then be expressed by *Kostant's multiplicity formula* [**46**]

$$m_\lambda(\mu) = \sum_{\sigma \in \mathcal{W}} (-1)^{\ell(\sigma)} \nu(\sigma(\lambda+\rho) - (\mu+\rho)) \tag{4}$$

where $\nu(\mu)$ is *Kostant's partition function*, the number of ways μ can be written as a nonnegative integer linear combination of positive roots, and ρ is the distinguished dominant integral weight $\lambda_1 + \cdots + \lambda_r$. The dimension of $V(\lambda)$ is given by the *Weyl dimension formula* [**37**, Corollary 24.3]

$$\dim V(\lambda) = \prod_{\alpha \succ 0} \frac{\langle \lambda+\rho, \alpha\rangle}{\langle \rho, \alpha\rangle}, \tag{5}$$

and $M^\mu_{\lambda,\gamma}$, the multiplicity of $V(\mu)$ in the tensor product $V(\lambda) \otimes V(\gamma)$, is given by the *Racah-Speiser formula* [**29**, Equation 15.23]

$$M^\mu_{\lambda,\gamma} = \sum_{\sigma \in \mathcal{W}} (-1)^{\ell(\sigma)} m_\gamma(\sigma(\mu+\rho) - (\lambda+\rho)). \tag{6}$$

EXAMPLE 7 (Type B_2). From the Dynkin diagram of type B_2 (Section 1.4) we have two simple roots α_1, α_2 with $\|\alpha_1\| > \|\alpha_2\|$. One then computes [**37**, Section 9.4] the angle between α_1, α_2 is 135 degrees and the ratio of the lengths of α_1, α_2 is $\sqrt{2}$. One should verify the fundamental weights are $\lambda_1 := (1/2)\alpha_1 + \alpha_2$ and $\lambda_2 := \alpha_1 + \alpha_2$ (Figure 1a).

Consider $V := V(2\lambda_2)$. The nonzero weight spaces corresponding to dominant integral weights are $2\lambda_2$, $2\lambda_1$, λ_2, and 0, illustrated below as rectangular nodes with multiplicities inside computed using (4). The remainder of the weight spaces in circular nodes are finally determined by the Weyl group symmetries, illustrated with thin lines in Figure 1b.

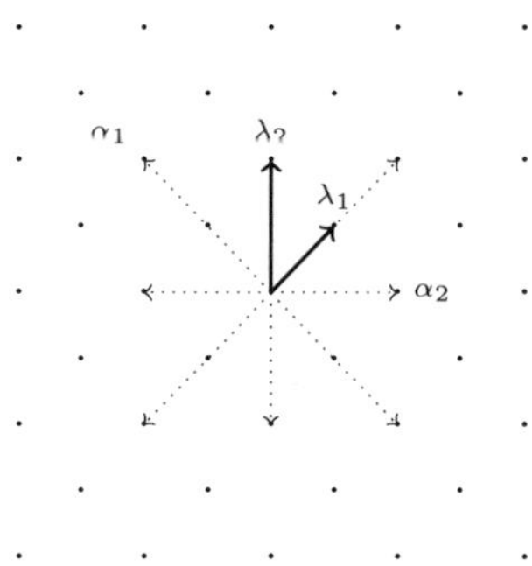

(A) Φ and Q

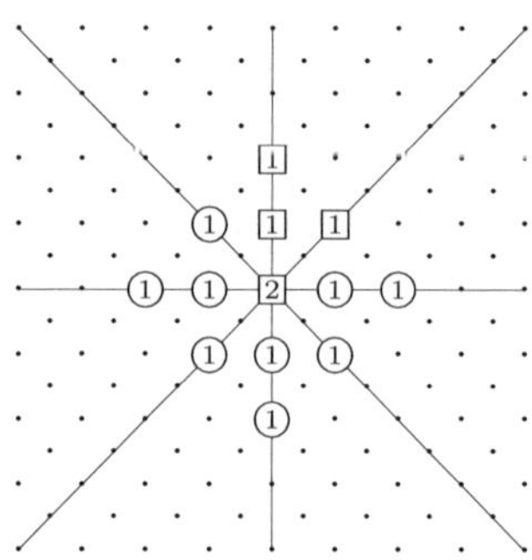

(B) Weight decomposition of $V(2\lambda_2)$

FIGURE 1. Example of representation theory for type B_2

3. Representation theory of quantized enveloping algebras

In 1986 at the International Congress of Mathematicians, Vladimir Drinfeld spoke on "recent works on Hopf algebras" motivated by mathematical physics. His summary which appeared the following year in the Proceedings of the ICM titled *Quantum Groups* [**21**] stands as one of the most oft-cited papers in the field. The process of quantization is there described as "something like replacing commutative algebras with noncommutative ones". We will briefly consider quantizations of the commutative algebra $\mathcal{U}(\mathfrak{g})$ in this sense with an ultimate goal of studying categories created from their representation theory.

Constructions of the *quantized enveloping algebra* $\mathcal{U}_q(\mathfrak{g})$ of a simple Lie algebra $\mathfrak{g}$ are numerous with subtle differences based on the end-goal of the respective authors [**24**, **41**, **50**]. These constructions involve introduction of a formal parameter q into the defining Serre relations of $\mathfrak{g}$ [**37**, Section 18.1]. But with care, apparent in Lusztig's construction [**33**], q can be specialized to nonzero complex numbers.

3.1. Weyl modules. The choice of our parameter q stems from creating interesting representation theory. In the case $q \neq 0$ is not a root of unity, the category of finite-dimensional $\mathcal{U}_q(\mathfrak{g})$-modules with tractable weight decomposition (described in [**6**, Section 3.3]) has the same isomorphism classes of simple objects and fusion coefficients as the category of finite-dimensional $\mathcal{U}(\mathfrak{g})$-modules. In particular these categories have infinitely many isomorphism classes of simple objects and do not verbatim create our desired examples of fusion categories. With q a root of unity, the category of finite-dimensional $\mathcal{U}_q(\mathfrak{g})$-modules satisfying the aforementioned weight decomposition criteria is a ribbon category [**33**] but in particular is *not* semisimple [**6**, Exercise 3.3.12 (ii)] and still has infinitely many isomorphism classes of simple objects.

Once-and-for-all we describe the most generic roots of unity to be considered in this exposition that will alleviate the aforementioned undesirable characteristics. Their description depends on the largest absolute value of an off-diagonal entry of the Cartan matrix [**37**, Section 11.4], which we denote by m. For simply-laced Lie algebras $m = 1$, for types B_r, C_r, and F_4 $m = 2$, and for type G_2 $m = 3$. Now let q be a root of unity such that q^2 has order ℓ. Some aspects of the representation theory of $\mathcal{U}_q(\mathfrak{g})$ will depend on ℓ alone, but some will depend on both q and ℓ. For a fixed simple Lie algebra $\mathfrak{g}$, a root of unity q such that $m \mid \ell$ will be called *divisible* for $\mathfrak{g}$ (also called *uniform* in the literature). Figure 2 [**65**, Table 2], lists the lower bound for ℓ that will produce a modular tensor category by the construction described in Section 4.

Type	Divisible	$\ell \geq$
A_r	yes	$r+1$
B_r	no	$2r+1$
	yes	$4r-2$
C_r	no	$2r+1$
	yes	$2r+2$
D_r	yes	$2r-2$
E_6	yes	12
E_7	yes	18
E_8	yes	30
F_4	no	13
	yes	18
G_2	no	7
	yes	12

FIGURE 2. Restrictions on roots of unity

To construct a semisimple category we first consider *Weyl modules* labelled by weights $\lambda \in Q$ [**6**, Section 3.1]. We will continue a standard practice of referring to these modules only by their corresponding weight. Context will clearly differentiate references to weights versus their corresponding modules. For each $\lambda \in Q$ (Section 2.1) the categorical, or quantum dimension of the corresponding Weyl module is given by the *quantum Weyl dimension formula* analogous to the classical Weyl dimension formula in Equation (5):

$$\dim(\lambda) = \prod_{\alpha \succ 0} \frac{[\langle \alpha, \lambda + \rho \rangle]}{[\langle \alpha, \rho \rangle]}, \tag{7}$$

where $[n] = (q^n - q^{-n})/(q - q^{-1})$ is the *quantum integer* n with respect to q. When computing these formulas by hand note the cancellation that happens in the classical case is absent here. For example [**37**, Section 24.3] the classical formula for the dimension of the representation of highest weight $s\lambda_1 + t\lambda_2$ for type G_2 is

$$\frac{1}{5!}(s+1)(t+1)(s+t+2)(s+2t+3)(s+3t+4)(2s+3t+5) \tag{8}$$

while in the quantum case [**70**, Section 2.3.4] one computes the dimension to be

$$\frac{[s+1][3(t+1)][3(s+t+2)][3(s+2t+3)][s+3t+4][2s+3t+5]}{[1][3][6][9][4][5]}. \tag{9}$$

Example 8. ($\mathfrak{sl}_2$ dimensions) Having exactly one positive root α and one fundamental weight λ implies the Weyl module of weight $s\lambda$ satisfies $\dim(s\lambda) = [s+1]$ using (7) above. Figure 3 plots these dimensions for $0 \leq s \leq 20$ in the case $q = \exp(\pi i/10)$ (white nodes) and $q = \exp(2\pi i/9)$ (black nodes). Note in particular that some Weyl modules have dimension zero.

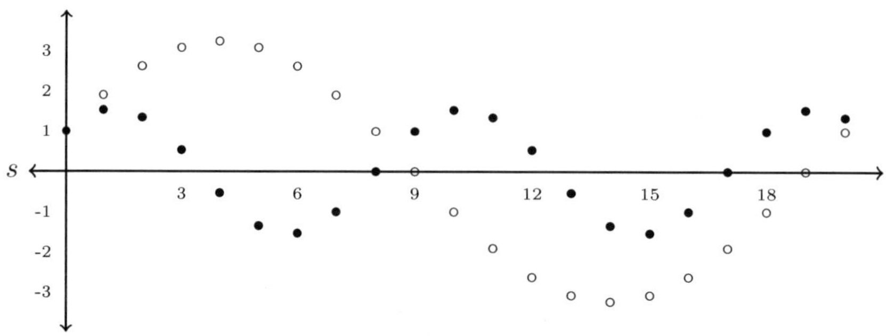

Figure 3. Dimension of $s\lambda$ when $q = \exp(\pi i/10)$ and $q = \exp(2\pi i/9)$

Figure 3 illustrates that computing quantum integers, and thus dimensions of Weyl modules, is an inherently trigonometric problem. Precisely when q is a root of unity with argument θ, $q^n - q^{-n} = 2i\sin(n\theta)$ and $q^n + q^{-n} = 2\cos(n\theta)$ which is useful if the long division $(q^n - q^n)/(q - q^{-1})$ is carried out, illustrated in Figure 4 when $\sin(n\theta) > 0$ with the center of the unit circle indicated with a white node.

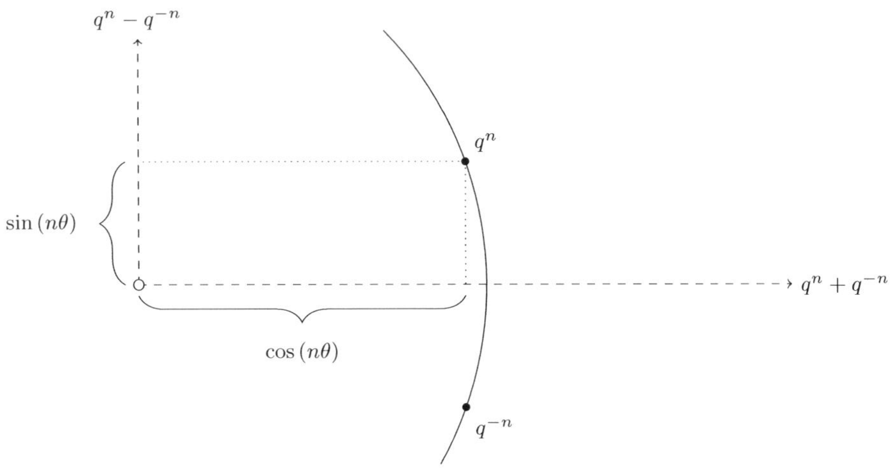

FIGURE 4. Trigonometric values for quantum integers

EXAMPLE 9 (Dimensions for B_2). If $q = e^{\theta i}$ for a root of unity q satisfying the criteria in Figure 2, then from Equation (7) with $\lambda := s\lambda_1 + t\lambda_2$ (see Example 7)

$$\dim(\lambda) = \frac{\sin((s+1)\theta)\sin(2(t+1)\theta)\sin(2(s+t+2)\theta)\sin((s+2t+3)\theta)}{\sin(\theta)\sin(2\theta)\sin(3\theta)\sin(4\theta)}.$$

3.2. Affine Weyl group. Recall that q is a root of unity such that q^2 has order ℓ. Let $\mathfrak{W}$, the *affine Weyl group*, be the group generated by the reflections τ_i through the corresponding hyperplanes

$$T_i := \{\lambda \in \mathfrak{t}^* : \langle \lambda + \rho, \alpha_i^\vee \rangle = 0\} \tag{10}$$

for all simple roots $\alpha_i \in \Delta$, and the single reflection through the hyperplane

$$T_0 := \{\lambda \in \mathfrak{t}^* : \langle \lambda + \rho, \beta^\vee \rangle = \ell\} \tag{11}$$

where β is the longest root if $m \mid \ell$ and β is the shortest root if $m \nmid \ell$. Note that the hyperplanes in (10) are the reflections generating the classical Weyl group $\mathcal{W}$ shifted by $-\rho$. The weights $\lambda \in Q$ strictly bounded by the hyperplanes T_i will be referred to as the *Weyl alcove* and be denoted Λ_0 and accordingly the hyperplanes T_i are the *walls* of the Weyl alcove.

EXAMPLE 10. (affine Weyl group of $\mathfrak{sl}_3$) Continuing from Example 6 we consider the two fundamental weights λ_1, λ_2 such that $\langle \lambda_1, \lambda_1 \rangle = \langle \lambda_2, \lambda_2 \rangle = 2/3$ and $\langle \lambda_1, \lambda_2 \rangle = 1/3$. One then computes $\lambda := s\lambda_1 + t\lambda_2$ lies on the hyperplane T_1 if and only if

$$\begin{aligned} & \langle (s+1)\lambda_1 + (t+1)\lambda_2, \lambda_1 - (1/2)\lambda_2 \rangle = 0 \\ \Leftrightarrow \quad & (2/3)(s+1) + (1/3)(t+1) - (1/6)(s+1) - (1/3)(t+1) = 0 \\ \Leftrightarrow \quad & s = -1. \end{aligned}$$

Similarly λ lies on T_2 if and only if $t = -1$, or lies on T_0 if and only if $s + t = \ell - 2$. Figure 5 illustrates $\mathfrak{W}$ in the case $\ell = 6$ with $-\rho$ indicated by a white node, 0 indicated by a black node, the reflections of $\mathfrak{W}$ indicated by dashed lines, and with the generating reflections T_0, T_1, T_2 emphasized.

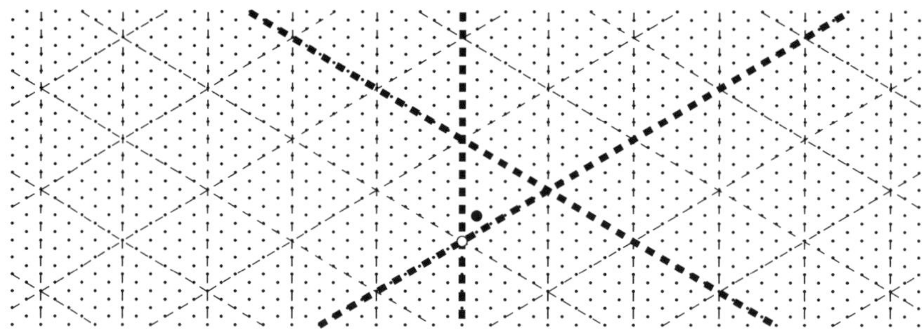

FIGURE 5. $\mathfrak{W}$ for $\mathfrak{sl}_3$ superimposed on Q

NOTE. In general $\mathfrak{W}$ is an example of an *affine Coxeter group* [**9**, Ch. VI, Sec. 4] which are classified by Dynkin diagrams akin to the classification of simple finite-dimensional Lie algebras in Section 1.4 (the classical Weyl groups are also Coxeter groups). For a complete table of the affine Dynkin diagrams corresponding to the affine Weyl groups defined above, refer to [**39**, page 54]. The connection between the representation theory of affine Lie algebras and the representation theory of quantum groups at roots of unity is a long but historically important part of this story. From a computational standpoint we have little reason to elaborate in this direction due to the equivalence of categories [**6**, Theorem 7.0.2] attributed to M. Finkelberg [**25**] based on the work of Kazhdan and Lusztig [**42**, **43**].

The affine Weyl group $\mathfrak{W}$ acts *anti*-symmetrically on the dimensions of Weyl modules. That is to say if μ is $\mathfrak{W}$-conjugate to $\lambda \in \Lambda_0$ by an element $\tau \in \mathfrak{W}$ then $\dim(\mu) = (-1)^{\ell(\tau)} \dim(\lambda)$ where $\ell(\tau)$ is the *length* of τ, i.e. the length of a shortest expression of τ in terms of the generating reflections τ_i. Hence the quantum dimensions of Weyl modules $\lambda \in \Lambda_0$ determine the quantum dimensions of all other Weyl modules. One may then conclude if λ lies on any hyperplane of reflection arising from $\mathfrak{W}$ then $\dim(\lambda) = 0$; the converse is also true.

EXAMPLE 11. ($\mathfrak{sl}_2$ continued) The weight lattice of $\mathfrak{sl}_2$ is the $\mathbb{Z}$-linear span of the unique fundamental weight λ and if for some $s \in \mathbb{Z}_{\geq 0}$ and $\ell \in \mathbb{Z}_{\geq 2}$, $\langle (s+1)\lambda, 2\lambda \rangle = \ell$ implies $s = \ell - 1$ since $\langle \lambda, \lambda \rangle = 1/2$. Similarly $\langle s\lambda, 2\lambda \rangle = 0$ implies $s = -1$. Thus the affine Weyl group has elements which are reflections through $-1 + j\ell$ for $j \in \mathbb{Z}$. The antisymmetric action of $\mathfrak{W}$ can be visually confirmed in Example 8.

4. The categories $\mathcal{C}(\mathfrak{g}, \ell, q)$

Section 3.2 hints to which Weyl modules would be feasible to consider when constructing a premodular category from the representation theory of quantum groups at roots of unity. Dimensions vanish along all hyperplanes of reflection in the affine Weyl group $\mathfrak{W}$, but inside a fundamental domain of the action of $\mathfrak{W}$, the Weyl alcove Λ_0, all dimensions are nonzero and all Weyl modules are irreducible. To disregard all other modules and achieve semisimplicity while retaining the familiar ribbon structure one uses a quotient construction which has proven to be applicable in a more general setting [**4**, **7**], but was originally concieved in the work of Andersen [**1**] and collaborators.

Roughly speaking the desired modules are *tilting modules* which coincide with Weyl modules inside Λ_0 and have zero dimension on the walls of Λ_0 [**67**, Lemma 9(a)].

Tilting modules are closed under fusion, quotients, etc. [**2**] but their collection as a category is not semisimple, nor even abelian. This can be rectified by quotienting out the ideal of the category of tilting modules by negligible morphisms [**22**, Exercise 8.18.9], leaving a ribbon category whose simple objects are enumerated by weights in Λ_0 up to isomorphism. We denote this category by $\mathcal{C}(\mathfrak{g}, \ell, q)$ where q^2 is a root of unity of order ℓ.

The quantum Weyl dimension formula is superficially impervious to the root of unity considered but the geometry of the Weyl alcove based on the order of q^2 greatly alters the remaining topics of discussion in this section and beyond. For $k \in \mathbb{Z}_{\geq 1}$, if $q = \exp(\pi i/\ell)$ where $\ell = m(k + h^\vee)$ ($h^\vee$ is the dual Coxeter number of a given $\mathfrak{g}$), we denote $\mathcal{C}(\mathfrak{g}, \ell, q)$ by $\mathcal{C}(\mathfrak{g}, k)$ instead. These cases are referred to as *positive integer levels* in the literature. The dual Coxeter numbers for the finite Dynkin diagrams are included in Figure 6 for ease of reference. Positive integer levels are the case of most interest in mathematical physics and the categories which have the most structure, which explains their preference. In particular, all $\mathcal{C}(\mathfrak{g}, k)$ are *pseudounitary* [**62**, Section 8.4], the major computational benefit being the quantum dimensions of all objects are positive. Unfortunately there are numerous instances of authors making statements about "arbitrary" roots of unity when, in fact, the categories in question are those at positive integer levels. One can extend the idea of "level" to all roots of unity by setting the level $k := \ell/m - h^\vee$, which may be negative or even fractional. Here we make no claim to cover arbitrary roots of unity (see Figure 2) as the cases when ℓ is small compared to $h^\vee$ have not been critically examined and in the author's opinion this is an open area of research.

Type	A_r	B_r	C_r	D_r	E_6	E_7	E_8	F_4	G_2
$h^\vee$	$r+1$	$2r-1$	$r+1$	$2r-2$	12	18	30	9	4

FIGURE 6. Dual Coxeter numbers

EXAMPLE 12. (simple objects of $\mathcal{C}(\mathfrak{g}_2, \ell, q)$) Let q be a root of unity such that q^2 has order ℓ (refer to Figure 2). In [**65**, Example 4.1.1] Rowell describes a generating function for the cardinality of Λ_0 with the two examples of $\ell = 14$ ($k = 2/3$) and $\ell = 27$ ($k = 5$). We illustrate these examples using black nodes for the weights in Λ_0, a single white node for $-\rho$, and dashed lines for the walls of the Weyl alcove.

(A) $\ell = 14$ ($k = 2/3$) (B) $\ell = 27$ ($k = 5$)

FIGURE 7. Λ_0 for $\mathcal{C}(\mathfrak{g}_2, \ell, q)$

5. Fusion

5.1. Fusion rules. As early as Andersen and Paradowski's seminal work [**2**, Proposition 2.10] the fusion rules of $\mathcal{C}(\mathfrak{g},\ell,q)$ have been expressed as a truncated version of the classical Racah-Speiser formula (see Equation (4)) for the decomposition of tensor products of $\mathcal{U}(\mathfrak{g})$-modules. The supplied proof by the aforementioned authors applies only to fusion coefficients between weights in the root lattice (see Section 5.2). A rigorous proof of the full *Quantum Racah Formula* was later given by Sawin [**68**, Corollary 8] which states that for all $\lambda,\gamma,\mu\in\Lambda_0$,

$$N_{\lambda,\gamma}^{\mu}=\sum_{\tau\in\mathfrak{W}}(-1)^{\ell(\tau)}m_\lambda(\tau(\mu)-\gamma). \tag{12}$$

Example 13. ($\mathfrak{sl}_2$ continued) With the description of $\mathfrak{W}$ for $\mathfrak{sl}_2$ from Example 11 we now derive an explicit formula for the fusion rules of $\mathcal{C}(\mathfrak{sl}_2,\ell,q)$. To this end let $s,t\in\mathbb{Z}_{\geq 0}$ (without loss of generality assume $s\geq t$ since the fusion rules are symmetric) and consider the product $s\lambda\otimes t\lambda$. First let $s+t<\ell-1$ and note $\tau_0(u\lambda)=2(\ell-1)-u$, with $2(\ell-1)-u-t\geq s$. Thus the terms in (12) are zero except that which comes from the identity. This implies $N_{s\lambda,t\lambda}^{u\lambda}=m_{s\lambda}((u-t)\lambda)$ and moreover $N_{s\lambda,t\lambda}^{u\lambda}=1$ when $u-t=s-2j$ for some $0\leq j\leq t$ and $N_{s\lambda,t\lambda}^{u\lambda}=0$ otherwise. Summarily,

$$s\lambda\otimes t\lambda=\bigoplus_{j=0}^{t}(s+t-2j)\lambda \tag{13}$$

which coincides with the *Clebsch-Gordan formula* for tensor products of $\mathcal{U}(\mathfrak{sl}_2)$-modules in the classical case [**22**, Equation 5.16]. In general if ℓ is sufficiently large the classical and quantum Racah formulas coincide. If $s+t\geq\ell-1$ there are two (possibly) nontrivial terms in (12) and

$$N_{s\lambda,t\lambda}^{u\lambda}=m_{s\lambda}((u-t)\lambda)-m_{s\lambda}((2(\ell-1)-(u+t))\lambda). \tag{14}$$

The first term is nonzero exactly as in the first case, but is negated by the second term when $2(\ell-1)-(u+t)=s-2j$ for $0\leq j\leq t$. Hence

$$s\lambda\otimes t\lambda=\bigoplus_{j=s+t-(\ell-1)}^{t}(s+t-2j)\lambda. \tag{15}$$

One could feasibly compute explicit formulas for the fusion rules by hand for rank 2 Lie algebras as there exist case-by-case formulas for Kostant's partition function [**71**, Table 1]. It is more constructive to see Equation (12) geometrically as we will illustrate in Example 14 and again in Example 16. The quantum Racah formula states that once $\lambda,\gamma\in\Lambda_0$ are fixed, for any $\mu\in\Lambda_0$, $N_{\lambda,\gamma}^{\mu}$ is computed by determining the classical weight multiplicities of $V(\lambda)$, shifting this weight diagram so it is centered at γ, then conjugating the weight multiplicities by the generating reflections $\tau_i\in\mathfrak{W}$ until they all lie within Λ_0, keeping track of the parity of the number of reflections required to achieve this.

Example 14. (Quantum Racah for $\mathcal{C}(\mathfrak{sl}_3,8)$) We will compute the fusion rules for $4\lambda_1\otimes 4\lambda_2\in\mathcal{C}(\mathfrak{sl}_3,8)$ using the above geometric interpretation of (12). To this end note that the convex hull of the weight diagram for $V(4\lambda_1)$ is an equilateral triangle with vertices $4\lambda_1$, $4(\lambda_2-\lambda_1)$, and $-4\lambda_2$. The weight-space multiplicities for generic $V(s\lambda_1+t\lambda_2)$ for $\mathfrak{sl}_3$ are easily computed [**5**], forming concentric "layers" of weight multiplicities beginning with the one-dimensional highest weight space, and increasing by 1 toward the weight 0, λ_1, or λ_2 depending on s and t. In the case of $V(4\lambda_1)$ there are only 2 layers. Figure 8b illustrates the process of conjugating these classical weight multiplicities (still transparently displayed) by generating reflections of $\mathfrak{W}$ after being shifted by $4\lambda_2$. Once contained in Λ_0, the weight multiplicities are summed in ovoid nodes with positive contribution if an even number of reflections was used, and negative contribution if an odd number of reflections was used.

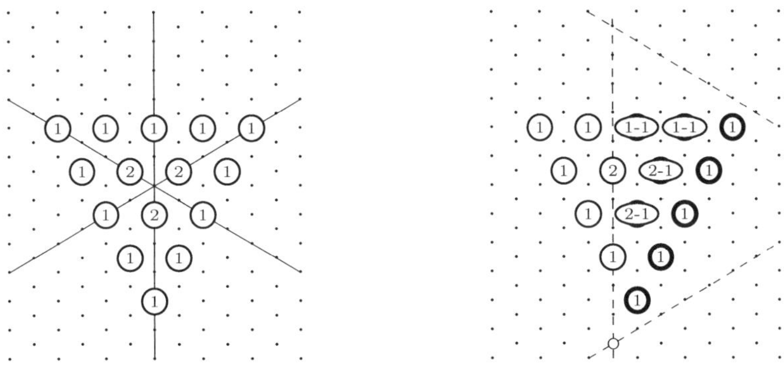

(A) Weight diagram of $V(4\lambda_1)$ (B) $N_{4\lambda_1,4\lambda_2}^{\lambda}$ for $\lambda \in \Lambda_0$ using (12)

FIGURE 8. $4\lambda_1 \otimes 4\lambda_2 \in \mathcal{C}(\mathfrak{sl}_3, 8)$

Numerous examples of this geometric interpretation can be found in Sections 4-6 of [**70**] for the rank 2 Lie algebras. In rank greater than 2 this task is substantially tedious and one may be satisfied deriving a coarser set of conclusions from the quantum Racah formula, as an explicit expression akin to Example 13 is unrealistic at this time. It should be noted that Kazhdan and Wenzl [**44**] have characterized all possible monoidal categories with fusion rings isomorphic to that of $\mathcal{C}(\mathfrak{sl}_n, \ell, q)$ while this was partially extended to other classical Lie algebras (or subcategories thereof) in [**72**] with the assumption that the given category is braided.

5.2. Fusion subcategories. Fusion subcategories of $\mathcal{C}(\mathfrak{g}, \ell, q)$ are rare. One construction for fusion subcategories of an arbitrary fusion category $\mathcal{C}$ is the *pointed* fusion subcategory generated by invertible objects, $\mathcal{C}_{\text{pt}}$ [**22**, Section 2.11].

EXAMPLE 15 (Pointed subcategories of $\mathcal{C}(\mathfrak{g}, k)$ and centers). Each simple Lie algebra $\mathfrak{g}$ has a corresponding simply-connected compact Lie group G [**12**, Chapters II-III]. Theorem 3 of [**66**] describes how elements of the center $Z(G)$ are in one-to-one correspondence with invertible objects in $\mathcal{C}(\mathfrak{g}, k)$ (originally classified by Fuchs [**28**]) with one exception: $\mathcal{C}(E_8, 2)_{\text{pt}}$ is rank 2 despite the simply-connected compact Lie group of type E_8 having trivial center. These centers are tabulated in Figure 9.

Type	A_r	$B_r/C_r/E_7$	D_{2r}	D_{2r+1}	E_6	$E_8/F_4/G_2$
$Z(G)$	$\mathbb{Z}/(r+1)\mathbb{Z}$	$\mathbb{Z}/2\mathbb{Z}$	$\mathbb{Z}/2\mathbb{Z} \oplus \mathbb{Z}/2\mathbb{Z}$	$\mathbb{Z}/4\mathbb{Z}$	$\mathbb{Z}/3\mathbb{Z}$	trivial

FIGURE 9. Centers of simply-connected compact Lie groups

The isomorphism classes of simple invertible objects of $\mathcal{C}(\mathfrak{g}, k)$ form an abelian group (with tensor unit $\mathbb{1}$), and the correspondence described in [**66**, Theorem 3] is a group homomorphism, hence any subgroup of $Z(G)$ corresponds to a pointed subcategory of $\mathcal{C}(\mathfrak{g}, k)$. Subcategories arising in this manner (along with the aforementioned E_8 exception) describe all pointed subcategories of $\mathcal{C}(\mathfrak{g}, k)$.

A systematic study of fusion subcategories under the name *closed subsets* was undertaken by Sawin [**67**, Theorem 1] for positive integer levels k. In these cases there are three types of nontrivial proper fusion subcategories which occur: the subcategories generated by weights in the root lattice P (Section 1.3), the pointed subcategories described in Example 15, and five exceptional cases occuring at level 2 for types B, D, and E_7. The first two types of fusion subcategories exist in the case of general q and one expects there to be a small but distinct list of exceptional subcategories that do not appear in the classification for positive integer levels. In the simply-laced case the geometry of Λ_0 does not depend on the root of unity q (only on ℓ) and thus Sawin's classification is complete for arbitrary roots of unity.

EXAMPLE 16. (An exceptional fusion subcategory in type G_2) By Sawin's classification for divisible q there are no proper nontrivial fusion categories in $\mathcal{C}(\mathfrak{g}_2, \ell, q)$ when $3 \mid \ell$. But let $q := \exp(\pi i/10)$ $(k = -2/3)$. The simple object indexed by λ_2 is the adjoint representation of $\mathfrak{g}_2$ so all of the classical weight-space multiplicities are 1 except $m_{\lambda_2}(0) = 2$ (Section 2.1). Using the quantum Racah formula (visualization in Figure 10) we compute $N_{\lambda_2,\lambda_2}^{\mathbb{1}} = 1$, $N_{\lambda_2,\lambda_2}^{\lambda_1} = 0$, $N_{\lambda_2,\lambda_2}^{\lambda_2} = 1$, $N_{\lambda_2,\lambda_2}^{2\lambda_1} = 0$. Hence $\lambda_2 \otimes \lambda_2 = \mathbb{1} \oplus \lambda_2$ and moreover λ_2 generates a fusion subcategory with the "Fibonacci" fusion rules. One can verify using (9) that $\dim(\lambda_2) = (1 - \sqrt{5})/2$ and so this subcategory is a Galois conjugate of the fusion subcategory of $\mathcal{C}(\mathfrak{sl}_2, 3)$ corresponding to the root lattice (with simple objects $\mathbb{1}$, 2λ) using Ostrik's classification of fusion categories of rank 2 [**57**, Section 2.5].

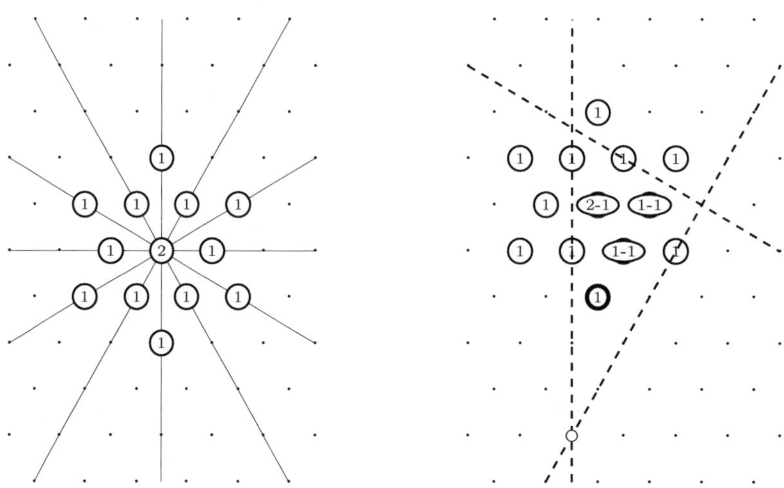

(A) Weight diagram of $V(\lambda_2)$ (B) $N_{\lambda_2,\lambda_2}^{\lambda}$ for $\lambda \in \Lambda_0$ using (12)

FIGURE 10. $\lambda_2 \otimes \lambda_2 \in \mathcal{C}\left(\mathfrak{g}_2, 10, e^{\pi i/10}\right)$

Using the decomposition method of Müger [**54**] and Bruguières [**10**], if $\mathcal{C}(\mathfrak{g}, \ell, q)$ and $\mathcal{C}(\mathfrak{g}, \ell, q)_{\mathrm{pt}}$ are nondegenerate, one may factor

$$\mathcal{C}(\mathfrak{g}, \ell, q) \simeq \mathcal{C}(\mathfrak{g}, \ell, q)_{\mathrm{pt}} \boxtimes \mathcal{C}(\mathfrak{g}, \ell, q)'_{\mathrm{pt}} \tag{16}$$

where the second factor is the centralizer [**54**, Definition 2.6] of the first. But this centralizer must also be a fusion subcategory and so it is, in general, the subcategory corresponding to the root lattice. Factorizing in this way is trivial in the cases $\mathcal{C}(\mathfrak{g}, \ell, q)$ is pointed (as is $\mathcal{C}(\mathfrak{g}, 1)$ for simply-laced $\mathfrak{g}$ [**27**]), or unpointed. The technique in (16) is applicable for all modular categories but in the case of $\mathcal{C}(\mathfrak{g}, k)$ (up to factorizations of $\mathcal{C}(\mathfrak{g}, k)_{\mathrm{pt}}$ and exceptional cases) these factors must be *simple* by Sawin's classification of closed subsets of Λ_0.

EXAMPLE 17. (Simple factorizations of $\mathcal{C}(\mathfrak{sl}_p, k)$) For primes p, the factorization in (16) into simple components is easily described for $\mathcal{C}(\mathfrak{sl}_p, k)$ when $p \nmid k$. In these cases the pointed subcategory has nontrivial simple objects $k\lambda_i$ for $1 \leq i \leq p-1$ whose fusion rules have the structure of the cyclic group $\mathbb{Z}/p\mathbb{Z}$. The quadratic form [**22**, Section 8.4] of the pointed subcategory is determined by the twists $\theta_{k\lambda_i}$ (see (17) below) which imply the form is degenerate if and only if $p \mid k$. As p is prime, $\mathcal{C}(\mathfrak{sl}_p, k)_{\mathrm{pt}}$ is simple, and by Sawin's classification of fusion subcategories $\mathcal{C}(\mathfrak{sl}_p, k)'_{\mathrm{pt}}$ is simple as well. When $p \mid k$, $\theta_{k\lambda_i} = 1$ for all $1 \leq i \leq n-1$ and $\mathcal{C}(\mathfrak{g}, k)_{\mathrm{pt}} \simeq \mathrm{Rep}(\mathbb{Z}/p\mathbb{Z})$ which is symmetrically braided.

6. Modular data

6.1. ***S* and *T* matrices.** One of the strongest available numerical invariants of a modular tensor category is the pair known as the S-matrix and T-matrix containing the traces of the double braids between simple objects, and the full twists of the simple objects, respectively. We note that there are various normalizations of S and the normalization we will consider ensures that the first row/column consists of the dimensions of the simple objects. Together these matrices comprise the *modular data* of a modular tensor category and are subject to a small list of compatibility conditions [**22**, Definition 8.17.1]. Modular tensor categories of very low rank are determined up to ribbon equivalence by their modular data [**64**, Section 5.3], and at the completion of this paper, examples (infinite families) of modular tensor categories are just now being proposed which are nonequivalent with identical modular data [**52**] coming from the representation theory of finite groups. One should also note the S and T-matrices of a premodular category have been referred to as "modular data" in many papers, though the category itself is not modular.

The formula for the entries of the S-matrix of $\mathcal{C}(\mathfrak{g}, \ell, q)$ is well-known, corresponding (up to scaling) to the Kac-Peterson formula for the modular transformations of characters of affine Lie algebras [**40**]. Finding a formula for the modular data of $\mathcal{C}(\mathfrak{g}, \ell, q)$ which does not rely on summing over the Weyl group, like for the quantum Racah formula is too complex to be expected in all but low-rank examples. For any $\lambda, \mu \in \Lambda_0$,

$$S_{\lambda,\mu} = \frac{\sum\limits_{\sigma \in \mathcal{W}} (-1)^{\ell(\sigma)} q^{2\langle \lambda+\rho, \sigma(\mu+\rho)\rangle}}{\sum\limits_{\sigma \in \mathcal{W}} (-1)^{\ell(\sigma)} q^{2\langle \rho, \sigma(\rho)\rangle}}, \qquad \text{and} \qquad \theta_\lambda = q^{\langle \lambda, \lambda+2\rho\rangle}. \tag{17}$$

EXAMPLE 18 (Modular data of $\mathcal{C}(\mathfrak{sl}_2,\ell,q)$). The Weyl group $\mathcal{W}$ is isomorphic to $\mathbb{Z}/2\mathbb{Z}$ so by (17) we have $\theta_{s\lambda}=q^{s(s+2)/2}$ and

$$S_{s\lambda,t\lambda}=\frac{q^{(s+1)(t+1)}-q^{-(s+1)(t+1)}}{q-q^{-1}}=[(s+1)(t+1)].$$

With $q:=e^{\pi i/3}$ we have the S,T-matrices for $\mathcal{C}(\mathfrak{sl}_2,3,q)$ and $\mathcal{C}(\mathfrak{sl}_2,3,q^2)$, respectively:

$$S,T=\begin{bmatrix}1&1\\1&-1\end{bmatrix},\begin{bmatrix}1&0\\0&i\end{bmatrix}\qquad\qquad S,T=\begin{bmatrix}1&-1\\-1&1\end{bmatrix},\begin{bmatrix}1&0\\0&-1\end{bmatrix}\tag{18}$$

Thus $\mathcal{C}(\mathfrak{sl}_2,3,q)$ is modular while $\mathcal{C}(\mathfrak{sl}_2,3,q^2)$ is not. This demonstrates ℓ is not sufficient to characterize the degeneracy/non-degeneracy of the braidings for $\mathcal{C}(\mathfrak{g},\ell,q)$.

The question of modularity has been answered in the affirmative for positive integer levels. The categories $\mathcal{C}(\mathfrak{g},k)$ are modular for all simple Lie algebras $\mathfrak{g}$ and positive integer levels $k\in\mathbb{Z}_{\geq1}$ [**6**,**40**,**73**], while other roots of unity can be analyzed using the techniques of [**74**]. For a detailed exposition on this line of reasoning refer to Section 4 of [**65**], where the question of unitarizability is discussed as well. To the extent of the author's knowledge, the question of modularity is still open in a small number of cases [**65**, Section 4.5-4.6] while the study of unitarizability was completed by Rowell in [**63**].

NOTE. A distinct lack of fusion subcategories outlined in Section 5.2 suggests that nondegeneracy of the braidings for general $\mathcal{C}(\mathfrak{g},\ell,q)$ is easily determined. In particular the collection of degenerate objects forms a fusion subcategory. For categories such as $\mathcal{C}(\mathfrak{g}_2,\ell,q)$ which are unpointed, and the root lattice coincides with the weight lattice, the only possible conclusions are that *every* simple object is degenerate, or the category is modular (the former conclusion being preposterous except in trivial examples).

6.2. The Galois action. The work of Anderson-Moore [**3**], and Vafa [**75**] (see also [**22**, Corollary 8.18.2]) implies that for any premodular category the full twists are roots of unity which is clear for the categories $\mathcal{C}(\mathfrak{g},\ell,q)$ by the formula in Section 6.1. What is less clear is that if such a category is also modular over $\mathbb{C}$, the entries of the S-matrix are contained in a cyclotomic extension of $\mathbb{Q}$ [**16**]. Furthermore [**11**, Section 2.1.4] if a modular category $\mathcal{C}$ is defined over $\mathbb{Q}(\xi)$ for some root of unity ξ, then each $\pi\in\mathrm{Gal}(\mathbb{Q}(\xi)/\mathbb{Q})$ induces a permutation $\hat{\pi}$ of $\mathcal{O}(\mathcal{C})$, the set of isomorphism classes of simple objects of $\mathcal{C}$, such that

$$\pi\left(\frac{S_{i,j}}{S_{0,j}}\right)=\frac{S_{i,\hat{\pi}(j)}}{S_{0,\hat{\pi}(j)}}.\tag{19}$$

As a result, $S_{i,j}=\epsilon_\pi(i)\epsilon_{\pi^{-1}}(j)S_{\hat{\pi}(i),\hat{\pi}^{-1}(j)}$ for a sign function $\epsilon_\pi:\mathcal{O}(\mathcal{C})\to\{\pm1\}$ depending on π. The details in the case of the categories $\mathcal{C}(\mathfrak{g},\ell,q)$ can be found in [**16**, Section 4], attributed to [**20**], which we summarize here with examples and geometric interpretation, limiting ourselves to positive integer levels to ensure modularity.

A cyclotomic field containing the entries of the S and T-matrices for $\mathcal{C}(\mathfrak{g},k)$ is $\mathbb{Q}(\xi)$ where $\xi=\exp(2\pi i/L)$, $L:=2\ell$, and the Galois group of this field extension is $\mathrm{Gal}(\mathbb{Q}(\xi)/\mathbb{Q})\simeq(\mathbb{Z}/L\mathbb{Z})^\times$. To define the sign functions of the Galois permutations on $\mathcal{O}(\mathcal{C}(\mathfrak{g},k))$ let $p\in(\mathbb{Z}/L\mathbb{Z})^\times$ correspond to $\pi\in\mathrm{Gal}(\mathbb{Q}(\xi)/\mathbb{Q})$ in the above isomorphism, i.e. $\pi(\xi)=\xi^p$. For each $\lambda\in\Lambda_0$ we define the unique weight $\lambda_\pi\in\Lambda_0$

to be the weight in Λ_0 conjugate to $p(\lambda+\rho)-\rho$ under some permutation $\tau_\lambda \in \mathfrak{W}$. The permutation $\hat{\pi}$ on $\mathcal{O}(\mathcal{C}(\mathfrak{g},k))$ is then defined by $\lambda \mapsto \lambda_\pi$ and the sign function $\epsilon_\pi : \Lambda_0 \to \{\pm 1\}$ is given by $\lambda \mapsto (-1)^{\ell(\tau_\lambda)}$.

EXAMPLE 19. In [**16**, Section 5] there is an example given for $\mathcal{C}(\mathfrak{sl}_2, 3)$. We have $L = 10$ in this example and the Galois automorphisms are represented by $\{1,3,7,9\} \subset \mathbb{Z}/10\mathbb{Z}$. Illustrated in Figure 11 is the geometric computation for the permutation ($\hat{\pi}_3$) and sign functions (ϵ_3) associated with $3 \in (\mathbb{Z}/10\mathbb{Z})^\times$ for the simple objects 2λ and 3λ. The weight $-\rho$ is identified with a white node, the map $\lambda \mapsto 3(\lambda+\rho)-\rho$ with dotted lines, the reflections in $\mathfrak{W}$ with dashed lines, and the reflections τ_λ for $\lambda \in \Lambda_0$ with solid lines.

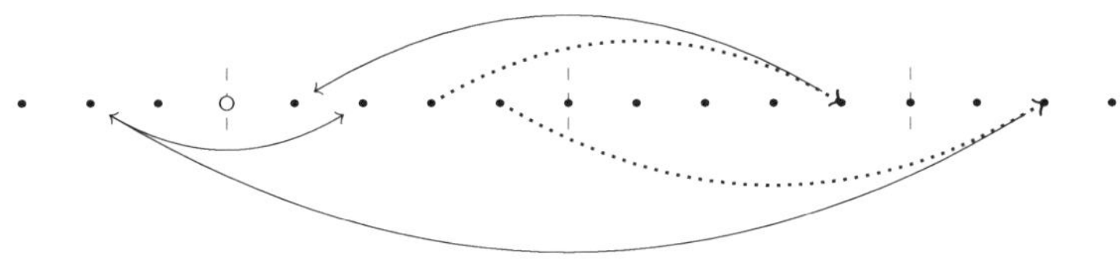

FIGURE 11. $\hat{\pi}_3 : \Lambda_0 \to \Lambda_0$ for $\mathcal{C}(\mathfrak{sl}_2, 3)$

In particular $\hat{\pi}_3(3\lambda) = \mathbb{1}$, $\hat{\pi}_3(4\lambda) = \lambda$, $\epsilon_3(3\lambda) = -1$, and $\epsilon_3(4\lambda) = 1$. The reader can verify that $\hat{\pi}_3(\mathbb{1}) = 2\lambda$, $\hat{\pi}_3(\lambda) = 3\lambda$, $\epsilon_3(\mathbb{1}) = 1$, and $\epsilon_3(\lambda) = -1$.

The Galois action for arbitrary modular tensor categories has been used extensively in the classification program for modular tensor categories by rank [**11**, **64**], as well as the classification of modular invariant partition functions in the work of Gannon [**30**, **31**].

EXAMPLE 20 (Permutation of the root lattice in B_2). The constant $L = 20$ for Type B_2 at level 4 hence the Galois group is $(\mathbb{Z}/20\mathbb{Z})^\times = \{1,3,7,9,11,13,17,19\}$. Since the action of $\mathfrak{W}$ and the map $\lambda \mapsto p(\lambda+\rho)-\rho$ fixes the root lattice P (see Example 7) for all $p \in (\mathbb{Z}/20\mathbb{Z})^\times$, for space limitations we compute $\hat{\pi}_3$ on $P \cap \Lambda_0$. The geometric symmetry of $\hat{\pi}_3$ is visualized in Figure 12 with the ambiguous central arrows representing $\hat{\pi}_3(2\lambda_2) = \lambda_1 + \lambda_2$, $\hat{\pi}_3(\lambda_1+\lambda_2) = 2\lambda_1$, and $\hat{\pi}_3(2\lambda_1) = 2\lambda_2$.

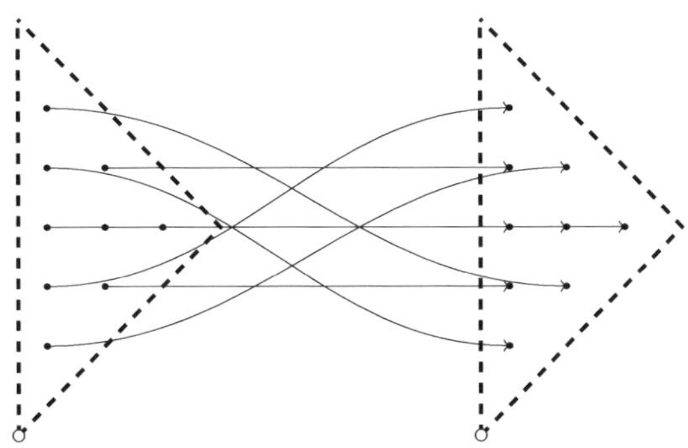

FIGURE 12. Visualization of $\hat{\pi}_3$ for the root lattice in $\mathcal{C}(\mathfrak{so}_5, 4)$

7. Symmetries and quantum subgroups

7.1. Fusion symmetries and autoequivalences. Section 6.2 describes one family of permutations of the isomorphism classes of simple objects of $\mathcal{C}(\mathfrak{g},\ell,q)$. These are not (in general) *fusion ring symmetries*, i.e. a permutation π of Λ_0 such that $N_{\lambda,\gamma}^{\mu}=N_{\pi\lambda,\pi\gamma}^{\pi\mu}$ for all $\lambda,\gamma,\mu\in\Lambda_0$. These permutations are not *automorphisms* of the fusion ring (in general) either, as we stipulate an automorphism be unital as well as multiplicative. A large class of fusion ring symmetries correspond to the action of tensoring with an invertible object (or *simple current*). Symmetries of this form were described in terms of pointed subcategories in Example 15, and it should be evident that the fusion symmetry from tensoring with any nontrivial invertible object cannot be an automorphism of the fusion ring by our definition.

NOTE. Tensoring with any invertible object (simple current) gives rise to a symmetry of the affine Dynkin diagram described in Section 3.2 except the non-trivial invertible object in $\mathcal{C}(E_8,2)$ (see Example 15).

EXAMPLE 21 ($\mathcal{C}(\mathfrak{so}_5,k)$ fusion symmetry). The only nontrivial invertible object in $\mathcal{C}(\mathfrak{so}_5,k)$ is $k\lambda_2$. The permutation on Λ_0 induced by tensoring with $k\lambda_2$ can be computed explicitly (using the quantum Racah formula (12) and the formulas for Kostant's partition function found in [**71**]) as $s\lambda_1+t\lambda_2\mapsto s\lambda_1+(k-s-t)\lambda_2$. Geometrically this is illustrated in Figure 13 as the reflection about the dotted line through $\{s\lambda+t\lambda\in Q : s+t=k/2\}$, along with the corresponding affine Dynkin diagram symmetry where α_0 is the imaginary root.

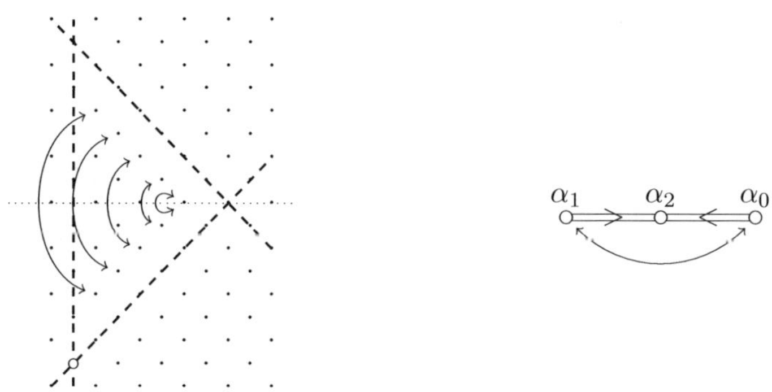

(A) Tensoring with $4\lambda_2$ in $\mathcal{C}(\mathfrak{so}_5,4)$ (B) Affine type B_2 symmetry

FIGURE 13. $\mathcal{C}(\mathfrak{so}_5,k)$ fusion symmetry

Symmetries of the finite Dynkin diagram of $\mathfrak{g}$ (Section 1.4) also correspond to fusion symmetries of $\mathcal{C}(\mathfrak{g},\ell,q)$. These permutations (called *conjugations* in the literature), unlike those from affine Dynkin diagram symmetries, are automorphisms. The conjugations induced from the duality endofunctor of $\mathcal{C}(\mathfrak{g},k)$ represent non-trivial fusion ring automorphisms in the case of A_n for $n\geq 2$, D_{2n+1} for $n\geq 1$ and E_6, while D_{2n} for $n\geq 1$ has a non-trivial conjugation which does not come from duality (representations of Type D_{2n} are self-dual). The Lie algebra D_4 is unique in that it has *triality*: an order 3 Dynkin diagram symmetry (See Example 23). Gannon [**32**] has classified all fusion symmetries for $\mathcal{C}(\mathfrak{g},k)$, hence fusion ring automorphisms, as well as the instances when the fusion rings of $\mathcal{C}(\mathfrak{g},k)$ coincide.

EXAMPLE 22 (Duality in $\mathcal{C}(\mathfrak{sl}_n, \ell, q)$). The fundamental weights $\lambda_1, \ldots, \lambda_{n-1}$ were computed in Example 6 which allows us to label simple objects of $\mathcal{C}(\mathfrak{sl}_n, \ell, q)$ by nonnegative integer tuples $(w_1, \ldots, w_{n-1})$. The duality endofunctor can then be described on objects as a reflection $(w_1, \ldots, w_n) \mapsto (w_n, \ldots, w_1)$. For $n = 1$ this is trivial, but for $n > 1$ this is the fusion ring automorphism corresponding to the Dynkin diagram automorphism given by reflection about a vertical axis. We illustrate the duality permutation in Figure 14a on the root lattice (to prevent clutter) of $\mathcal{C}(\mathfrak{sl}_3, 6)$, and in Figure 14b on the Dynkin diagram A_r for arbitary $r \geq 2$.

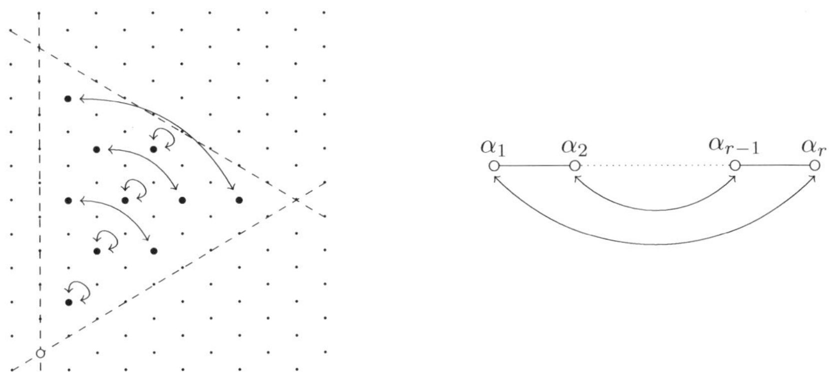

(A) Duality in $\mathcal{C}(\mathfrak{sl}_3, 6)$ (B) Type A_r diagram automorphism

FIGURE 14. Duality of $\mathcal{C}(\mathfrak{sl}_n, \ell, q)$

In addition one may ask which fusion symmetries give rise to a tensor autoequivalence, or braided tensor autoequivalence of the category $\mathcal{C}(\mathfrak{g}, k)$. As $\mathcal{C}(\mathfrak{g}, k)$ are nondegenerate, braided tensor autoequivalences may be used to compute the *Picard group* of $\mathcal{C}(\mathfrak{g}, k)$ [**23**, Section 4.4][**51**] consisting of equivalence classes of invertible module categories over $\mathcal{C}(\mathfrak{g}, k)$ (see Section 7.2). Note this is ostensibly a great increase in complexity as given a fusion ring automorphism describing the autoequivalence F on objects, one must also define a tensor functor structure on F which is braided (recall that a tensor functor being braided is a property, not an additional structure).

One motivating result in support of a clean solution for $\mathcal{C}(\mathfrak{g}, k)$ is that of Neshveyev and Tuset [**55**, Theorem 5] which states that tensor autoequivalences of the category of $\mathcal{U}_q(\mathfrak{g})$-modules when $q \neq 0$ is *not* a root of unity are determined by (i) automorphisms of the Dynkin diagram corresponding to $\mathfrak{g}$, and (ii) a very limited set of cohomological data [**55**, Theorem 1] corresponding to tensor structures on an endofunctor fixing all objects. This general flavor of classification was recently applied to the representation categories of *small quantum groups* $\mathfrak{u}_q(\mathfrak{g})$ at roots of unity [**48, 49**] by Davydov, Etingof, and Nikshych [**17**].

Similarly the problem of classifying (braided) tensor autoequivalences of $\mathcal{C}(\mathfrak{g}, \ell, q)$ can be thought of in two halves: one combinatorial and the other categorical. Combinatorially one needs to choose a fusion ring automorphism as discussed above and determine whether there is a (braided) autoequivalence realizing it (or many). As in the proof of [**55**, Theorem 5], what remains to be described is all possible tensor structures that can be equipped to an endofunctor which fixes the objects (so-called *gauge automorphisms* [**47**, Definition 4.1.8]). This is a problem of great interest and is still open to the author's knowledge.

7.2. Quantum subgroups. If one considers fusion categories as a categorical analog of rings, then considering *module categories* over a fusion category is a natural progression in abstraction to considering a "representation theory" of fusion categories [**22**, Chapter 7]. Furthermore, module categories offer yet another construction of infinite families of fusion and modular tensor categories coming from quantum groups as we will see below. The work of Ostrik [**59**, Section 3.2] demonstrates that the study of the category of modules over a fusion category can be done *internally* to the category itself. Specifically, each module category over a fusion category $\mathcal{C}$ is equivalent to the category of modules $\mathcal{C}_A$ over an algebra $A \in \mathcal{C}$. For each fusion category there are finitely many *simple* module categories up to equivalence coming from algebras which are *connected* and *separable*.

Categories $\mathcal{C}(\mathfrak{g}, \ell, q)$ are also braided which allows for a sensible notion of a commutative algebra. Algebras which are commutative and separable (called *étale*), and connected are those of interest in this setting as such an algebra A implies $\mathcal{C}_A$ in turn has the structure of a fusion category. The category $\mathcal{C}_A$ does not have an obvious braiding but the full subcategory $\mathcal{C}_A^0$ of *dyslectic* modules [**18**, Definition 3.12], has a braiding which is nondegenerate provided $\mathcal{C}$ is nondegenerately braided. Dyslectic modules where introduced by Pareigis [**61**] and can be alternatively described as the intersection of the images of the α-induction functors [**59**, Section 5.1]. Thus the study of connected étale algebras provides a construction of new infinite families of modular tensor categories. Connected étale algebras (simple module categories) are often referred to as *quantum subgroups* as connected étale algebras in fusion categories arising from the representation theory of a finite group G correspond to subgroups $H \subset G$ with additional cohomological data [**58**, Theorem 3.1].

EXAMPLE 23. (Quantum subgroups of $\mathcal{C}(\mathfrak{sl}_2, k)$) A succinct description of the classification of quantum subgroups of $\mathcal{C}(\mathfrak{sl}_2, k)$, or the *quantum McKay correspondence*, can be found in [**45**] in categorical language. Each quantum subgroup (connected étale algebra) which appears in $\mathcal{C}(\mathfrak{sl}_2, k)$ corresponds to one of the Dynkin diagrams of type A_n, D_{2n}, E_6, or E_8 with Coxeter number $k+2$. If A is a connected étale algebra in $\mathcal{C}(\mathfrak{sl}_2, k)$ with free module functor $F : \mathcal{C}(\mathfrak{sl}_2, k) \to \mathcal{C}(\mathfrak{sl}_2, k)_A$, the corresponding Dynkin diagram is the fusion graph of $F(\lambda)$. For example the quantum subgroup corresponding to the Dynkin diagram of type D_4 (which implies $k+2 = 2(4)-2 = 6$) occurs in $\mathcal{C}(\mathfrak{sl}_2, 4)$. We will encode the isomorphism classes of simple objects by $(0), (1), (2), (3), (4)$ for brevity. There is a unique algebra structure on $(0) \oplus (4)$ which is connected étale (the regular algebra of $\mathcal{C}(\mathfrak{sl}_2, 4)_{\text{pt}}$). The dyslectic module subcategory is indicated in Figure 15 with black nodes.

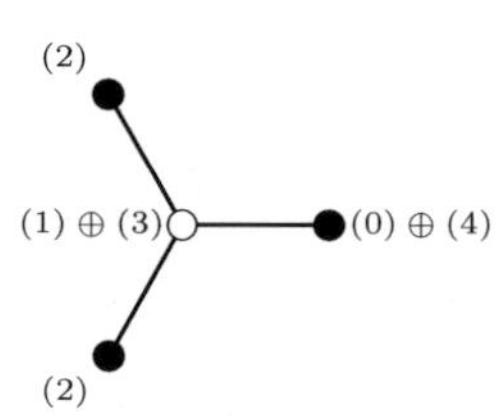

FIGURE 15. Quantum subgroup of type D_4

There is an alternate construction for the dyslectic module category $\mathcal{C}(\mathfrak{sl}_2,4)_A^0$ that displays the symmetry associated with algebras arising from the pointed subcategory in this way. Recall $\mathcal{C}(\mathfrak{sl}_2,4)_{\mathrm{pt}} \simeq \mathrm{Rep}(\mathbb{Z}/2\mathbb{Z})$. By the quantum Clebsch-Gordan rules derived in Example 13, tensoring with simple object $k\lambda$ corresponds to reflection through $k/2$, i.e. $s\lambda \mapsto (k-s)\lambda$. Moreover this defines a categorical $\mathbb{Z}/2\mathbb{Z}$ group action on the subcategory of $\mathcal{C}(\mathfrak{sl}_2,4)$ corresponding to the root lattice. The *de-equivariantization* [**22**, Section 8.23] of the root lattice subcategory with respect to the $\mathbb{Z}/2\mathbb{Z}$-action recovers $\mathcal{C}(\mathfrak{sl}_2,4)_A^0$.

The ability to classify quantum subgroups of fusion categories is still quite rare. There are few results known in this direction beside the classifications for group-theoretical categories, the infinite families of $\mathcal{C}(\mathfrak{sl}_2,k)$ and $\mathcal{C}(\mathfrak{sl}_3,k)$ categories (the latter following from the classification of modular invariant partition functions for affine $SU(3)$ due to Gannon [**31**]), and the Haagerup fusion categories [**34**]. Some additional examples of quantum subgroups arise from conformal embeddings of vertex operator algebras [**14,15**][**18**, Section 6.2], while Ocneanu [**56**] has proposed a complete classification of quantum subgroups for $\mathcal{C}(\mathfrak{sl}_4,k)$, the rigorous details of which are still being worked out by many researchers. In most researched cases the number of nontrivial quantum subgroups is few or, as in the case of the Fibonacci categories [**8**] and many others, zero.

EXAMPLE 24 (Witt group relations). One application of classifying quantum subgroups is to classify relations in the *Witt group* of nondegenerate braided fusion categories [**18**], which organizes nondegenerate braided fusion categories by identifying them up to Drinfeld centers. A characterization of this equivalence relation is that two non-degenerate braided fusion categories $\mathcal{C},\mathcal{D}$ are Witt equivalent (denoted $[\mathcal{C}]=[\mathcal{D}]$) if there exist connected étale algebras A,B such that $\mathcal{C}_A^0\simeq\mathcal{D}_B^0$ is a braided equivalence.

For example there is an infinite family of Witt group relations coming from *rank-level duality* [**26**] which was translated in terms of connected étale algebras for type A in [**60**, Theorem 5.1]. Specifically, for $n,m\in\mathbb{Z}_{\geq 2}$, $[\mathcal{C}(\mathfrak{sl}_n,m)]=[\mathcal{C}(\mathfrak{sl}_m,n)]$. A table of conformal embeddings, and thus many more Witt group relations, can be found in the Appendix of [**18**]. From a lack of a general classification of quantum subgroups, Witt group relations generated by the infinite families $\mathcal{C}(\mathfrak{g},k)$ have only been classified for $\mathfrak{sl}_2$ [**19**, Section 5.5] and $\mathfrak{sl}_3$ [**69**, Section 4.2].

NOTE. Each Witt group relation $[\mathcal{C}]=[\mathcal{D}]$ posits the existence of an unknown fusion category $\mathcal{A}$ such that $\mathcal{C}\boxtimes\mathcal{D}^{\mathrm{rev}}\simeq\mathcal{Z}(\mathcal{A})$, the Drinfeld center of $\mathcal{A}$. For Witt group relations of the form $[\mathcal{C}]=[\mathcal{C}_A^0]$ for some connected étale algebra A, the fusion category $\mathcal{A}=\mathcal{C}_A$ [**18**, Corollary 3.30] but for arbitrary Witt group relations the categories $\mathcal{A}$ have not been identified and it is unclear whether they produce novel fusion categories. This is an important open question and one can see an example of this process in [**13**, Appendix A].

7.3. Structure of module categories over $\mathcal{C}(\mathfrak{g},k)$. Once connected étale algebras $A\in\mathcal{C}$ have been classified, it still remains to compute the structure of the module category $\mathcal{C}_A$ (or even $\mathcal{C}_A^0$) to understand which fusion and modular tensor categories arise from this construction. The fusion rules for the module categories $\mathcal{C}(\mathfrak{sl}_2,k)_A$ where A is the algebra of type D_{2n} (Example 23) were computed in [**45**, Section 7]. In particular these categories are simple.

EXAMPLE 25 ($\mathcal{C}(\mathfrak{sl}_3,3)_A^0$). We know the isomorphism classes of simple objects of $\mathcal{C}(\mathfrak{sl}_3,3)_{\text{pt}}$ have the abelian group structure of $\mathbb{Z}/3\mathbb{Z}$ (Section 5.2) so we may consider the regular connected étale algebra $A := \mathbb{1}\oplus 3\lambda_1 \oplus 3\lambda_2$. The simple dyslectic A-modules (as objects of $\mathcal{C}(\mathfrak{sl}_3,3)$) are the trivial module A, and three copies of $\rho = \lambda_1+\lambda_2$ whose A-module structures are parameterized by third roots of unity. Using the quantum Weyl dimension formula and [**45**, Theorem 1.18], $\mathcal{C}(\mathfrak{sl}_3,3)$ is pointed and thus equivalent to $\mathcal{C}(A,q)$ where A is an abelian group of order 4, which is either cyclic or the Klein-4 group. But the automorphism of the simple dyslectic A-modules given by tensoring with $3\lambda_1$ or $3\lambda_2$ has order three so we must have $\mathcal{C}(\mathfrak{sl}_3,3)_A^0 \simeq \mathcal{C}(\mathbb{Z}/2\oplus\mathbb{Z}/2\mathbb{Z},q)$ with quadratic form $q:\mathbb{Z}/2\mathbb{Z}\oplus\mathbb{Z}/2\mathbb{Z}\longrightarrow\mathbb{C}^\times$ which is 1 on the unit object and -1 on the nontrivial simple objects. Note this category is not simple.

Another example (the fusion rules of $\mathcal{C}(\mathfrak{sl}_3,6)_A^0$ where A is the regular algebra of $\mathcal{C}(\mathfrak{sl}_3,6)_{\text{pt}}$) was computed in [**69**, Example 3.5.2], but the fusion rules of other dyslectic module categories are still unknown (though these "type D" dyslectic module categories for $\mathcal{C}(\mathfrak{sl}_3,k)$ have been shown to be simple aside from the example above [**69**, Theorem 1]).

NOTE. One application of identifying these fusion rules is that factoring $\mathcal{C}(\mathfrak{g},k)_A^0$ is a necessary process for classifying Witt group relations [**18**, Section 5.4].

Tensor autoequivalences of the module categories $\mathcal{C}(\mathfrak{g},k)_A^0$ (and $\mathcal{C}(\mathfrak{g},k)_A$) for a connected étale algebra A are less understood than those of $\mathcal{C}(\mathfrak{g},k)$. Upon first glance it seems that $\mathcal{C}(\mathfrak{g},k)_A^0$ would have less symmetry than $\mathcal{C}(\mathfrak{g},k)$ by looking at the examples where passing to the dyslectic module category is the same as de-equivariantizing (collapsing) by a group action. But in some low-level cases the dyslectic module categories $\mathcal{C}(\mathfrak{g},k)_A^0$ have *more* symmetry than the original.

EXAMPLE 26 ($\mathcal{C}(\mathfrak{sl}_2,16)_A^0$). Let A be the regular algebra of $\mathcal{C}(\mathfrak{sl}_2,16)_{\text{pt}}$ (see Example 23). The dyslectic module category $\mathcal{C}(\mathfrak{sl}_2,16)_A^0$ is the de-quivariantization of the root lattice subcategory (the so-called *even part*) by the $\mathbb{Z}/2\mathbb{Z}$-action of tensoring with the invertible object (16) (abbreviated from 16λ for brevity). Figure 16 illustrates the simple dyslectic modules with dashed lines, which correspond to orbits of the $\mathbb{Z}/2\mathbb{Z}$-action plus two modules $(8)^\pm$ isomorphic to (8) as objects of $\mathcal{C}(\mathfrak{sl}_2,16)$, but with A-module structure twisted by -1.

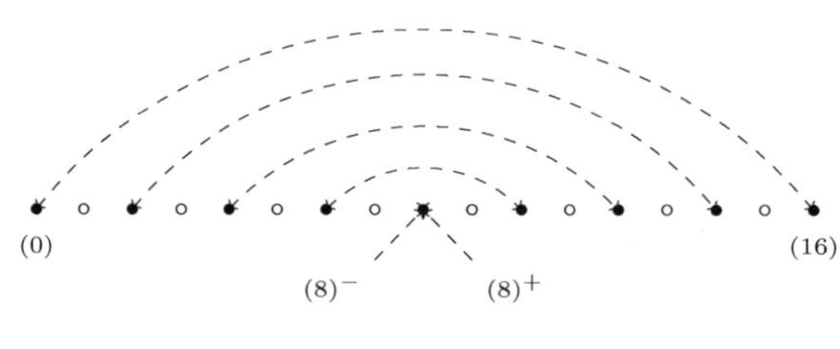

FIGURE 16. Λ_0 for $\mathcal{C}(\mathfrak{sl}_2,16)$, and simple objects of $\mathcal{C}(\mathfrak{sl}_2,16)_A^0$

Using the quantum Weyl dimension formula and [**45**, Theorem 1.18] we compute that $(8)^\pm$ and $(2)\oplus(14)$ have the same dimensions as A-modules. Morrison, Peters, and Snyder [**53**, Theorem 4.3] proved there is a tensor autoequivalence of

$\mathcal{C}(\mathfrak{sl}_2, 16)^0_A$, cyclically permuting $(8)^- \mapsto (8)^+ \mapsto (2) \oplus (14)$ using subfactor planar algebras, while $\mathcal{C}(\mathfrak{sl}_2, 16)$ has no tensor autoequivalences of order 3.

Another layer of complexity arising from Example 26 is that $\mathcal{C}(\mathfrak{sl}_2, 16)^0_A$ and $\mathcal{C}(\mathfrak{sl}_3, 6)^0_B$ [**69**, Example 3.5.2] are braided equivalent where B is the regular algebra of $\mathcal{C}(\mathfrak{sl}_3, 6)_{\text{pt}}$. This equivalence can also be seen via the conformal embedding of vertex operator algebras $A_{2,6} \times A_{1,16} \subset E_{8,1}$ which implies the Witt group relation

$$[\mathcal{C}(\mathfrak{sl}_3, 6)][\mathcal{C}(\mathfrak{sl}_2, 16)] = [\text{Vec}] \tag{20}$$

as noted in [**69**, Theorem 4]. But then we have $\mathcal{C}(\mathfrak{sl}_2, 16)$, $\mathcal{C}(\mathfrak{sl}_2, 16)^0_A$, and thus $\mathcal{C}(\mathfrak{sl}_3, 6)^0_B$ are self-dual, while $\mathcal{C}(\mathfrak{sl}_3, 6)$ is *not*. Conversely (see Example 23), $\mathcal{C}(\mathfrak{sl}_2, 6)$ is self-dual, while $\mathcal{C}(\mathfrak{sl}_2, 6)^0_A$ where A is the quantum subgroup of type D_4 is not. Computing tensor autoequivalences of dyslectic module categories $\mathcal{C}(\mathfrak{g}, k)^0_A$ is an open area of research, and one would hope to find some relationship that connects them to the symmetries discussed in Section 7.1.

References

[1] Henning Haahr Andersen, *Tensor products of quantized tilting modules*, Comm. Math. Phys. **149** (1992), no. 1, 149–159. MR1182414

[2] Henning Haahr Andersen and Jan Paradowski, *Fusion categories arising from semisimple Lie algebras*, Comm. Math. Phys. **169** (1995), no. 3, 563–588. MR1328736

[3] Greg Anderson and Greg Moore, *Rationality in conformal field theory*, Comm. Math. Phys. **117** (1988), no. 3, 441–450. MR953832

[4] N. Andruskiewitsch, I. Angiono, A. García Iglesias, B. Torrecillas, and C. Vay, *From Hopf algebras to tensor categories*, Conformal field theories and tensor categories, Math. Lect. Peking Univ., Springer, Heidelberg, 2014, pp. 1–31. MR3585364

[5] J.-P. Antoine and D. Speiser, *Characters of irreducible representations of the simple groups. I. General theory*, J. Mathematical Phys. **5** (1964), 1226–1234, DOI 10.1063/1.1704230. MR168699

[6] Bojko Bakalov and Alexander Kirillov Jr., *Lectures on tensor categories and modular functors*, University Lecture Series, vol. 21, American Mathematical Society, Providence, RI, 2001. MR1797619

[7] John W. Barrett and Bruce W. Westbury, *Spherical categories*, Adv. Math. **143** (1999), no. 2, 357–375, DOI 10.1006/aima.1998.1800. MR1686423

[8] Thomas Booker and Alexei Davydov, *Commutative algebras in Fibonacci categories*, J. Algebra **355** (2012), 176–204, DOI 10.1016/j.jalgebra.2011.12.029. MR2889539

[9] Nicolas Bourbaki, *Groupes et algèbres de Lie: Chapitres 4, 5 et 6*, Éléments de mathématique, Springer Berlin Heidelberg, 2007.

[10] Alain Bruguières. Catègories prémodulaires, modularisations et invariants des variétés de dimension 3. *Mathematische Annalen*, 316, 2000.

[11] Paul Bruillard, Siu-Hung Ng, Eric C. Rowell, and Zhenghan Wang, *On classification of modular categories by rank*, Int. Math. Res. Not. IMRN **24** (2016), 7546–7588, DOI 10.1093/imrn/rnw020. MR3632091

[12] Daniel Bump, *Lie groups*, 2nd ed., Graduate Texts in Mathematics, vol. 225, Springer, New York, 2013. MR3136522

[13] Frank Calegari, Scott Morrison, and Noah Snyder, *Cyclotomic integers, fusion categories, and subfactors*, Comm. Math. Phys. **303** (2011), no. 3, 845–896, DOI 10.1007/s00220-010-1136-2. MR2786219

[14] R. Coquereaux, R. Rais, and E. H. Tahri, *Exceptional quantum subgroups for the rank two Lie algebras B_2 and G_2*, J. Math. Phys. **51** (2010), no. 9, 092302, 34, DOI 10.1063/1.3476319. MR2742806

[15] Robert Coquereaux and Gil Schieber, *Quantum symmetries for exceptional* SU(4) *modular invariants associated with conformal embeddings*, SIGMA Symmetry Integrability Geom. Methods Appl. **5** (2009), Paper 044, 31, DOI 10.3842/SIGMA.2009.044. MR2506168

[16] A. Coste and T. Gannon, *Remarks on Galois symmetry in rational conformal field theories*, Phys. Lett. B **323** (1994), no. 3-4, 316–321, DOI 10.1016/0370-2693(94)91226-2. MR1266785

[17] Alexei Davydov, Pavel Etingof, and Dmitri Nikshych, *Autoequivalences of tensor categories attached to quantum groups at roots of 1*, Lie groups, geometry, and representation theory, Progr. Math., vol. 326, Birkhäuser/Springer, Cham, 2018, pp. 109–136. MR3890207

[18] Alexei Davydov, Michael Müger, Dmitri Nikshych, and Victor Ostrik, *The Witt group of non-degenerate braided fusion categories*, J. Reine Angew. Math. **677** (2013), 135–177. MR3039775

[19] Alexei Davydov, Dmitri Nikshych, and Victor Ostrik, *On the structure of the Witt group of braided fusion categories*, Selecta Math. (N.S.) **19** (2013), no. 1, 237–269, DOI 10.1007/s00029-012-0093-3. MR3022755

[20] Jan de Boer and Jacob Goeree, *Markov traces and* II_1 *factors in conformal field theory*, Comm. Math. Phys. **139** (1991), no. 2, 267–304. MR1120140

[21] V. G. Drinfel′d, *Quantum groups*, Proceedings of the International Congress of Mathematicians, Vol. 1, 2 (Berkeley, Calif., 1986), Amer. Math. Soc., Providence, RI, 1987, pp. 798–820. MR934283

[22] Pavel Etingof, Shlomo Gelaki, Dmitri Nikshych, and Victor Ostrik, *Tensor categories*, Mathematical Surveys and Monographs, vol. 205, American Mathematical Society, Providence, RI, 2015. MR3242743

[23] Pavel Etingof, Dmitri Nikshych, and Victor Ostrik, *Fusion categories and homotopy theory*, Quantum Topol. **1** (2010), no. 3, 209–273, DOI 10.4171/QT/6. With an appendix by Ehud Meir. MR2677836

[24] Pavel Etingof and Olivier Schiffmann, *Lectures on quantum groups*, Lectures in Mathematical Physics, International Press, Boston, MA, 1998. MR1698405

[25] M. Finkelberg, *An equivalence of fusion categories*, Geom. Funct. Anal. **6** (1996), no. 2, 249–267, DOI 10.1007/BF02247887. MR1384612

[26] I. B. Frenkel, *Representations of affine Lie algebras, Hecke modular forms and Korteweg-de Vries type equations*, Lie algebras and related topics (New Brunswick, N.J., 1981), Lecture Notes in Math., vol. 933, Springer, Berlin-New York, 1982, pp. 71–110. MR675108

[27] I. B. Frenkel and V. G. Kac, *Basic representations of affine Lie algebras and dual resonance models*, Invent. Math. **62** (1980/81), no. 1, 23–66, DOI 10.1007/BF01391662. MR595581

[28] Jürgen Fuchs, *Simple WZW currents*, Comm. Math. Phys. **136** (1991), no. 2, 345–356. MR1096120

[29] Jürgen Fuchs and Christoph Schweigert, *Symmetries, Lie algebras and representations*, Cambridge Monographs on Mathematical Physics, Cambridge University Press, Cambridge, 1997. A graduate course for physicists. MR1473220

[30] Terry Gannon, *WZW commutants, lattices and level-one partition functions*, Nuclear Phys. B **396** (1993), no. 2-3, 708–736, DOI 10.1016/0550-3213(93)90660-G. MR1218796

[31] Terry Gannon, *The classification of affine* SU(3) *modular invariant partition functions*, Comm. Math. Phys. **161** (1994), no. 2, 233–263. MR1266482

[32] Terry Gannon, *The automorphisms of affine fusion rings*, Adv. Math. **165** (2002), no. 2, 165–193, DOI 10.1006/aima.2001.2006. MR1887583

[33] George Lusztig, *Introduction to quantum groups*, Modern Birkhäuser Classics, Birkhäuser/Springer, New York, 2010. Reprint of the 1994 edition. MR2759715

[34] Pinhas Grossman and Noah Snyder, *Quantum subgroups of the Haagerup fusion categories*, Comm. Math. Phys. **311** (2012), no. 3, 617–643, DOI 10.1007/s00220-012-1427-x. MR2909758

[35] Brian Hall, *Lie groups, Lie algebras, and representations*, 2nd ed., Graduate Texts in Mathematics, vol. 222, Springer, Cham, 2015. An elementary introduction. MR3331229

[36] G. Hochschild, *An addition to Ado's theorem*, Proc. Amer. Math. Soc. **17** (1966), 531–533, DOI 10.2307/2035206. MR194482

[37] James E. Humphreys, *Introduction to Lie algebras and representation theory*, Springer-Verlag, New York-Berlin, 1972. Graduate Texts in Mathematics, Vol. 9. MR0323842

[38] James E. Humphreys, *Reflection groups and Coxeter groups*, Cambridge Studies in Advanced Mathematics, vol. 29, Cambridge University Press, Cambridge, 1990. MR1066460

[39] Victor G. Kac, *Infinite-dimensional Lie algebras*, 2nd ed., Cambridge University Press, Cambridge, 1985. MR823672

[40] Victor G. Kac and Dale H. Peterson, *Infinite-dimensional Lie algebras, theta functions and modular forms*, Adv. in Math. **53** (1984), no. 2, 125–264, DOI 10.1016/0001-8708(84)90032-X. MR750341

[41] Christian Kassel, *Quantum groups*, Graduate Texts in Mathematics, vol. 155, Springer-Verlag, New York, 1995. MR1321145

[42] D. Kazhdan and G. Lusztig, *Tensor structures arising from affine Lie algebras. I, II*, J. Amer. Math. Soc. **6** (1993), no. 4, 905–947, 949–1011, DOI 10.2307/2152745. MR1186962

[43] D. Kazhdan and G. Lusztig, *Tensor structures arising from affine Lie algebras. III*, J. Amer. Math. Soc. **7** (1994), no. 2, 335–381, DOI 10.2307/2152762. MR1239506

[44] David Kazhdan and Hans Wenzl, *Reconstructing monoidal categories*, I. M. Gel′fand Seminar, Adv. Soviet Math., vol. 16, Amer. Math. Soc., Providence, RI, 1993, pp. 111–136. MR1237835

[45] Alexander Kirillov Jr. and Viktor Ostrik, *On a q-analogue of the McKay correspondence and the ADE classification of* $\mathfrak{sl}_2$ *conformal field theories*, Adv. Math. **171** (2002), no. 2, 183–227, DOI 10.1006/aima.2002.2072. MR1936496

[46] Bertram Kostant, *A formula for the multiplicity of a weight*, Proc. Nat. Acad. Sci. U.S.A. **44** (1958), 588–589, DOI 10.1073/pnas.44.6.588. MR0099387

[47] Jesse Liptrap, *From hypergroups to anyonic twines*, ProQuest LLC, Ann Arbor, MI, 2010. Thesis (Ph.D.)–University of California, Santa Barbara. MR2941611

[48] George Lusztig, *Finite-dimensional Hopf algebras arising from quantized universal enveloping algebra*, J. Amer. Math. Soc. **3** (1990), no. 1, 257–296, DOI 10.2307/1990988. MR1013053

[49] George Lusztig, *Quantum groups at roots of* 1, Geom. Dedicata **35** (1990), no. 1-3, 89–113, DOI 10.1007/BF00147341. MR1066560

[50] Shahn Majid, *Foundations of quantum group theory*, Cambridge University Press, Cambridge, 1995. MR1381692

[51] Ian Marshall and Dmitri Nikshych, *On the Brauer-Picard groups of fusion categories*, Math. Z. **288** (2018), no. 3-4, 689–712, DOI 10.1007/s00209-017-1907-y. MR3778972

[52] Michael Mignard and Peter Schauenburg. Modular categories are not determined by their modular data. *ArXiv e-prints*, August 2017.

[53] Scott Morrison, Emily Peters, and Noah Snyder, *Knot polynomial identities and quantum group coincidences*, Quantum Topol. **2** (2011), no. 2, 101–156, DOI 10.4171/QT/16. MR2783128

[54] Michael Müger, *On the structure of modular categories*, Proc. London Math. Soc. (3) **87** (2003), no. 2, 291–308, DOI 10.1112/S0024611503014187. MR1990929

[55] Sergey Neshveyev and Lars Tuset, *Autoequivalences of the tensor category of* $U_q\mathfrak{g}$*-modules*, Int. Math. Res. Not. IMRN **15** (2012), 3498–3508, DOI 10.1093/imrn/rnr151. MR2959039

[56] Adrian Ocneanu, *The classification of subgroups of quantum* SU(N), Quantum symmetries in theoretical physics and mathematics (Bariloche, 2000), Contemp. Math., vol. 294, Amer. Math. Soc., Providence, RI, 2002, pp. 133–159, DOI 10.1090/conm/294/04972. MR1907188

[57] Viktor Ostrik, *Fusion categories of rank 2*, Math. Res. Lett. **10** (2003), no. 2-3, 177–183, DOI 10.4310/MRL.2003.v10.n2.a5. MR1981895

[58] Viktor Ostrik, *Module categories over the Drinfeld double of a finite group*, Int. Math. Res. Not. **27** (2003), 1507–1520, DOI 10.1155/S1073792803205079. MR1976233

[59] Victor Ostrik, *Module categories, weak Hopf algebras and modular invariants*, Transform. Groups **8** (2003), no. 2, 177–206, DOI 10.1007/s00031-003-0515-6. MR1976459

[60] Victor Ostrik and Michael Sun, *Level-rank duality via tensor categories*, Comm. Math. Phys. **326** (2014), no. 1, 49–61, DOI 10.1007/s00220-013-1869-9. MR3162483

[61] Bodo Pareigis, *On braiding and dyslexia*, J. Algebra **171** (1995), no. 2, 413–425, DOI 10.1006/jabr.1995.1019. MR1315904

[62] Pavel Etingof, Dmitri Nikshych, and Viktor Ostrik, *On fusion categories*, Ann. of Math. (2) **162** (2005), no. 2, 581–642, DOI 10.4007/annals.2005.162.581. MR2183279

[63] Eric C. Rowell, *Unitarizability of premodular categories*, J. Pure Appl. Algebra **212** (2008), no. 8, 1878–1887, DOI 10.1016/j.jpaa.2007.11.004. MR2414692

[64] Eric Rowell, Richard Stong, and Zhenghan Wang, *On classification of modular tensor categories*, Comm. Math. Phys. **292** (2009), no. 2, 343–389, DOI 10.1007/s00220-009-0908-z. MR2544735

[65] Eric C. Rowell, *From quantum groups to unitary modular tensor categories*, Representations of algebraic groups, quantum groups, and Lie algebras, Contemp. Math., vol. 413, Amer. Math. Soc., Providence, RI, 2006, pp. 215–230, DOI 10.1090/conm/413/07848. MR2263097

[66] Stephen F. Sawin, *Jones-Witten invariants for nonsimply connected Lie groups and the geometry of the Weyl alcove*, Adv. Math. **165** (2002), no. 1, 1–34, DOI 10.1006/aima.1999.1910. MR1880320

[67] Stephen F. Sawin, *Closed subsets of the Weyl alcove and TQFTs*, Pacific J. Math. **228** (2006), no. 2, 305–324, DOI 10.2140/pjm.2006.228.305. MR2274523

[68] Stephen F. Sawin, *Quantum groups at roots of unity and modularity*, J. Knot Theory Ramifications **15** (2006), no. 10, 1245–1277, DOI 10.1142/S0218216506005160. MR2286123

[69] Andrew Schopieray, *Classification of $\mathfrak{sl}_3$ relations in the Witt group of nondegenerate braided fusion categories*, Comm. Math. Phys. **353** (2017), no. 3, 1103–1127, DOI 10.1007/s00220-017-2831-z. MR3652484

[70] Andrew Schopieray, *Level bounds for exceptional quantum subgroups in rank two*, Internat. J. Math. **29** (2018), no. 5, 1850034, 33, DOI 10.1142/S0129167X18500349. MR3808050

[71] Jan Tarski, *Partition function for certain simple Lie algebras*, J. Mathematical Phys. **4** (1963), 569–574, DOI 10.1063/1.1703992. MR147584

[72] Imre Tuba and Hans Wenzl, *On braided tensor categories of type BCD*, J. Reine Angew. Math. **581** (2005), 31–69, DOI 10.1515/crll.2005.2005.581.31. MR2132671

[73] V. Turaev and H. Wenzl, *Quantum invariants of 3-manifolds associated with classical simple Lie algebras*, Internat. J. Math. **4** (1993), no. 2, 323–358, DOI 10.1142/S0129167X93000170. MR1217386

[74] Vladimir Turaev and Hans Wenzl, *Semisimple and modular categories from link invariants*, Math. Ann. **309** (1997), no. 3, 411–461, DOI 10.1007/s002080050120. MR1474200

[75] Cumrun Vafa, *Toward classification of conformal theories*, Phys. Lett. B **206** (1988), no. 3, 421–426, DOI 10.1016/0370-2693(88)91603-6. MR944264

School of Mathematics and Statistics, University of New South Wales
Email address: szopraj@gmail.com

Contemporary Mathematics
Volume **747**, 2020
https://doi.org/10.1090/conm/747/15037

Entanglement and the Temperley-Lieb category

Michael Brannan and Benoît Collins

Abstract. We survey some recent results from [**BrCo16b**], where a class of highly entangled subspaces of bipartite quantum systems is described, which arises from unitary fiber functors on the Temperley-Lieb category associated to the representation theory of free orthogonal quantum groups. By exploiting the rich structure of the Temperley-Lieb category and this particular fiber functor, we are able to precisely determine the largest singular values for these subspaces and obtain lower bounds for the minimum output entropy of the corresponding quantum channels. Future research directions and some open problems are also discussed.

1. Introduction

Entanglement is a fundamental notion in quantum mechanics that does not have an analogue in the classical world. Within the framework of quantum computation and quantum information, entanglement in bipartite or multipartite systems produces, on the one hand, many counterintuitive phenomena, while on the other hand, it can be used to design new communications protocols which admit no classical analogues [**Eis06, Gro96, CLSZ95, EJ96**].

Throughout this paper we will focus on entanglement in bipartite quantum systems. Within the formalism of quantum mechanics, a quantum mechanical system is described by a complex Hilbert space H. The (pure) states of the system are described by unit norm vectors $\xi \in H$, taken up to a complex phase factor. (In this paper, all Hilbert spaces are taken to be finite-dimensional, unless otherwise specified.) Equivalenty, a pure state of the system can be described by the rank one projector $\rho = |\xi\rangle\langle\xi|$ onto the subspace $\mathbb{C}\xi \subset H$. The (closed) convex hull of pure states (viewed as rank-one projectors on H) is denoted by $\mathcal{D}(H)$, and elements $\rho \in \mathcal{D}(H)$ are called *mixed states*. If we fix a basis of H and identify $\mathcal{B}(H) \cong M_n(\mathbb{C})$ ($n = \dim H$), then the convex set $\mathcal{D}(H)$ is nothing more than the collection of all trace-one positive semidefinite matrices, and the extreme points of $\mathcal{D}(H)$ are the precisely the rank one projectors, i.e., pure states on H.

In the quantum context, one often has to deal with *bipartite systems* AB, built from subsystems A, B. Mathematically, such a bipartitie system is modeled by the

2010 *Mathematics Subject Classification.* Primary 46N50, 47L90, 81P45, 20G42.

The first author's research was supported by NSF grant DMS-1700267.

The second author's research was supported by NSERC discovery and accelerator grants, ANR-14-CE25-0003, JSPS Challenging Research grant 17K18734, JSPS Fund for the Promotion of Joint International Research grant 15KK0162, and JSPS wakate A grant 17H04823.

©2020 American Mathematical Society

tensor product Hilbert space $H = H_A \otimes H_B$, where the Hilbert spaces H_A and H_B describe the states of systems A and B, respectively. Given such a bipartite system modeled on $H = H_A \otimes H_B$, a mixed state $\rho \in \mathcal{D}(H)$ is said to be *separable* if it belongs to the convex hull of the set of product states $\rho = \rho_A \otimes \rho_B$, where $\rho_A \in \mathcal{D}(H_A)$ and $\rho_B \in \mathcal{D}(H_B)$. A state ρ is called *entangled* if it is not separable. We shall call a Hilbert subspace $H_0 \subset H_A \otimes H_B$ an *entangled subspace* if all of its associated pure states are entangled. In this paper, we are concerned with studying non-trivial examples of highly-entangled subspaces $H_0 \subset H = H_A \otimes H_B$. By highly-entangled, we shall mean that the set of pure states on H associated to the subspace H_0 are uniformly "far away" from the set of product states $\rho_A \otimes \rho_B \in \mathcal{D}(H)$ with respect to some suitable measure of distance. The choice of "distance" here is not unique, and our choice is based on the largest singular value of pure states - precise details will be given in the next section.

What are some explicit examples of highly entangled subspaces? Well, one trivial example that we always have access to is the one-dimensional subspace $H_0 = \mathbb{C}\xi \subset H_A \otimes H_B$ spanned by a maximally entangled (Bell) state ξ. That is, a state ξ of the form $\xi = d^{-1/2} \sum_{i=1}^d e_i \otimes f_i$, where $d = \min\{\dim H_A, \dim H_B\}$ and $(e_i)_i \subset H_A$, $(f_i)_i \subset H_B$ are orthonormal systems. Another important example that has been well-studied in the literature (see for example [**GHP10**]) is the anti-symmetric subspsace $H \wedge H \subset H \otimes H$ associated to the tensor square of any Hilbert space H. As one might expect, as the relative dimension of the subspace $H_0 \subseteq H$ grows, the more challenging it becomes to find explicit examples of (highly) entangled subspaces. However, in recent years it has become a very important problem in Quantum Information Theory (QIT) and Quantum Computing (QC) to develop means to *construct subspaces H_0 of large relative dimension in a tensor product $H = H_A \otimes H_B$ such that all states are highly entangled.* One major reason for the importance of this problem is for applications to entanglement-assisted communication protocols [**NC00, LB13**], and also to the construction of counterexamples to additivity questions related to capacities of quantum channels [**Has09**].

As is the case for many mathematical problems, the question of the *existence* of a rich supply of highly entangled subspaces with large relative dimension can be settled using probabilistic techniques. The idea of studying random subspaces of tensor products dates back to the work of Hayden, Leung, Shor, Winter, Hastings [**HLSW04, HW08, HLW06, Has09**], among others, and it was explored in great detail by Aubrun, Belinschi, Collins, Fukuda, King, Nechita, Szarek, Werner [**ASW11, ASY14, BCN12, FK10**], and others. The whole theory of random subspaces of tensor products is intimately connected with random matrix theory and free probability theory (see [**CN16**] for a good survey on this), and has led to many fruitful interactions between these communities. The general outcome of these works was the conclusion that (at least in certain asymptotic dimension regimes) highly entangled subspaces of large relative dimension are ubiquitous: with high probability, a randomly selected subspace of a tensor product will be highly entangled. These random constructions have had a profound impact on the field, solving several open problems, most notably the minimum output entropy additivity problem [**Has09, ASW11, BCN16**]. The downside to these highly random techniques is that they provide no information on finding concrete examples that are predicted to exist by these methods. In fact, there seems to be embarrassingly few known examples of such subspaces (beyond the ones already mentioned above). Thus, there is a need for a systematic development of non-random examples of highly entangled subspaces.

The purpose of this survey is to promote our belief that one natural and fruitful place to search for *deterministic* examples of highly entangled subspaces is within the framework of *representation theory.* For example, if we are given a (compact) group G and a pair of (irreducible) unitary representations H_π, H_σ of G, then we can form their tensor product representation $H_\pi \otimes H_\sigma$, and attempt to quantify the entanglement of the irreducible subrepresentations $H_\nu \subset H_\pi \otimes H_\sigma$ that arise in the decomposition of $H_\pi \otimes H_\sigma$ into irreducibles. In this way, we have a natural playground of examples of subspaces of Hilbert space tensor products, and we shall see that understanding the entanglement (=relative position) of these subspaces is essentially equivalent to understanding the representation theory of G.

Our idea here is of course not new. For example, a first attempt was made in this direction by M. Al Nuwairan [**AN13, AN14**], by studying the entanglement of subrepresentations of tensor products of irreducible representations of the group $SU(2)$. Here, Al Nuwairan showed that entanglement is always achieved for subrepresentations of tensor products of $SU(2)$-irreducibles (except when one takes the highest weight subrepresentation). However, as is evidenced by the results in [**AN13**, Section 3], a high degree of entanglement is unfortunately not achieved when working with $SU(2)$.

In order to use representation theory to obtain examples of entangled subspaces exhibiting a higher level of entanglement, there are two natural approaches. The first approach would be to consider more complicated examples of compact groups and their representations. The significant downside of this approach is that for most examples of groups G, one lacks the complete understanding of the representation category Rep(G) that one has for $SU(2)$. The second approach, which we follow in this paper, is to instead consider "q-deformations" of the representation category $\mathrm{Rep}(SU(2))$ arising from certain quantum group constructions.

Perhaps the most well-known examples of such deformations are the canonical realizations of (i.e., unitary fiber functors on) the Temperley-Lieb Categories that are associated to the Drinfeld-Jimbo-Woronowicz q-deformations of $SU(2)$ [**Dri87, Jim85, Wor88, Wor87**]. In this paper, we consider a very different fiber functor on the Temperley-Lieb categories which act on higher dimensional spaces, and come from another class of quantum groups (more closely linked with operator algebra theory and free probability theory), called *free orthogonal quantum groups.*

Given an integer $N \geq 2$, the free orthogonal quatum group O_N^+ is the (compact) quantum group whose Hopf $*$-algebra of polynomial functions $\mathcal{O}(O_N^+)$ is given as a certain natural non-commutative (or free) version of the algebra of polynomial functions on the classical $N \times N$ orthogonal matrix group O_N. In the context of C*-algebraic compact quantum groups, O_N^+ was first introduced and studied by Wang [**Wan95**]. Shortly after Wang's original paper was published, Banica [**Ban96**] studied the representation category $\mathrm{Rep}(O_N^+)$ and showed that there is a natural unitary fiber functor on the Temperley-Lieb Category $\mathrm{TL}(N)$, which concretely realizes the representation category $\mathrm{Rep}(O_N^+)$. In Banica's fiber functor, the generating object of $\mathrm{TL}(N)$ is given by the N-dimensional fundamental representation space $\mathbb{C}^N$ (in contrast to $\mathbb{C}^2$ associated to the usual q-deformation of $SU(2)$). It is the entanglement phenomena associated to this "higher dimensional" unitary fiber functor on the Temperley-Lieb category that we study here.

Our motivation to study entanglement in the context of $\mathrm{Rep}(O_N^+) \cong \mathrm{TL}(N)$ comes from the pioneering work of Vergnioux [**Ver07**] on the seemingly unrelated *property of rapid decay (propety RD)* for quantum groups. The property of rapid decay is a geometric-analytic property posessed by certain (quantum) groups and corresponds the existence of polynomial bounds relating non-commutative L^∞-norms of polynomial functions on quantum groups to their (much easier to calculate) L^2-norms. The operator algebraic notion of property RD has its origins in the groundbreaking work of Haagerup [**Haa79**] on approximation properties of free group C*-algebras. Unlike in the case of ordinary groups, where property RD is connected to the combinatorial geometry of a discrete group G, in the quantum world, property RD was observed by Vergnioux to be intrinsically connected to the geometry of the relative position of a subrepresentation of a tensor product of irreducible representations of a given quantum group. More precisely, Vergnioux [**Ver07**, Section 4] points out that property RD for a given quantum group $\mathbb{G}$ is related to the following geometric requirement: *Given any pair of irreducible representations H_A, H_B of $\mathbb{G}$, all multiplicity-free irreducible subrepresentations $H_0 \subset H_A \otimes H_B$ must be asymptotically far from the cone of decomposable tensors in $H_A \otimes H_B$.*

The work [**BC18b**] that we survey here is largely an in-depth exploration of this passing remark of Vergnioux [**Ver07**], and our goal is to show how a rather modest understanding of the structure of the Temperley-Lieb category can be extremely fruitful when analyzing the entanglement problem for $\mathrm{Rep}(O_N^+)$. In this context, we show that one can describe very precisely the largest singular values of states that appear in irreducible subrepresentations of tensor product representations (see Theorem 3.3). This construction produces a new non-random class of subspaces of tensor products with the property of being highly entangled and of large relative dimension. As easy applications of these entanglement results, we establish some interesting properties for a class of quantum channels associated to these subspaces. We compute explicitly the $\mathcal{S}_1 \to \mathcal{S}_\infty$ norms of these channels, and obtain lower bounds on their minimum output entropies (see Section 4). We also show in Section 5 how one can use "planar diagrammatic" arguments to study further properties of our quantum channels, including their entanglement breaking property, and constructing positive maps on matrix algebras which are not completely positive. It is our hope that this survey will inspire others to view quantum groups/symmetries and their associated tensor categories as a new, rich source of entangled subspaces with interesting geometric properties.

Acknowledgements. The authors are indebted to Vern Paulsen and the referees for helpful comments.

2. Preliminaries

We refer to [**NC00**] for the basics on entangled subspaces and quantum channels. For the reader who is interested in learning more about quantum groups and their representation categories, see [**NT13, Tim08**]. In the following, we use standard notations and conventions from operator algebra theory: Hilbert spaces are typically denoted by the letter H (possibly with some additional subscripts), and are always assumed to be finite-dimensional. The inner product on H is always linear in the right variable, $\mathcal{B}(H)$ denotes the unital $*$-algebra of (automatically bounded) linear operators on H, and $\mathcal{D}(H) \subseteq \mathcal{B}(H)$ denotes the collection of *mixed states on H*. That is, $\rho \in \mathcal{D}(H)$ iff ρ is positive semidefinite (written $\rho \geq 0$) and has unit trace, $\mathrm{Tr}(\rho) = 1$, where Tr denotes the canonical trace on $\mathcal{B}(H)$ satisfying $\mathrm{Tr}(1) = \dim H$. A state $\rho \in \mathcal{D}(H)$ is called a *pure state* if there exists a unit vector $\xi \in H$ so that ρ is given by the rank-one projector $\rho_\xi = |\xi\rangle\langle\xi|$. We denote by $\mathcal{S}_1(H)$ the Banach algebra $\mathcal{B}(H)$, equipped with the *trace norm* $\|\rho\|_{\mathcal{S}_1(H)} = \mathrm{Tr}(|\rho|)$, where $|\rho| := (\rho^*\rho)^{1/2}$. At times, we will also write $\mathcal{S}_\infty(H)$ for the space $\mathcal{B}(H)$ equipped with the operator norm $x \mapsto \|x\|_\infty = \sup_{0 \neq \xi \in H} \frac{\|x\xi\|}{\|\xi\|}$.

2.1. Entangled vectors and subspaces. Consider a pair of finite-dimensional complex Hilbert spaces H_A and H_B with $\dim H_A, \dim H_B \geq 2$. Our first goal is to define what it means for a unit vector $\xi \in H_A \otimes H_B$ to be entangled, and how one can quantify the amount of entanglement that ξ has. These goals are best achieved using the *singular value decomposition (SVD)*. Namely, any unit vector $\xi \in H_A \otimes H_B$ admits a representation of the following form:

$$\xi = \sum_{i=1}^{d} \sqrt{\lambda_i} e_i \otimes f_i,$$

where $(e_i)_{i=1}^d \subset H_A$ and $(f_i)_{i=1}^d \subset H_B$ are orthonormal systems, $d = \min\{\dim H_A, \dim H_B\}$, and $\lambda_1 \geq \lambda_2 \geq \ldots \lambda_d \geq 0$ satisfy $\sum_{i=1}^d \lambda_i = \|\xi\|^2 = 1$. Although the orthonormal systems defining the singular value decomposition of ξ above are never unique, it turns out the sequence of numbers $(\lambda_i)_i$ is uniquely determined (as a multi-set) by the vector ξ and these numbers are called the *singular values (or Schmidt coefficients)* of ξ.

Thus, to any unit vector $\xi \in H_A \otimes H_B$, we have an essentially unique probability distribution $(\lambda_i)_i$ corresponding to its string of singular values. With this in mind, we shall call $\xi \in H_A \otimes H_B$ *separable* if the corresponding distribution is *deterministic*: $(\lambda_i)_i = (1, 0, 0, \ldots, 0)$. Otherwise, we shall call ξ *entangled*. It is easy to see that $\xi \in H_A \otimes H_B$ is entangled if and only if it cannot be expressed as a simple tensor $\xi = \eta \otimes \zeta$ for $\eta \in H_A$, $\zeta \in H_B$. The most fundamental example of an entangled vector is a *Bell vector* (or maximally entangled state), which is a unit vector $\xi_{\mathrm{Bell}} \in H_A \otimes H_B$ with SVD

$$\xi_{\mathrm{Bell}} = \frac{1}{\sqrt{d}} \sum_{i=1}^{d} e_i \otimes f_i.$$

In this case, the singular values of ξ_{Bell} are given by the uniform distribution $\left(\frac{1}{d}, \frac{1}{d}, \ldots, \frac{1}{d}\right)$, which is in many senses "far" from the detereministic distribution $(1, 0, 0, \ldots, 0)$ associated to a separable vector. In fact, the above remark leads to a rigorous measurement of how entangled a unit vector $\xi \in H_a \otimes H_B$ is, namely the so-called *entanglement entropy*:

$$H(\xi) = -\sum_{i=1}^{d} \lambda_i \log \lambda_i,$$

where $(\lambda_i)_{i=1}^d$ are the singular values of ξ. In other words, $H(\xi)$ is simply the Shannon entropy of the string of singular values associated to ξ, and it follows from the basic properties of the Shannon entropy functional that $0 \leq H(\xi) \leq \log d$, and $H(\xi)$ is minimized (resp. maximized) if and only if ξ is separable (resp. maximally entangled). In general, $H(\xi)$ gives a measure of how entangled ξ is. Another (generally less precise) measure of the entanglement of a unit vector $\xi \in H_a \otimes H_B$

that we shall primarily use in this paper is given by the size of the largest singular value λ_1. Note that geometrically, λ_1 can be interpreted as the square of the cosine of the smallest angle between ξ and any product vector $\eta \otimes \zeta$. In particular, we always have

$$\frac{1}{d} \leq \lambda_1 = \sup_{\|\eta\|_{H_A} = \|\zeta\|_{H_B} = 1} |\langle \xi | \eta \otimes \zeta \rangle|^2 \leq 1,$$

with ξ being entangled (resp. maximally entangled) if and only if $\lambda_1 < 1$ (resp. $\lambda_1 = \frac{1}{d}$).

Suppose now that we are given a linear subspace $H_0 \subseteq H_A \otimes H_B$. We will call H_0 a *separable subspace* (resp. *entangled subspace*) if H_0 contains (resp. does not contain) separable vectors. Using our measure of entanglement coming from the largest singular values for unit vectors in H_0, we can call $H_0 \subseteq H_A \otimes H_B$ *highly entangled* if the supremum of all maximal Schmidt coefficients associated to all unit vectors in H_0 is bounded away from one. That is, the quantity

$$\lambda_1(H_0) := \sup_{\xi \in H_0,\ \|\xi\| = 1} \lambda_1(\xi) = \sup_{\|\xi\|_{H_0} = \|\eta\|_{H_A} = \|\zeta\|_{H_B} = 1} |\langle \xi | \eta \otimes \zeta \rangle|^2 << 1.$$

Of course, in general we have $\frac{1}{d} \leq \lambda_1(H_0) \leq 1$ with $d = \min\{\dim H_a, \dim H_B\}$.

REMARK 1. For the sake of comparison, we note here that current random techniques prove the existence of highly entangled subspaces $H_0 \subseteq H_A \otimes H_B$ with $\lambda_1(H_0) \sim t$ for any $t \in (0,1)$, where $t = \frac{\dim H_0}{\dim(H_A \otimes H_B)}$ in certain asymptotically large dimensional regimes. See [**BCN12, BCN16**] and the references therein. On the other hand, in Section 3 we shall exhibit *deterministic examples* of $H_0 \subseteq H_A \otimes H_B$ with $\lambda_1(H_0) \sim \sqrt{t}$ where $t = \frac{\dim H_0}{\dim(H_A \otimes H_B)} \in (0,1)$.

2.2. Quantum channels. Given two Hilbert spaces H_A and H_B, a *quantum channel* is a mathematical model that describes the transmission of mixed states on H_A to mixed states on H_B. More precisely, a quantum channel is a linear, completely positive and trace-preserving map (CPTP map) $\Phi : \mathcal{B}(H_A) \to \mathcal{B}(H_B)$ [**NC00**]. By definition, we have $\Phi(\mathcal{D}(H_A)) \subseteq \mathcal{D}(H_B)$ for any quantum channel Φ. The perspective on quantum channels we take here is that they are intimately connected to the geometry subspaces of Hilbert space tensor products. Indeed, suppose we are given a triple of finite dimensional Hilbert spaces (H_A, H_B, H_C) and an isometric linear map $\alpha_A^{B,C} : H_A \to H_B \otimes H_C$, we can then form a pair of quantum channels

$$\Phi_A^{\overline{B},C} : \mathcal{B}(H_A) \to \mathcal{B}(H_C); \quad \Phi_A^{\overline{B},C}(\rho) = (\mathrm{Tr}_{H_B} \otimes \iota)(\alpha_A^{B,C} \rho (\alpha_A^{B,C})^*)$$
$$\Phi_A^{B,\overline{C}} : \mathcal{B}(H_A) \to \mathcal{B}(H_B); \quad \Phi_A^{B,\overline{C}}(\rho) = (\iota \otimes \mathrm{Tr}_{H_C})(\alpha_A^{B,C} \rho (\alpha_A^{B,C})^*).$$

In other words, associated to the subspace $\alpha(H_A) \subseteq H_B \otimes H_C$, we have two channels $\Phi_A^{\overline{B},C}$ and $\Phi_A^{B,\overline{C}}$. Note that in the literature the channel $\Phi_A^{B,\overline{C}}$ is called the *complement* of $\Phi_A^{\overline{B},C}$.

It is a remarkable fact that every quantum channel in fact arises from the above construction. This fact is a special case of the Stinespring dilation theorem for completely positive maps (see [**HW08**]). Stinespring's theorem basically says that if we are given any quantum channel $\Phi : \mathcal{B}(H_A) \to \mathcal{B}(H_B)$, then there exists an essentially unique *Stinespring pair* $(H_C, \alpha_A^{B,C})$, where H_C is an auxiliary *environment* Hilbert space $\alpha_A^{B,C} : H_A \to H_B \otimes H_C$ is a linear isometry, and $\Phi = \Phi_A^{B,\overline{C}}$ in the above notation.

The main quantity associated to quantum channel $\Phi : \mathcal{B}(H_A) \to \mathcal{B}(H_B)$ that we will be interested in this work is called the *minimum output entropy (MOE)*, $H_{\min}(\Phi)$, which is defined by

$$H_{\min}(\Phi) = \min_{\xi \in H_A,\ \|\xi\|=1} H(\Phi(|\xi\rangle\langle\xi|)).$$

where $H(\cdot)$ denotes the *von Neumann entropy* of a state: $H(\rho) = -\mathrm{Tr}(\rho \log \rho)$. Note that by functional calculus, we have $H(\rho) = -\sum_i \lambda_i \log \lambda_i$, where $(\lambda_i)_i \subset [0,\infty)$ denotes the spectrum of ρ. In other words, $H(\rho)$ is nothing but the Shannon entropy of the probability vector $(\lambda_i)_i$ corresponding to the eigenvalues of ρ.

To get a better handle on what exactly $H_{\min}(\Phi)$ is, let us suppose that $\Phi = \Phi_A^{B,\overline{C}} = (\iota \otimes \mathrm{Tr}_{H_C})(\alpha_A^{B,C}(\cdot)(\alpha_A^{B,C})^*)$ is a Stinespring representation for Φ, where $\alpha_A^{B,C} : H_A \to H_B \otimes H_C$ is our Stinespring isometry. We claim that the MOE $H_{\min}(\Phi)$ only depends on the geometry of the set of singular values associated to unit vectors in the subspace $\alpha_A^{B,C}(H_A) \subseteq H_B \otimes H_C$. Indeed, if we fix a unit vector $\xi \in H_A$ and write down the corresponding SVD $\alpha_A^{B,C}(\xi) = \sum_i \sqrt{\lambda_i} e_i \otimes f_i$, then one readily sees that

$$\Phi(|\xi\rangle\langle\xi|) = (\iota \otimes \mathrm{Tr}_{H_C})(|\alpha_A^{B,C}(\xi)\rangle\langle\alpha_A^{B,C}(\xi)|) = \sum_i \lambda_i |e_i\rangle\langle e_i|,$$

and thus

$$H(\Phi(|\xi\rangle\langle\xi|)) = -\sum_i \lambda_i \log \lambda_i = H(\alpha_A^{B,C}(\xi)).$$

In particular, computing $H_{\min}(\Phi)$ amounts to minimizing the entanglement entropy of unit vectors in the subspace $\alpha_A^{B,C}(H_A) \subseteq H_B \otimes H_C$. Namely,

$$H_{\min}(\Phi) = \min_{\xi \in H_A,\ \|\xi\|=1} H(\alpha_A^{B,C}(\xi)).$$

It follows from this calculation that $H_{\min}(\Phi)$ is zero (resp. large) if and only if $\alpha_A^{B,C}(H_A) \subseteq H_B \otimes H_C$ is a separable subspace (resp. highly entangled subspace).

2.3. Free orthogonal quantum groups, their representations, and the Temperley-Lieb Category. We now come to the main algebraic objects of study for us – quantum groups and their representation categories. In fact we will only consider one class of quantum groups here, called the free orthogonal quantum groups, and explain how their representations are connected to the Temperley-Lieb category, which is itself an amazing tensor category that is ubiquitous across many branches of mathematics (e.g., subfactors [**Jon83**], quantum computation [**Abr08**], knot theory [**Jon85**], and mathematical physics [**TL71**].)

2.3.1. *The notion of a quantum group.* The theory of quantum groups was initiated in the second half of the 20th century by several people, including Kac, Vainerman, Enock, Schwartz, Drinfeld, Jimbo, Woronowicz, Kustermans and Vaes [**ES92, Dri87, Jim85, Wor98, KV00**]. The perspective taken by each author here varies quite significantly, but the guiding principle is always the same: if we start with, say, a compact matrix group $G \subset U_N$ ($U_N \subseteq M_N(\mathbb{C})$ being the unitary group), then we can encode the entire structure of G in terms of the commutative $*$-algebra

$\mathcal{O}(G)$ of polynomial functions on G, which itself is generated as a $*$-algebra by the N^2 coordinate functions $u_{ij} : G \to \mathbb{C}$ defining the embedding $G \subseteq M_N(\mathbb{C})$. For example, the multiplication $G \times G \to G$, the inversion $G \to G$, the unit $e \in G$ are all encoded at the level of $\mathcal{O}(G)$ in terms of the following unital algebra homomorphisms

$$\Delta : \mathcal{O}(G) \to \mathcal{O}(G) \otimes \mathcal{O}(G); \qquad \Delta(u_{ij}) = \sum_{k=1}^{N} u_{ik} \otimes u_{kj} \qquad \text{(comultiplication)}$$
$$S : \mathcal{O}(G) \to \mathcal{O}(G); \qquad S(u_{ij}) = u_{ji}^* \qquad \text{(coinverse)}$$
$$\epsilon : \mathcal{O}(G) \to \mathbb{C}; \qquad \epsilon(u_{ij}) = \delta_{ij} \qquad \text{(counit).}$$

From these defintions, one can readily check that the Hopf identities $(\epsilon \otimes \iota)\Delta = \iota$, $m(S \otimes \iota)\Delta = m(\iota \otimes S)\Delta = \epsilon(\cdot)1$ are satisfied (where $m : \mathcal{O}(G) \otimes \mathcal{O}(G) \to \mathcal{O}(G)$ is the multiplication map), and what one obtains is the structure of a *commutative Hopf $*$-algebra* $(\mathcal{O}(G), \Delta, S, \epsilon)$.

Very loosely speaking, a quantum group is then given by a (possibly) *noncommutative* Hopf $*$-algebra $\mathbb{G} := (\mathcal{O}(\mathbb{G}), \Delta, S, \epsilon)$. Of course, when our $*$-algebra $\mathcal{O}(\mathbb{G})$ is not commutative, there is no longer anything like an underlying group representing $\mathcal{O}(\mathbb{G})$ as an algebra of coordinate functions. Nonetheless, if one places "reasonable" assumptions on $\mathcal{O}(\mathbb{G})$, then one obtains an algebraic/analytic structure that deserves to be considered group like (e.g., the existence of a "Haar measure", a rich finite-dimensional unitary representation theory, a Peter-Weyl theorem, and so on). See [**Tim08, Wor98, DK94**] for more details. Without going into any of the gory details, we assure the readers that the free orthogonal quantum groups defined below satisfy the reasonable assumptions alluded to above.

First some notation and terminology. Let $N \geq 2$, let A be a unital $*$-algebra over $\mathbb{C}$, and let $u = [u_{ij}]_{1 \leq i,j \leq N} \in M_N(A)$ be a matrix with entries in A. We will write $u^* = [u_{ji}^*] \in M_N(A)$ and $\bar{u} = [u_{ij}^*] \in M_N(A)$. We will call the matrix u an $N \times N$ *quantum orthogonal matrix* if u is invertible in $M_N(A)$, $u^* = u^{-1}$, and $\bar{u} = u$.

DEFINITION (Free Orthogonal Quantum Groups). The *free orthogonal quantum group* (of rank N) is given by the quadruple $O_N^+ := (\mathcal{O}(O_N^+), \Delta, S, \epsilon)$, where

(1) $\mathcal{O}(O_N^+)$ is the universal unital $*$-algebra (over $\mathbb{C}$) generated by the coefficients $(u_{ij})_{1 \leq i,j \leq N}$ of a quantum orthogonal matrix $u = [u_{ij}] \in M_N(\mathcal{O}(O_N^+))$. More precisely, $\mathcal{O}(O_N^+)$ is the universal unital $*$-algebra with generators $(u_{ij})_{1 \leq i,j \leq N}$ satisfing the relations $u_{ij} = u_{ij}^*$ and $\sum_{k=1}^N u_{ik}u_{jk} = \sum_{k=1}^N u_{ki}u_{kj} = \delta_{i,j}1$ for each $1 \leq i,j \leq N$.

(2) $\Delta : \mathcal{O}(O_N^+) \to \mathcal{O}(O_N^+) \otimes \mathcal{O}(O_N^+)$ is the unique unital $*$-algebra homomorphism, called the *coproduct*, given by
$$\Delta(u_{ij}) = \sum_{k=1}^{N} u_{ik} \otimes u_{kj} \qquad (1 \leq i,j \leq N).$$

(3) $S : \mathcal{O}(O_N^+) \to \mathcal{O}(O_N^+)$ is the $*$-antiautomorphism given by $S(u_{ij}) = u_{ji}$, $1 \leq i,j \leq N$.

(4) $\epsilon : \mathcal{O}(O_N^+) \to \mathbb{C}$ is the $*$-character given by $\epsilon(u_{ij}) = \delta_{ij}$.

REMARK 2. Of course, what we have defined above is a non-commutative Hopf $*$-algebra, which we like to interpret as a noncommutative analogue of the algebra of coordinate functions on O_N. To support this perspective, note that if we quotient $\mathcal{O}(O_N^+)$ by its commutator ideal, we obtain the abelianization of $\mathcal{O}(O_N^+)$, which is isomorphic to $\mathcal{O}(O_N)$, the Hopf $*$-algebra of polynomial functions on the real orthogonal group O_N. The map $\mathcal{O}(O_N^+) \to \mathcal{O}(O_N)$ is given by $u_{ij} \mapsto v_{ij}$, where $v = [v_{ij}] \in M_N(\mathcal{O}(O_N))$ forms the matrix of basic coordinate functions on O_N (a.k.a. the *fundamental representation* of O_N). In this context, the coproduct map Δ on $\mathcal{O}(O_N^+)$ factors through the quotient and induces the corresponding coproduct map Δ on $\mathcal{O}(O_N)$. In this sense, we are justified in calling the quantum group O_N^+ a "free analogue" of the classical orthogonal group O_N, and we can even view O_N as a "quantum subgroup" of O_N^+.

2.3.2. *Unitary representations of* O_N^+. A (finite-dimensional unitary) *representation* of O_N^+ is given by a finite dimensional Hilbert space H_v and unitary matrix $v \in \mathcal{O}(O_N^+) \otimes \mathcal{B}(H_v)$ satisfying

$$(\Delta \otimes \iota)v = v_{13}v_{23} \in \mathcal{O}(O_N^+) \otimes \mathcal{O}(O_N^+) \otimes \mathcal{B}(H_v),$$

where above we use the standard leg numbering notation for linear maps on tensor products. If we fix an orthonormal basis $(e_i)_{i=1}^d \subset H_v$, then we can write v as the matrix $[v_{ij}] \in M_d(\mathcal{O}(O_N^+))$ with respect to this basis, and the above formula translates to

$$\Delta v_{ij} = \sum_{k=1}^{d} v_{ik} \otimes v_{kj} \qquad (1 \leq i, j \leq d).$$

Observe that the above definition corresponds precisely to our usual notion of a unitary representation of a group if we were to assume that our Hopf $*$-algebra was commutative.

What are some examples of unitary representations of O_N^+? Based on the above formulas, it should be evident to the reader that we have immediate access to at least two distinct representations. The first one is the one-dimensional *trivial representation*, which we denote by $v^0 := 1 \in \mathcal{O}(O_N^+) = M_1(\mathcal{O}(O_N^+))$. The second example is the N-dimensional *fundamental representation* $v^1 := u = [u_{ij}] \in M_N(\mathcal{O}(O_N^+))$ (which is simply the matrix of generators for $\mathcal{O}(O_N^+)$).

In order to generate more examples of unitary representations we use our intuition from group theory and try to build more representations from v^0, v^1 via the operations of direct sum, tensor product, and compression to subrepresentations. Let us recall these notions. Given two representations $v = [v_{ij}]$ and $w = [w_{kl}]$, we can naturally form their *direct sum* $v \oplus w \in \mathcal{O}(O_N^+) \otimes \mathcal{B}(H_v \oplus H_w)$ and their *tensor product* $v \otimes w = v_{12}w_{13} = [v_{ij}w_{kl}] \in \mathcal{O}(O_N^+) \otimes \mathcal{B}(H_v \otimes H_w)$ to obtain new examples of representations from known ones. From a unitary representation $v = [v_{ij}]$, we may also form the *contragredient representation* $\bar{v} := [v_{ij}^*] \in \mathcal{O}(O_N^+) \otimes \mathcal{B}(\overline{H_v})$. Finally, if $p = p^2 = p^* \in \mathcal{B}(H_v)$ satisfies $(1 \otimes p)v = v(1 \otimes p)$, we can form the *subrepresentation* $v' := (1 \otimes p)v(1 \otimes p) \in \mathcal{O}(O_N^+) \otimes \mathcal{B}(pH)$ of v.

In order to fully understand the structure of the unitary representations of O_N^+, we need to study intertwiner spaces between representations. Given two representations u and v of O_N^+, we define the space of *intertwiners* between u and v as

$$\mathrm{Hom}(u, v) = \{T \in \mathcal{B}(H_u, H_v) : (\iota \otimes T)u = v(\iota \otimes T).\}$$

Two representations u, v are called *equivalent* (written $u \cong v$) if $\mathrm{Hom}(u, v)$ contains an invertible operator, and a representation u is called *irreducible* if $\mathrm{Hom}(u, u) = \mathbb{C}1$.

It is a consequence of a general fact about compact quantum groups that every unitary representation of O_N^+ is equivalent to a direct sum of irreducible unitary representations [**Wor87, Wor98**].

2.3.3. *Fusion rules for O_N^+-irreducibles.* As is the case for any (compact) quantum group, a problem of fundamental importance concerning O_N^+ following its introduction by Wang [**Wan95**] was to classify its irreducible unitary representations up to unitary equivalence. This problem was solved in the groundbreaking work of Banica [**Ban96**] where he showed that there exists a complete list of irreducible unitary representations of O_N^+, $(v^k)_{k\in\mathbb{N}_0}$, (taken up to unitary equivalence) such that $v^0 = 1$ (the trivial representation), $v^1 = u$ (the fundamental representation), each v^k is unitarily equivalent to its conjugate $\overline{v^k}$, and moreover the following fusion rules hold:

$$v^l \otimes v^m \cong v^{|l-m|} \oplus v^{|l-m|+2} \oplus \ldots \oplus v^{l+m} = \bigoplus_{0\le r\le \min\{l,m\}} v^{l+m-2r}. \tag{1}$$

Note that the above labeling of irreducible representations and fusion rules is exactly the same as those for $SU(2)$. The main difference here is that dimensions of the corresponding representation spaces are larger. More precisely, if we denote by H_k the Hilbert space associated to v^k, then the fusion rules $v^1 \otimes v^k \cong v^{k+1} \oplus v^{k-1}$ dictate that the dimension relation $\dim H_1 \dim H_k = \dim H_{k+1} + \dim H_{k-1}$ must hold for all $k \ge 1$. Taken together with the initial conditions $H_0 = \mathbb{C}$ and $H_1 = \mathbb{C}^N$, one obtains that the dimensions $\dim H_k$ are given by the *quantum integers*

$$\dim H_k = [k+1]_q := q^{-k}\Big(\frac{1-q^{2k+2}}{1-q^2}\Big) \qquad (N \ge 3),$$

where $q \in (0,1)$ is given by $q + q^{-1} = N$. (When $N = 2$, we have $q = 1$, and then $\dim H_k = k+1 = \lim_{q\to 1^-}[k+1]_q$. However, in the sequel we will always take $N \ge 3$.)

2.3.4. *The connection to Temperley-Lieb.* The striking similarity between the fusion rules for the irreducible representations of O_N^+ to those of $SU(2)$ is no coincidence. This turns out to be a consequence of the fact (observed by Banica) that both representation categories are described in terms of certain unitary fiber functors on Temperley-Lieb categories [**TL71**]. Let $d \ge 2$. Recall that the *Temperley-Lieb Category* $\mathrm{TL}(d)$ is the strict tensor category with duals generated by two simple objects $\{0,1\}$, where 0 denotes the unit object for the tensor category, and $1 \ne 0$ is a self-dual simple object with the property that the morphism spaces $\mathrm{TL}_{k,l}(d) := \mathrm{Hom}(1^{\otimes k}, 1^{\otimes l})$ $(k,l \in \mathbb{N})$ are generated by the identity map $\iota \in \mathrm{Hom}(1,1)$ together with a unique morphism $\cup \in \mathrm{Hom}(0, 1\otimes 1)$ satisfying $\cap \circ \cup = d \in \mathrm{Hom}(0,0) = \mathbb{C}$. Here $\cap := \cup^* \in \mathrm{Hom}(1\otimes 1, 0)$. The Temperley-Lieb category admits a nice diagrammatic presentation [**KL94**] in terms of the so-called Kauffman (or Temperley-Lieb) diagrams. Let $k,l \in \mathbb{N}$ and $d \in \mathbb{C}\backslash\{0\}$ be as above. If $k+l$ is odd, we have $\mathrm{TL}_{k,l}(d) = 0$. Otherwise we plot the set $[k+l] = \{1,\ldots,k+l\}$ on a rectangle clockwise with $\{1,\ldots,k\}$ on the top edge and $\{k+l,\ldots,k+1\}$ on the bottom edge. Next, we consider the set $NC_2(k+l)$ of non-crossing pairings of these $k+l$ points on the boundary of our rectangle (see [**NS06**] for more details). Geometrically, any $p \in NC_2(k+l)$ corresponds to a partition of the set $[k+l]$ into $\frac{k+l}{2}$ pairs with the property that if we connect the $k+l$ boundary points on our rectangle that are paired off by p with smooth curves lying inside of our rectangle, then these curves can all be arranged so that none of them cross. The resulting non-crossing diagram is our Kauffman diagram, and is denoted by D_p. The collection of all Kauffman diagrams $(D_p)_{p \in NC_2(k+l)}$ (taken up to equivalence by planar isotopy) spans a basis for the vector space $\mathrm{TL}_{k,l}(d)$. For example, when $k = l = 3$ there are $|NC_2(6)| = 5$ Kauffman diagrams spanning $\mathrm{TL}_{k,k}(d)$:

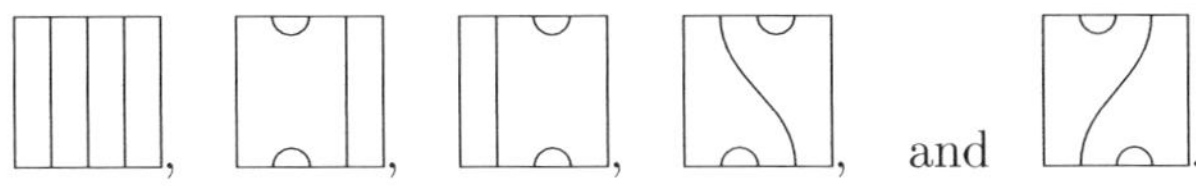

In the diagrammatic description of the morphism spaces $\mathrm{TL}_{k,l}(d)$, the composition $D_p D_q$ of diagrams $D_p \in \mathrm{TL}_{k,l}(d)$ and $D_q \in \mathrm{TL}_{m,k}(d)$ is obtained by first stacking the diagram D_p on top of D_q, connecting the bottom row of k points on D_p to the top row of k points on D_q. The result is a new planar diagram, which may have a certain number, c, of internal loops. By removing these loops, we obtain a new Kauffman diagram $D_r \in \mathrm{TL}_{m,l}(d)$, corresponding to some $r \in NC_2(m+l)$ (which is unique up to planar isotopy). The composition $D_p D_q$ is then defined to be $d^c D_r$. For example, we have

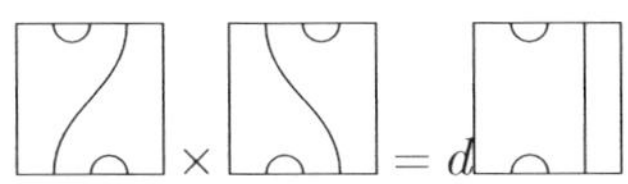

As for the tensor structure on $\mathrm{TL}(d)$, this is simply diagrammatically represented by horizontal concatenation of Kauffman diagrams. We leave it to the reader to verify how each of the above diagrams is obtained from sequences of the the basic operations of tensoring and composing the basic maps $\cup, \cap$, and ι.

Returning now to the connection with $\mathrm{Rep}(O_N^+)$ – observe that we can produce a natural unitary tensor functor $\mathrm{TL}(N) \to \mathrm{Rep}(O_N^+)$ given by $\iota \in \mathrm{TL}_{1,1}(N) \mapsto \mathrm{id}_{\mathbb{C}^N} \in \mathrm{Hom}(u,u)$ and $\cup \in \mathrm{TL}_{0,2}(N) \mapsto \sum_{i=1}^N e_i \otimes e_i \in \mathrm{Hom}(1, u \otimes u)$, where $(e_i)_{i=1}^N$ is an orthonormal basis for $\mathbb{C}^N$. The key point here is that the universal properties of O_N^+ guarantee that this functor is both injective and surjective (in the sense of [**ENO05**]). More precisely, we have the following theorem of Banica.

THEOREM 2.1 (Banica [**Ban96**]). *The above functor is in fact a unitary fiber functor $TL(N) \to Rep(O_N^+)$.*

With the above connection between $\mathrm{TL}(N)$ and $\mathrm{Rep}(O_N^+)$, an explicit construction of the irreducible representation spaces $(H_k)_{k \in \mathbb{N}_0}$ of O_N^+ can now proceed as follows [**Ban96, VV07, BDRV06**]. Denote by $(e_i)_{i=1}^N$ a fixed orthonormal basis for $H_1 := \mathbb{C}^N$, and as above, put $\cup = \sum_{i=1}^N e_i \otimes e_i \in \mathrm{Hom}(1, u \otimes u)$. (I.e., $u^{\otimes 2}(1 \otimes \cup) = (1 \otimes \cup)$.) Next, we consider the intertwiner space $\mathrm{Hom}(u^{\otimes k}, u^{\otimes k}) \subseteq \mathcal{B}((\mathbb{C}^N)^{\otimes k})$, which can be shown (using its identification with $\mathrm{TL}_{k,k}(N)$) to contain a unique non-zero self-adjoint projection p_k (*the Jones-Wenzl projection*) [**Wen87**] with the defining property that

$$(\iota_{H_1^{\otimes i-1}} \otimes \cup \cup^* \otimes \iota_{H_1^{\otimes k-i-1}}) p_k = 0 \qquad (1 \le i \le k-1).$$

The projections p_k are known to satisfy the *Wenzl recursion*

$$p_1 = \iota_{H_1}, \quad p_k = \iota_{H_1} \otimes p_{k-1} - \frac{[k-1]_q}{[k]_q}(\iota_{H_1} \otimes p_{k-1})(\cup \cup^* \otimes \iota_{H_1^{\otimes k-2}})(\iota_{H_1} \otimes p_{k-1}) \qquad (k \geq 2),$$

which can be used to determine p_k. In passing, we point out that the problem of obtaining explicit formulas for Jones-Wenzl projections (beyond the above recursion) has attracted a lot of attention over the years from various mathematical communities. See [**BC18a, Mor15, FK97, Rez07, Rez02**] and the references therein.

We conclude this section with a description of the non-empty intertwiner spaces $\mathrm{Hom}(v^k, v^l \otimes v^m)$ that arise from the fusion rules (1). To begin, let us call a triple $(k,l,m) \in \mathbb{N}_0^3$ *admissible* if there exists an integer $0 \leq r \leq \min\{l,m\}$ such that $k = l + m - 2r$. In other words, $(k,l,m) \in \mathbb{N}_0^3$ is admissible if and only if the tensor product representation $v^l \otimes v^m$ contains a (multiplicity-free) subrepresentation equivalent to v^k. It is easy to see that the set of admissible triples is invariant under coordinate permutations: (k_1,k_2,k_3) is admissible **iff** $(k_{\sigma(1)}, k_{\sigma(2)}, k_{\sigma(3)})$ is admissible for all $\sigma \in S_3$. Fix an admissible triple $(k,l,m) \in \mathbb{N}_0^3$. Then $\mathrm{Hom}(v^k, v^l \otimes v^m) \subseteq \mathcal{B}(H_k, H_l \otimes H_m) \subseteq \mathcal{B}(H_1^{\otimes k}, H_1^{\otimes l} \otimes H_1^{\otimes m})$ is one-dimensional and is spanned by the following canonical non-zero intertwiner

$$A_k^{l,m} = (p_l \otimes p_m)\Big(\iota_{H_{l-r}} \otimes \cup_r \otimes \iota_{m-r}\Big)p_k, \tag{2}$$

where $\cup_r \in \mathrm{Hom}(1, u^{\otimes 2r})$ is defined recursively from $\cup_1 := \sum_{i=1}^N e_i \otimes e_i$ via $\cup_r = (\iota_{H_1} \otimes \cup_1 \otimes \iota_{H_1})\cup_{r-1}$. In terms of the planar diagrammatics, $\cup_r$ is simply r nested cups, viewed as an element of $\mathrm{TL}_{0,2r}(N)$. The maps $A_k^{l,m}$ are well studied in the Temperley-Lieb recoupling theory [**KL94**], and are known there as *three-vertices*. See also [**EMM17**]. A three-vertex is typically diagrammatically represented as follows:

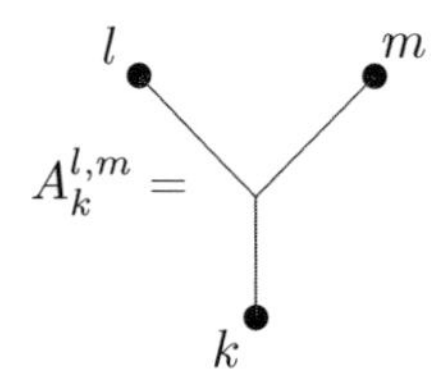

Here, the solid dots at the vertices are meant to depict the Jones-Wenzl projectors at the inputs/outputs. In the following we will simply omit these solid dots in our pictures, and simply draw the three-vertex as

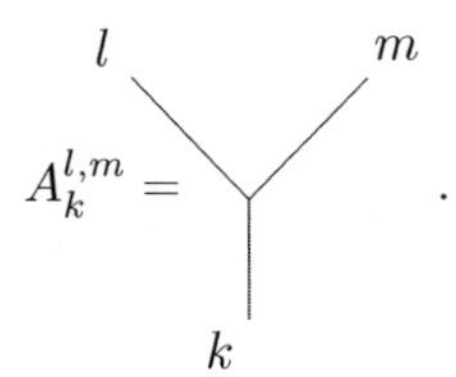

In order to find the unique O_N^+-equivariant *isometry* $\alpha_k^{l,m} : H_k \to H_l \otimes H_m$ (up to multiplication by $\mathbb{T}$), we simply have to renormalize $A_k^{l,m}$, which amounts to computing the norm of $A_k^{l,m}$. To do this, we define (following the terminology and

diagrammatics from [**KL94**]) the *θ-net*

$$\theta_q(k,l,m) = \mathrm{Tr}_{H_k}((A_k^{l,m})^* A_k^{l,m}) =$$

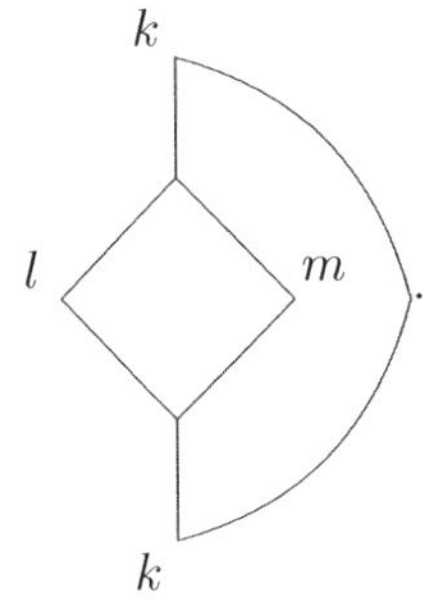

Note that the trace on $\mathcal{B}(H_k)$ corresponds to the usual Markov trace on $\mathrm{TL}(N)$ [**KL94, Ban96**].

Now, since $A_k^{l,m}$ is a multiple of an isometry, it easily follows that $\|A_k^{l,m}\|^2[k+1]_q = \theta_q(k,l,m)$. θ-net evaluations are well known [**KL94, Ver05, VV07**], and are given by

$$\theta_q(k,l,m) := \frac{[r]_q![l-r]_q![m-r]_q![k+r+1]_q!}{[l]_q![m]_q![k]_q!}, \tag{3}$$

where $k = l+m-2r$ and $[x]_q! = [x]_q[x-1]_q \dots [2]_q[1]_q$ denotes the quantum factorial. We thus arrive at the following formula for our isometry $\alpha_k^{l,m}$:

$$\alpha_k^{l,m} = \|A_k^{l,m}\|^{-1} A_k^{l,m} = \Big(\frac{[k+1]_q}{\theta_q(k,l,m)}\Big)^{1/2} A_k^{l,m}. \tag{4}$$

Pictorially, we have

$$\alpha_k^{l,m} = \Big(\frac{[k+1]_q}{\theta_q(k,l,m)}\Big)^{1/2}$$

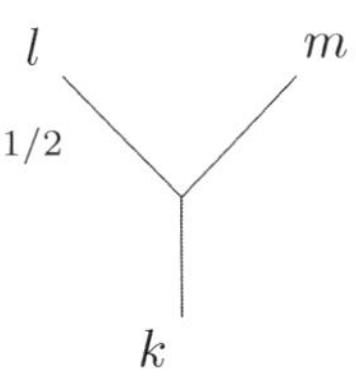

3. Entanglement analysis

In this section we begin our study of the entanglement geometry of irreducible subrepresentations of tensor products of irreducible representations of O_N^+. The general setup we will consider is a fixed $N \geq 3$ and an admissible triple $(k,l,m) \in \mathbb{N}_0^3$. This corresponds to irreducible representations (v^k, v^l, v^m) of O_N^+ with corresponding representation Hilbert spaces (H_k, H_l, H_m), and a O_N^+-equivariant isometry $\alpha_k^{l,m} : H_k \to H_l \otimes H_m$ as constructed in the previous section. Recall that we set $q = \frac{1}{N}\Big(\frac{2}{1+\sqrt{1-4/N^2}}\Big) \in (0,1)$. Our main interest is to study the entanglement of the subspace $\alpha_k^{l,m}(H_k) \subseteq H_l \otimes H_m$, and the following proposition yields a measure of this.

Proposition 3.1 ([**BC18b**]). *Fix $N \geq 3$ and let $(k,l,m) \in \mathbb{N}_0^3$ be an admissible triple. Then for any unit vectors $\xi \in H_k, \eta \in H_l, \zeta \in H_m$, we have*

$$|\langle \alpha_k^{l,m}(\xi) | \eta \otimes \zeta \rangle| \leq \Big(\frac{[k+1]_q}{\theta_q(k,l,m)}\Big)^{1/2} \leq C(q) q^{\frac{l+m-k}{4}},$$

where

$$C(q) = (1-q^2)^{-1/2}\Big(\prod_{s=1}^{\infty} \frac{1}{1-q^{2s}}\Big)^{3/2}$$

REMARK 3. We note that the bound $C(q)q^{\frac{l+m-k}{4}}$ appearing in Proposition 3.1 is equivalent, as N is large, to the fourth root of the relative dimension, $\Big(\frac{\dim H_k}{\dim H_l \dim H_m}\Big)^{1/4}$.

Proposition 3.1 can be interpreted as giving a general upper bound on the largest Schmidt coefficient of a unit vector belonging to the subspace $\alpha_k^{l,m}(H_k) \subseteq H_l \otimes H_m$. That is, if $\xi \in H_k$ is a unit vector and $\alpha_k^{l,m}(\xi)$ is represented by its singular value decomposition

$$\alpha_k^{l,m}(\xi) = \sum_i \sqrt{\lambda_i} e_i \otimes f_i,$$

with $(e_i)_i \subset H_l, (f_i)_i \subset H_m$ orthonormal systems, and $\lambda_1 \geq \lambda_2 \geq \ldots \geq 0$ satisfy $\sum_i \lambda_i = 1$, then

$$\lambda_1 \leq C(q)^2 q^{\frac{l+m-k}{2}}. \tag{5}$$

Since the above quantity is much smaller than 1 when $k < l+m$, we conclude that $\alpha_k^{l,m}(H_k)$ is "far" from containing separable unit vectors of the form $\eta\otimes\zeta \in H_l\otimes H_m$. That is, $\alpha_k^{l,m}(H_k) \subset H_l \otimes H_m$ is highly entangled. We summarize this in the following theorem.

THEOREM 3.2 ([**BC18b**]). *For k, l, m as above, the subspaces $\alpha_k^{l,m}(H_k) \subseteq H_l \otimes H_m$ are (highly) entangled provided $k < l+m$. When $k = l+m$, the highest weight subspace $\alpha_{l+m}(H_{l+m}) \subset H_l \otimes H_k$ is a separable subspace.*

PROOF. The first statement follows from the previous proposition and the remarks that follow. The second statement follows from the observation that if one considers the elementary (separable) tensors

$$(\ldots \xi \otimes \eta \otimes \xi \otimes \eta) \otimes (\xi \otimes \eta \otimes \xi \otimes \ldots) \in (\mathbb{C}^N)^{\otimes l} \otimes (\mathbb{C}^N)^{\otimes m} \qquad (\xi \perp \eta),$$

then they always lie in the subspace $\alpha_{l+m}^{l,m}(H_{l+m}) \subset H_l \otimes H_m$ (thanks to the algebraic properties of the Jones-Wenzl projections!). See [**BC18b**] for details. □

In fact it turns out that one can say quite a lot more about the largest possible Schmidt coefficients for irreducible subspaces of tensor products than what is said in Proposition 3.1. The following theorem shows that the bound given above is in fact optimal in a very strong sense: For any $d \in \mathbb{N}$, we can find a unit vector $\xi \in H_k$ (provided N is sufficiently large) with the property that $\alpha_k^{l,m}(\xi)$ admits at least d Schmidt coefficients with the same magnitude as that predicted by (5).

THEOREM 3.3 ([**BC18b**]). *Let $(k, l, m) \in \mathbb{N}_0^3$ be an admissible triple, $N \geq 3$, and $d \leq (N-2)(N-1)^{\frac{l+m-k-2}{2}}$. Then there exists a unit vector $\xi \in H_k$ such that $\alpha_k^{l,m}(\xi)$ has a singular value decomposition $\alpha_k^{l,m}(\xi) = \sum_i \sqrt{\lambda_i} e_i \otimes f_i$ with $\lambda_1 \geq \lambda_2 \geq \ldots$ satisfying*

$$\lambda_1 = \lambda_2 = \ldots = \lambda_d = \frac{[k+1]_q}{\theta_q(k,l,m)} \geq q^{\frac{l+m-k}{2}}.$$

REMARK 4. For various applications of the above theorem, it is of critical importance to understand if the above result is optimal in the sense that the number d of maximal Schmidt coefficients that is obtainable is indeed given by the above bound. At this stage, we are unable to fully answer this question. However, we can show that the upper bound $d(N) := (N-2)(N-1)^{\frac{m+l-k-2}{2}}$ of maximal Schmidt coefficients $\lambda_{\max} = \frac{[k+1]_q}{\theta_q(k,l,m)}$ is *asymptotically maximal* in the sense that

$$\lim_{N\to\infty} d(N)\frac{[k+1]_q}{\theta_q(k,l,m)} = 1.$$

This shows that in the limit as $N \to \infty$, the vector $\xi \in H_k$ which is asserted to exist by Theorem 3.3 becomes maximally entangled, with the bulk of its Schmidt coefficients equaling the maximal value $\lambda_{\max}$ allowed by Proposition 3.1.

4. O_N^+-equivariant quantum channels and minimum output entropy estimates

In this section we consider some applications of the entanglement results of the preceding section to study the outputs of the canonical quantum channels related to our subspaces.

Following Section 2, we form, for any admissible triple $(k,l,m) \in \mathbb{N}_0^3$, the complementary pair of quantum channels

$$\Phi_k^{\overline{l},m} : \mathcal{B}(H_k) \to \mathcal{B}(H_m); \qquad \rho \mapsto (\mathrm{Tr}\otimes\iota)(\alpha_k^{l,m}\rho(\alpha_k^{l,m})^*),$$

$$\Phi_k^{l,\overline{m}} : \mathcal{B}(H_k) \to \mathcal{B}(H_l); \qquad \rho \mapsto (\iota\otimes\mathrm{Tr})(\alpha_k^{l,m}\rho(\alpha_k^{l,m})^*).$$

In terms of the planar diagramatics of the Temperley-Lieb category, we have

$\Phi_k^{l,\overline{m}}(\rho) = \frac{[k+1]_q}{\theta_q(k,l,m)}$ 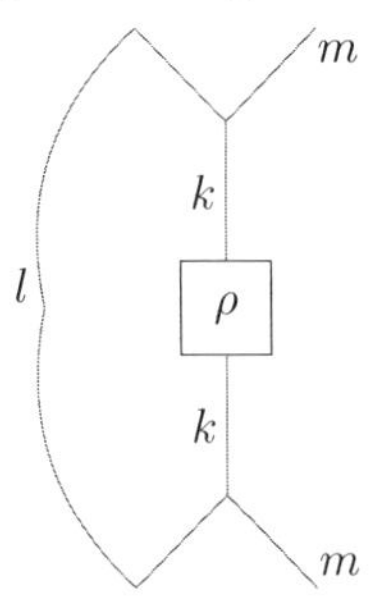

and $\Phi_k^{l,\overline{m}}(\rho) = \frac{[k+1]_q}{\theta_q(k,l,m)}$ 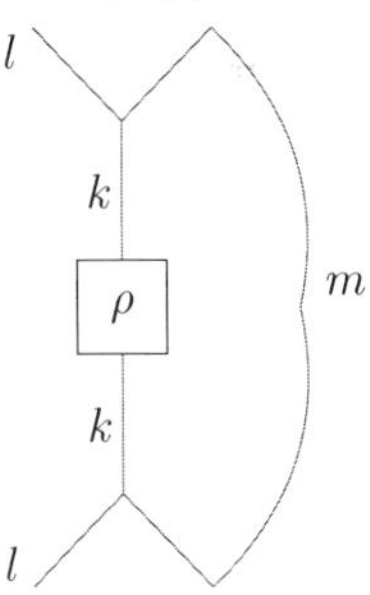

.

We then have the following proposition concerning the $\mathcal{S}_1 \to \mathcal{S}_\infty$ behavior of these channels.

PROPOSITION 4.1 ([**BC18b**]). *Given any admissible triple* $(k,l,m) \in \mathbb{N}_0^3$ *and* $N \geq 3$, *we have*

$$\begin{aligned}\|\Phi_k^{\overline{l},m}\|_{\mathcal{S}_1(H_k)\to\mathcal{S}_\infty(H_m)} &= \|\Phi_k^{l,\overline{m}}\|_{\mathcal{S}_1(H_k)\to\mathcal{S}_\infty(H_l)} \\ &= \frac{[k+1]_q}{\theta_q(k,l,m)} \in \big[q^{\frac{l+m-k}{2}}, C(q)^2 q^{\frac{l+m-k}{2}}\big].\end{aligned}$$

PROOF. We shall only consider $\Phi_k^{\overline{l},m}$ as the proof of the other case is identical. To prove the upper bound $\|\Phi_k^{\overline{l},m}\|_{\mathcal{S}_1(H_k)\to\mathcal{S}_\infty(H_m)} \leq \frac{[k+1]_q}{\theta_q(k,l,m)}$, note that by complete positivity, convexity and the triangle inequality, it suffices to consider a pure state

$\rho = |\xi\rangle\langle\xi| \in \mathcal{D}(H_k)$ and show that $\|\Phi_k^{\overline{l},m}(\rho)\|_{\mathcal{S}_\infty(H_m)} \le \frac{[k+1]_q}{\theta_q(k,l,m)}$. But in this case, we have

$$\Phi_k^{\overline{l},m}(\rho) = (\mathrm{Tr} \otimes \iota)(|\alpha_k^{l,m}\xi\rangle\langle\alpha_k^{l,m}\xi|) = \sum_i \lambda_i |f_i\rangle\langle f_i|,$$

where $\alpha_k^{l,m}(\xi) = \sum_i \sqrt{\lambda_i} e_i \otimes f_i$ is the corresponding singular value decomposition. In particular, $\|\Phi_k^{\overline{l},m}(\rho)\|_{\mathcal{S}_\infty(H_m)} = \max_i \lambda_i$, which by Proposition 3.1 is bounded above by $\frac{[k+1]_q}{\theta_q(k,l,m)}$. This upper bound is obtained by taking $\rho = |\xi\rangle\langle\xi|$, where ξ satisfies the hypotheses of Theorem 3.3. □

The preceeding norm computation for the channels $\Phi_k^{\overline{l},m}, \Phi_k^{l,\overline{m}}$ allows for an easy estimate of a lower bound on their minimum output entropies.

COROLLARY 4.2. [**BC18b**] *Given any admissible triple* $(k,l,m) \in \mathbb{N}_0^3$ *and* $N \ge 3$, *we have*

$$H_{\min}(\Phi_k^{\overline{l},m}), H_{\min}(\Phi_k^{l,\overline{m}}) \ge \log\Big(\frac{\theta_q(k,l,m)}{[k+1]_q}\Big) \ge -\Big(\frac{l+m-k}{2}\Big)\log(q) - 2\log(C(q)).$$

PROOF. Given a quantum channel $\Phi : \mathcal{B}(H) \to \mathcal{B}(K)$ and $\rho \in \mathcal{D}(H)$, we note that $H(\Phi(\rho)) = -\sum_i \lambda_i \log \lambda_i$, where $(\lambda_i)_i$ is the spectrum of $\Phi(\rho)$. In particular, we have the estimate

$$H(\Phi(\rho)) \ge -\log\Big(\max_i \lambda_i\Big) = -\log\|\Phi(\rho)\|_{\mathcal{B}(K)} \ge -\log\|\Phi\|_{\mathcal{S}_1(H)\to\mathcal{B}(K)}.$$

The first inequality in the corollary now follows immediately from Proposition 4.1. The second inequality is just a consequence of the inequality $\frac{[k+1]_q}{\theta_q(k,l,m)} \le C(q)^2 q^{\frac{l+m-k}{2}}$. □

REMARK 5. The above estimates show that for N large and $k < l+m$ fixed, the minimum output entropy of the channels is quite large and grows logarithmically in N.

On the other hand, if we fix $N \ge 3$ and consider, for example, the sequence of channels $\big(\Phi_{k-1}^{k,\overline{1}} : \mathcal{B}(H_{k-1}) \to \mathcal{B}(H_k)\big)_{k\in\mathbb{N}}$, then Corollary 4.2 yields the uniform positive lower bound

$$H_{\min}(\Phi_{k-1}^{k,\overline{1}}) \ge -\log(q) - 2\log(C(q)) > 0 \qquad (k \in \mathbb{N}).$$

This phenomenon stands in sharp contrast to what happens in the case of the $SU(2)$-equivariant quantum channels studied by Al Nuwairan in [**AN13**, Section 2]. Indeed, in the corresponding $SU(2)$ setting one has $H_{\min}(\Phi_{k-1}^{k,\overline{1}}) \approx \frac{\log(k+1)}{k+1} \to 0$ as $k \to \infty$.

In the case where $k = l+m$ (the highest weight case), we note that

$$H_{\min}(\Phi_k^{\overline{l},m}) = H_{\min}(\Phi_k^{l,\overline{m}}) = 0,$$

which follows from the fact that $\alpha_k^{l,m}(H_k) \subseteq H_l \otimes H_m$ is a separable subspace (cf. Theorem 3.2).

REMARK 6. We expect that the lower bound for the minimum output entropies given in Corollary 4.2 to be asymptotically optimal as $N \to \infty$, at least in some cases (e.g. m fixed). Evidence for this is provided by Theorem 3.3 and Remark 4, which shows that $\alpha_k^{l,m}(H_k)$ contains unit vectors which are asymptotically maximally entangled with the bulk of their Schmidt coefficients equal to $\frac{[k+1]_q}{\theta_q(k,l,m)}$.

5. The Choi map and Planar Isotopy

In this final section we indicate how the planar structure of our representation theoretic model for highly entangled subspaces can be used to easily describe the Choi maps associated to our quantum channels. As applications of this description, we construct non-random examples of d-positive maps between matrix algebras that fail to be completely positive, and we also study the entanglement breaking property for our channels.

First we recall the definition of the Choi map associated to a linear map $\Phi : \mathcal{B}(H_A) \to \mathcal{B}(H_B)$. Let $(e_i)_{i\in I}$, $(f_i)_{i\in I}$ be two fixed orthonomal bases for H_A, and let $(e_{ij})_{i,j\in I}$, $(f_{ij})_{i,j\in I}$ be the corresponding matrix units in $\mathcal{B}(H_A)$. Then the Choi map is the operator $C_\Phi \in \mathcal{B}(H_A \otimes H_B)$ given by

$$C_\Phi = \sum_{i,j\in I} \Phi(e_{ij}) \otimes f_{ij} = (\Phi \otimes \iota)(|\psi\rangle\langle\psi|), \tag{6}$$

where $\psi = \sum_{i\in I} e_i \otimes f_i \in H_A \otimes H_A$ (which is an unnormalized Bell state in $H_A \otimes H_A$). Of course, C_Φ is only defined uniquely up to the choice of matrix units e_{ij} and f_{ij}. Moreover, one could also define a "right-handed" version of $\tilde{C}_\Phi$ of C_Φ given by $\tilde{C}_\Phi = (\iota \otimes \Phi)|\psi\rangle\langle\psi|$ (i.e., slicing on the right instead of the left). However, for our purposes, the relevant properties of C_Φ (e.g., entanglement, positivity, etc.) do not depend on the choice of matrix units or side of the tensor product on which one slices $|\psi\rangle\langle\psi|$ by Φ. We also note the obvious fact that the map $\Phi \mapsto C_\Phi$ is linear in Φ.

Turning back to our representation category $\mathrm{Rep}(O_N^+)$ and our quantum channels $\Phi_k^{\bar{l},m} : \mathcal{B}(H_k) \to \mathcal{B}(H_m)$ ($(k,l,m) \in \mathbb{N}_0^3$ admissible), we judiciously choose orthonormal bases $(e_i)_i$ and $(f_i)_i$ of H_k so that the unnormalized Bell vector $\psi_k = \sum_i e_i \otimes f_i \in H_k \otimes H_k$ belongs to the one-dimensional Hom-space $\mathrm{Hom}(u^0, u^k \otimes u^k)$ (this is always possible, thanks to the fact that O_N^+ is a compact quantum group of Kac type. See for example [**Ver07**]). Using our identification $\mathrm{Rep}(O_N^+) \cong \mathrm{TL}(N)$, we can depict ψ_k (in terms of planar diagrams) as a three-vertex corresponding to the admissible triple $(0,k,k)$, which is explicitly given by $(p_k \otimes p_k) \circ \cup_k \in \mathrm{TL}_{0,2k}(N)$, where p_k is the kth Jones-Wenzl projector. Considering the projection $|\psi_k\rangle\langle\psi_k|$, we have

$|\psi_k\rangle\langle\psi_k| =$ 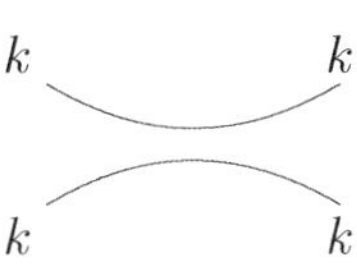

Then we can compute the corresponding Choi map $C_{\Phi_k^{\bar{l},m}} = (\Phi_k^{\bar{l},m} \otimes \iota)(|\psi_k\rangle\langle\psi_k|)$ diagrammatically by

$$\frac{\theta_q(k,l,m)}{[k+1]_q} C_{\Phi_k^{\bar{l},m}} =$$

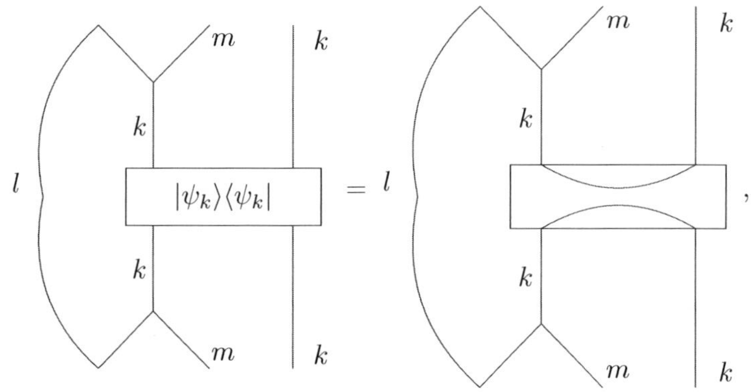

Since the linear map defined by the above planar tangle is invariant under planar isotopy (by construction it belongs to the Temperley-Lieb category!), we see that $\frac{\theta_q(k,l,m)}{[k+1]_q} C_{\Phi_k^{\bar{l},m}}$ also corresponds to the following planar tangle:

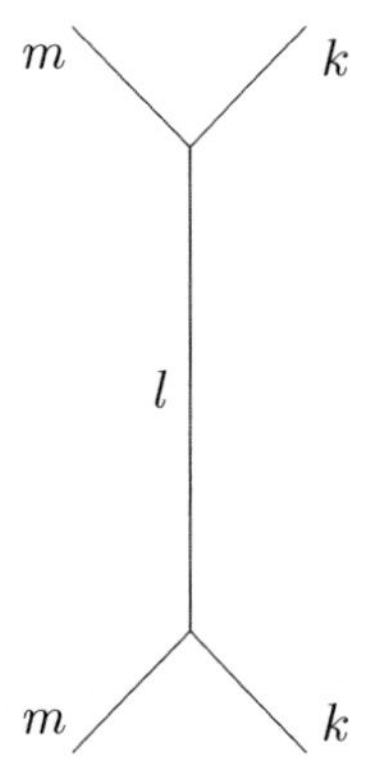

$$= \frac{\theta_q(k,l,m)}{[l+1]_q} \alpha_l^{m,k} (\alpha_l^{m,k})^*,$$

Note here that $\alpha_l^{m,k}(\alpha_l^{m,k})^*$ is simply the orthogonal equivariant projection from $H_m \otimes H_k$ onto the unique subspace equivalent to H_l. We have therefore arrived at the following theorem.

THEOREM 5.1. *For the O_N^+-equivariant quantum channel $\Phi_k^{\bar{l},m} : \mathcal{B}(H_k) \to \mathcal{B}(H_m)$, we have*

$$C_{\Phi_k^{\bar{l},m}} = \frac{[k+1]_q}{[l+1]_q} \alpha_l^{m,k} (\alpha_l^{m,k})^*. \tag{7}$$

A similar argument for the complementary channel $\Phi_k^{l,\bar{m}} : \mathcal{B}(H_k) \to \mathcal{B}(H_l)$, yields

$$\tilde{C}_{\Phi_k^{l,\bar{m}}} = \frac{[k+1]_q}{[m+1]_q} \alpha_m^{k,l} (\alpha_m^{k,l})^*. \tag{8}$$

In the following subsections, we show the utility of Theorem 5.1.

5.1. Examples of positive but not completely positive maps. A crucial property of the Choi map C_Φ associated to a linear map $\Phi : \mathcal{B}(H_A) \to \mathcal{B}(H_B)$ is that it can be used to detect positivity properties of Φ. More precisely, we have that Φ is completely positive if and only if C_Φ is positive semidefinite [**Cho75**]. More generally, C_Φ can be used to detect whether or not Φ is d-positive for any $d \in \mathbb{N}$ [**HLP$^+$12**]: Φ is d-positive if and only if

$$\langle C_\Phi x | x \rangle \geq 0$$

for all $x \in H_A \otimes H_B$ with a *Schmidt rank* of at most d. (That is, x admits a singular value decomposition $x = \sum_{i=1}^s \sqrt{\lambda_i} e_i \otimes f_i$ with $\min_i \lambda_i > 0$ and $s \leq d$).

Let us now return to our usual setup of an admissible triple $(k, l, m) \in \mathbb{N}_0^3$ corresponding to a non-highest-weight inclusion $\alpha_k^{l,m} : H_k \hookrightarrow H_l \otimes H_m$ of irreducible representations of O_N^+, $N \geq 3$. For each $t \geq 0$, we can consider the linear map $\Phi_t : \mathcal{B}(H_k) \to \mathcal{B}(H_m)$ given by

$$\Phi_t = \mathrm{Tr}_{H_k}(\cdot) 1_{\mathcal{B}(H_l)} - t \frac{[l+1]_q}{[k+1]_q} \Phi_k^{\bar{l},m}. \tag{9}$$

Using Theorem 5.1 together with the simple fact that the Choi map associated to $\mathcal{B}(H_k) \ni \rho \mapsto \mathrm{Tr}_{H_k}(\rho) 1_{\mathcal{B}(H_l)}$ is given by $1_{\mathcal{B}(H_m \otimes H_k)}$, we conclude that the Choi map of Φ_t is given by

$$C_{\Phi_t} = 1_{\mathcal{B}(H_m \otimes H_k)} - t \alpha_l^{m,k} (\alpha_l^{m,k})^*. \tag{10}$$

From this expression for C_{Φ_t}, it is clear that Φ_t is completely positive iff $C_{\Phi_t} \geq 0$ iff $t \leq 1$. On the other hand, we can prove the following result on d-positivity of Φ_t.

THEOREM 5.2 ([**BC18b**]). *Fix $N \geq 3$ and $(k, l, m) \in \mathbb{N}_0^3$, and fix a natural number $d \leq (N-2)(N-1)^{\frac{k+m-l-2}{2}}$. Then the map $\Phi_t : \mathcal{B}(H_k) \to \mathcal{B}(H_m)$ is d-positive (but not completely positive) if and only if*

$$1 < t \leq \frac{\theta_q(k,l,m)}{d[l+1]_q} \leq C(q)^{-2} q^{-\frac{k+m-l}{2}} d^{-1}.$$

PROOF. We have already observed that Φ_t is not completely positive when $t > 1$. Now fix $d \in \mathbb{N}$ and $x = \sum_{i=1}^s \sqrt{\lambda_i} e_i \otimes f_i \in H_m \otimes H_k$ with Schmidt-rank at most d. Using the inequality of Proposition 3.1, the triangle inequality, and the Cauchy-Schwarz inequality, we have

$$\begin{aligned}
\langle C_{\Phi_t} x | x \rangle &= \|x\|^2 - t \langle \alpha_l^{m,k} (\alpha_l^{m,k})^*(x) | x \rangle \\
&\geq \|x\|^2 - t \frac{[l+1]_q}{\theta_q(k,l,m)} \Big(\sum_{1 \leq i \leq s} \sqrt{\lambda_i} \|e_i\| \|f_i\| \Big)^2 \\
&\geq \|x\|^2 - t \frac{[l+1]_q}{\theta_q(k,l,m)} \Big(\sum_{1 \leq i \leq s} \sqrt{\lambda_i} \Big)^2 \\
&\geq \|x\|^2 - t \frac{[l+1]_q}{\theta_q(k,l,m)} s \|x\|^2 \\
&\geq \|x\|^2 \Big(1 - td \frac{[l+1]_q}{\theta_q(k,l,m)} \Big).
\end{aligned}$$

From this inequality, we obtain d-positivity of Φ_t provided $1 - td \frac{[l+1]_q}{\theta_q(k,l,m)} \geq 0$, as claimed.

To show failure of d-positivity when $t > \frac{\theta_q(k,l,m)}{d[l+1]_q}$, one has to find $x = \sum_{i=1}^d \eta_i \otimes \zeta_i \in H_l \otimes H_m$ with Schmidt rank d satisfying $\langle C_{\Phi_t} x | x \rangle < 0$. It turns out that such an x can be canonically constructed – see [**BC18b**] for details. □

REMARK 7. The above theorem can readily be used to construct maps on matrix algebras that are d positive but not $d+1$ positive. Indeed, one just has to choose $t > 1$, $N \geq 3$ and an admissible triple $(k,l,m) \in \mathbb{N}_0^3$ so that

$$\frac{\theta_q(k,l,m)}{(d+1)[l+1]_q} < t \leq \frac{\theta_q(k,l,m)}{d[l+1]_q}.$$

Then the corresponding Φ_t will do the job.

5.2. Entanglement breaking channels. We now turn to another application of Theorem 5.1, to the entanglement breaking property of our quantum channels $\Phi_k^{\bar{l},m}$.

DEFINITION. A quantum channel $\Phi : \mathcal{B}(H_A) \to \mathcal{B}(H_B)$ is called *entanglement breaking (or EBT)* if for any finite-dimensional auxiliary Hilbert space H_0, and any state $\rho \in \mathcal{D}(H_0 \otimes H_A)$, we have that $(\iota \otimes \Phi)(\rho) \in \mathcal{D}(H_0 \otimes H_B)$ is a separable state.

The class of EBT channels are precisely those which eliminate entanglement between the input states of composite systems. These channels form an important class which are amenable to analysis. For example, it is known that for EBT channels, both the minimum output entropy and the Holevo capacity (i.e., the capacity of a quantum channel used for classical communication with product inputs) is additive [**Hol01, Sho02**].

In order to detect whether or not a given quantum channel is EBT, it suffices to check whether or not the corresponding Choi map is a multiple of an entangled state. The following result is well known: see for example [**AN13**, Proposition 3.4].

PROPOSITION 5.3. *For a quantum channel $\Phi : \mathcal{B}(H_A) \to \mathcal{B}(H_B)$, the following conditions are equivalent.*

(1) *Φ is EBT.*
(2) *The state $\rho := \frac{1}{\dim H_A} C_\Phi \in \mathcal{D}(H_B \otimes H_A)$ is separable.*

Before coming to our main result of this section characterizing the EBT property for the channels $\Phi_k^{\bar{l},m}$, we first need an elementary lemma.

LEMMA 5.4. *Let H_A and H_B be finite dimensional Hilbert spaces, let $0 \neq p \in \mathcal{B}(H_B \otimes H_A)$ be an orthogonal projection, and let $H_0 \subseteq H_B \otimes H_A$ denote the range of p. If H_0 is an entangled subspace of $H_B \otimes H_A$, then the state $\rho := \frac{1}{\dim H_0} p$ is entangled.*

PROOF. We prove the contrapositive. If ρ is separable, then we can write

$$p = \sum_i |\xi_i\rangle\langle\xi_i| \otimes |\eta_i\rangle\langle\eta_i| \qquad (0 \neq \xi_i \in H_B,\ 0 \neq \eta_i \in H_A).$$

For each i put $x_i = |\xi_i\rangle\langle\xi_i| \otimes |\eta_i\rangle\langle\eta_i|$. Then since $x_i \leq p$ and p is a projection, it follows that $x_i = p x_i p$, which implies that the range of x_i is contained in the range of p. In particular, $\xi_i \otimes \eta_i \in H_0$, so H_0 is separable. □

THEOREM 5.5. *Let $(k,l,m) \in \mathbb{N}_0^3$ be an admissible triple. If $k \neq l - m$, then the quantum channel $\Phi_k^{\bar{l},m}$ is not EBT.*

PROOF. We have from Theorem 5.1 that $C_{\Phi_k^{\bar{l},m}} = \frac{[k+1]_q}{[l+1]_q}\alpha_l^{m,k}(\alpha_l^{m,k})^* \in \mathcal{B}(H_m \otimes H_k)$. Consider the orthogonal projection $p = \alpha_l^{m,k}(\alpha_l^{m,k})^*$. The range of p is the subrepresentation of $H_m \otimes H_k$ equivalent to H_l, and by Theorem 3.2 this subspace is entangled iff $l \neq k+m$. Applying Lemma 5.4 and Proposition 5.3, we conclude that $\Phi_k^{\bar{l},m}$ is not EBT whenever $k \neq l-m$. □

REMARK 8. We note that Theorem 5.5 leaves open whether or not the channels $\Phi_{l-m}^{\bar{l},m}$ are EBT. In this case, the corresponding Choi map is a multiple of a projection onto a separable subspace, and we do not know if this projection is a multiple of an entangled state.

6. Future work and open problems

We conclude this survey with a list of open problems and directions for future work.

(1) A major problem in QIT is to find explicit examples of quantum channels Φ, Ψ which are *strictly MOE-subbaditive*: $H_{\min}(\Phi \otimes \Psi) < H_{\min}(\Phi) + H_{\min}(\Psi)$. Such channels are known to exist with high probability [**Has09**, **ASW11**, **BCN16**], but no explicit examples are known. It is therefore tempting to wonder whether or not the channels considered in this work might be MOE subadditive. The first step in considering this question is to have an effective means of estimating the MOE of tensor products of our channels. In this context some computations are actually possible. In particular, if one takes one of our Temperley-Lieb chanels Φ, then it is always possible to explicitly compute the von Neumann entropy $H(\Phi \otimes \Phi^c)(\rho)$ of the output of a Bell state ρ, where Φ^c denotes the so-called *complementary channel* associated to Φ. It turns out that this computation involves the *quantum 6j-symbols* associated to the Temperley-Lieb category. This particular calculation is the topic of work in preparation [**BCLY18**]. At the present time, it seems that in order to have any hope of whitnessing strict MOE subadditivity in our channels, more tensor products beyond simply channels and their complements need to be studied, and at this time, a new idea is needed.

(2) Another important question related to our class of quantum channels is the problem of computing their classical and quantum capacities. This is another completely open and important research direction.

(3) As we have seen in this work, the Temperley-Lieb category provides a tractable concrete model for highly entangled subspaces. It is natural to wonder what other nice tensor categories or related structures give nice models of entanglement. Perhaps certain examples coming from planar algebras [**Jon99**] might give some interesting results?

(4) It would be interesting to make a further study of the family of d-positive maps Φ_t given here. The importance of such maps in QIT is for *entanglement detection* in bipartite systems: Positive maps that are not completely positive can be used to distinguish entangled states from separable ones. Of particular interest is the problem of detecting entangled states from the *positive partial transpose (PPT) states*. In this context, the relevant maps for entanglement detection are the *indecomposable maps*. I.e., positive maps Φ which are not of the form $\Phi = \Phi_1 + \Phi_2 \circ t$, where $\Phi_{1,2}$ are

completely positive, and t denotes the transpose map. In this context, we ask: *Are our families of maps Φ_t indecomposable?*

References

[Abr08] Samson Abramsky, *Temperley-Lieb algebra: from knot theory to logic and computation via quantum mechanics*, Mathematics of quantum computation and quantum technology, Chapman & Hall/CRC Appl. Math. Nonlinear Sci. Ser., Chapman & Hall/CRC, Boca Raton, FL, 2008, pp. 515–558. MR2422231

[AN13] Muneerah Al Nuwairan. Potential examples for non-additivity of the minimal output entropy. Preprint, arXiv:1312.2200, 2013.

[AN14] M. Al Nuwairan, *The extreme points of SU(2)-irreducibly covariant channels*, Internat. J. Math. **25** (2014), no. 6, 1450048, 30, DOI 10.1142/S0129167X14500487. MR3225575

[ASW11] Guillaume Aubrun, Stanisław Szarek, and Elisabeth Werner, *Hastings's additivity counterexample via Dvoretzky's theorem*, Comm. Math. Phys. **305** (2011), no. 1, 85–97, DOI 10.1007/s00220-010-1172-y. MR2802300

[ASY14] Guillaume Aubrun, Stanisław J. Szarek, and Deping Ye, *Entanglement thresholds for random induced states*, Comm. Pure Appl. Math. **67** (2014), no. 1, 129–171, DOI 10.1002/cpa.21460. MR3139428

[Ban96] Teodor Banica, *Théorie des représentations du groupe quantique compact libre* O(n) (French, with English and French summaries), C. R. Acad. Sci. Paris Sér. I Math. **322** (1996), no. 3, 241–244. MR1378260

[BC18a] Michael Brannan and Benoît Collins, *Dual bases in Temperley-Lieb algebras, quantum groups, and a question of Jones*, Quantum Topol. **9** (2018), no. 4, 715–748, DOI 10.4171/QT/118. MR3874001

[BC18b] Michael Brannan and Benoît Collins, *Highly entangled, non-random subspaces of tensor products from quantum groups*, Comm. Math. Phys. **358** (2018), no. 3, 1007–1025, DOI 10.1007/s00220-017-3023-6. MR3778349

[BCLY18] Michael Brannan, Benoît Collins, Hun-Hee Lee, and Sang-Gyun Youn. Temperley-Lieb quantum channels. In preparation, 2018.

[BCN12] Serban Belinschi, Benoît Collins, and Ion Nechita, *Eigenvectors and eigenvalues in a random subspace of a tensor product*, Invent. Math. **190** (2012), no. 3, 647–697, DOI 10.1007/s00222-012-0386-3. MR2995183

[BCN16] Serban T. Belinschi, Benoît Collins, and Ion Nechita, *Almost one bit violation for the additivity of the minimum output entropy*, Comm. Math. Phys. **341** (2016), no. 3, 885–909, DOI 10.1007/s00220-015-2561-z. MR3452274

[BDRV06] Julien Bichon, An De Rijdt, and Stefaan Vaes, *Ergodic coactions with large multiplicity and monoidal equivalence of quantum groups*, Comm. Math. Phys. **262** (2006), no. 3, 703–728, DOI 10.1007/s00220-005-1442-2. MR2202309

[Cho75] Man Duen Choi, *Completely positive linear maps on complex matrices*, Linear Algebra and Appl. **10** (1975), 285–290, DOI 10.1016/0024-3795(75)90075-0. MR0376726

[CLSZ95] I. L. Chuang, R. Laflamme, P. W. Shor, and W. H. Zurek, *Quantum computers, factoring, and decoherence*, Science **270** (1995), no. 5242, 1633–1635, DOI 10.1126/science.270.5242.1633. MR1362059

[CN16] Benoît Collins and Ion Nechita, *Random matrix techniques in quantum information theory*, J. Math. Phys. **57** (2016), no. 1, 015215, 34, DOI 10.1063/1.4936880. MR3432743

[DK94] Mathijs S. Dijkhuizen and Tom H. Koornwinder, *CQG algebras: a direct algebraic approach to compact quantum groups*, Lett. Math. Phys. **32** (1994), no. 4, 315–330, DOI 10.1007/BF00761142. MR1310296

[Dri87] V. G. Drinfel′d, *Quantum groups*, Proceedings of the International Congress of Mathematicians, Vol. 1, 2 (Berkeley, Calif., 1986), Amer. Math. Soc., Providence, RI, 1987, pp. 798–820. MR934283

[Eis06] Jens Eisert. Entanglement in quantum information theory. Ph.D. Thesis, arXiv:quant-ph/0610253v1, 2006.

[EJ96] Artur Ekert and Richard Jozsa, *Quantum computation and Shor's factoring algorithm*, Rev. Modern Phys. **68** (1996), no. 3, 733–753, DOI 10.1103/RevModPhys.68.733. MR1406773

[EMM17] Cain Edie-Michell and Scott Morrison. A field guide to categories with A_n fusion rules. Preprint, arXiv:1710.07362, 2017.

[ENO05] Pavel Etingof, Dmitri Nikshych, and Viktor Ostrik, *On fusion categories*, Ann. of Math. (2) **162** (2005), no. 2, 581–642, DOI 10.4007/annals.2005.162.581. MR2183279

[ES92] Michel Enock and Jean-Marie Schwartz, *Kac algebras and duality of locally compact groups*, Springer-Verlag, Berlin, 1992. With a preface by Alain Connes; With a postface by Adrian Ocneanu. MR1215933

[FK97] Igor B. Frenkel and Mikhail G. Khovanov, *Canonical bases in tensor products and graphical calculus for* $U_q(\mathfrak{sl}_2)$, Duke Math. J. **87** (1997), no. 3, 409–480, DOI 10.1215/S0012-7094-97-08715-9. MR1446615

[FK10] Motohisa Fukuda and Christopher King, *Entanglement of random subspaces via the Hastings bound*, J. Math. Phys. **51** (2010), no. 4, 042201, 19, DOI 10.1063/1.3309418. MR2662469

[GHP10] Andrzej Grudka, Michał Horodecki, and Łukasz Pankowski, *Constructive counterexamples to the additivity of the minimum output Rényi entropy of quantum channels for all* $p > 2$, J. Phys. A **43** (2010), no. 42, 425304, 7, DOI 10.1088/1751-8113/43/42/425304. MR2726721

[Gro96] Lov K. Grover, *A fast quantum mechanical algorithm for database search*, Proceedings of the Twenty-eighth Annual ACM Symposium on the Theory of Computing (Philadelphia, PA, 1996), ACM, New York, 1996, pp. 212–219, DOI 10.1145/237814.237866. MR1427516

[Haa79] Uffe Haagerup, *An example of a nonnuclear* C^**-algebra, which has the metric approximation property*, Invent. Math. **50** (1978/79), no. 3, 279–293, DOI 10.1007/BF01410082. MR520930

[Has09] Matthew B. Hastings. Superadditivity of communication capacity using entangled inputs. *Nature Physics*, 5:255–257, 2009.

[HLP+12] Jinchuan Hou, Chi-Kwong. Li, Yiu-Tong Poon, Xiaofei Qi, and Nung-Szing Sze. Criteria and new classes of k-positive maps. Preprint, arXiv:1211.0386, 2012.

[HLSW04] Patrick Hayden, Debbie Leung, Peter W. Shor, and Andreas Winter, *Randomizing quantum states: constructions and applications*, Comm. Math. Phys. **250** (2004), no. 2, 371–391, DOI 10.1007/s00220-004-1087-6. MR2094521

[HLW06] Patrick Hayden, Debbie W. Leung, and Andreas Winter, *Aspects of generic entanglement*, Comm. Math. Phys. **265** (2006), no. 1, 95–117, DOI 10.1007/s00220-006-1535-6. MR2217298

[Hol01] A. S. Holevo, *Coding theorems of quantum information theory*, XIIIth International Congress on Mathematical Physics (London, 2000), Int. Press, Boston, MA, 2001, pp. 415–422. MR1883334

[HW08] Patrick Hayden and Andreas Winter, *Counterexamples to the maximal p-norm multiplicity conjecture for all* $p > 1$, Comm. Math. Phys. **284** (2008), no. 1, 263–280, DOI 10.1007/s00220-008-0624-0. MR2443305

[Jim85] Michio Jimbo, *A q-difference analogue of* $U(\mathfrak{g})$ *and the Yang-Baxter equation*, Lett. Math. Phys. **10** (1985), no. 1, 63–69, DOI 10.1007/BF00704588. MR797001

[Jon83] V. F. R. Jones, *Index for subfactors*, Invent. Math. **72** (1983), no. 1, 1–25, DOI 10.1007/BF01389127. MR696688

[Jon85] Vaughan F. R. Jones, *A polynomial invariant for knots via von Neumann algebras*, Bull. Amer. Math. Soc. (N.S.) **12** (1985), no. 1, 103–111, DOI 10.1090/S0273-0979-1985-15304-2. MR766964

[Jon99] Vaughan Jones. Planar algebras 1. Preprint, arXiv:math/9909027, 1999.

[KL94] Louis H. Kauffman and Sóstenes L. Lins, *Temperley-Lieb recoupling theory and invariants of 3-manifolds*, Annals of Mathematics Studies, vol. 134, Princeton University Press, Princeton, NJ, 1994. MR1280463

[KV00] Johan Kustermans and Stefaan Vaes, *Locally compact quantum groups* (English, with English and French summaries), Ann. Sci. École Norm. Sup. (4) **33** (2000), no. 6, 837–934, DOI 10.1016/S0012-9593(00)01055-7. MR1832993

[LB13] Ching-Yi Lai and Todd A. Brun. Entanglement increases the error-correcting ability of quantum error-correcting codes. Phys. Rev. A 88, 012320 (2013), 2013.

[Mor15] Scott Morrison, *A formula for the Jones-Wenzl projections*, Proceedings of the 2014 Maui and 2015 Qinhuangdao conferences in honour of Vaughan F. R. Jones' 60th birthday, Proc. Centre Math. Appl. Austral. Nat. Univ., vol. 46, Austral. Nat. Univ., Canberra, 2017, pp. 367–378. MR3635675

[NC00] Michael A. Nielsen and Isaac L. Chuang, *Quantum computation and quantum information*, Cambridge University Press, Cambridge, 2000. MR1796805

[NS06] Alexandru Nica and Roland Speicher, *Lectures on the combinatorics of free probability*, London Mathematical Society Lecture Note Series, vol. 335, Cambridge University Press, Cambridge, 2006. MR2266879

[NT13] Sergey Neshveyev and Lars Tuset, *Compact quantum groups and their representation categories*, Cours Spécialisés [Specialized Courses], vol. 20, Société Mathématique de France, Paris, 2013. MR3204665

[Rez02] Sarah Anne Reznikoff, *Representations of the Temperley-Lieb planar algebra*, ProQuest LLC, Ann Arbor, MI, 2002. Thesis (Ph.D.)–University of California, Berkeley. MR2703792

[Rez07] Sarah A. Reznikoff, *Coefficients of the one- and two-gap boxes in the Jones-Wenzl idempotent*, Indiana Univ. Math. J. **56** (2007), no. 6, 3129–3150, DOI 10.1512/iumj.2007.56.3140. MR2375712

[Sho02] Peter W. Shor, *Additivity of the classical capacity of entanglement-breaking quantum channels*, J. Math. Phys. **43** (2002), no. 9, 4334–4340, DOI 10.1063/1.1498000. Quantum information theory. MR1924442

[Tim08] Thomas Timmermann, *An invitation to quantum groups and duality*, EMS Textbooks in Mathematics, European Mathematical Society (EMS), Zürich, 2008. From Hopf algebras to multiplicative unitaries and beyond. MR2397671

[TL71] H. N. V. Temperley and E. H. Lieb, *Relations between the "percolation" and "colouring" problem and other graph-theoretical problems associated with regular planar lattices: some exact results for the "percolation" problem*, Proc. Roy. Soc. London Ser. A **322** (1971), no. 1549, 251–280, DOI 10.1098/rspa.1971.0067. MR0498284

[Ver05] Roland Vergnioux, *Orientation of quantum Cayley trees and applications*, J. Reine Angew. Math. **580** (2005), 101–138, DOI 10.1515/crll.2005.2005.580.101. MR2130588

[Ver07] Roland Vergnioux, *The property of rapid decay for discrete quantum groups*, J. Operator Theory **57** (2007), no. 2, 303–324. MR2329000

[VV07] Stefaan Vaes and Roland Vergnioux, *The boundary of universal discrete quantum groups, exactness, and factoriality*, Duke Math. J. **140** (2007), no. 1, 35–84, DOI 10.1215/S0012-7094-07-14012-2. MR2355067

[Wan95] Shuzhou Wang, *Free products of compact quantum groups*, Comm. Math. Phys. **167** (1995), no. 3, 671–692. MR1316765

[Wen87] Hans Wenzl, *On sequences of projections*, C. R. Math. Rep. Acad. Sci. Canada **9** (1987), no. 1, 5–9. MR873400

[Wor87] S. L. Woronowicz, *Compact matrix pseudogroups*, Comm. Math. Phys. **111** (1987), no. 4, 613–665. MR901157

[Wor88] S. L. Woronowicz, *Tannaka-Kreĭn duality for compact matrix pseudogroups. Twisted* SU(N) *groups*, Invent. Math. **93** (1988), no. 1, 35–76, DOI 10.1007/BF01393687. MR943923

[Wor98] S. L. Woronowicz, *Compact quantum groups*, Symétries quantiques (Les Houches, 1995), North-Holland, Amsterdam, 1998, pp. 845–884. MR1616348

Michael Brannan, Department of Mathematics, Mailstop 3368, Texas A&M University, College Station, TX 77843-3368, USA

Email address: `mbrannan@math.tamu.edu`

Benoît Collins, Department of Mathematics, Kyoto University, and CNRS, Institut Camille Jordan Université Lyon 1, France

Email address: `collins@math.kyoto-u.ac.jp`

Contemporary Mathematics
Volume **747**, 2020
https://doi.org/10.1090/conm/747/15038

Lifting shadings on symmetrically self-dual subfactor planar algebras

Zhengwei Liu, Scott Morrison, and David Penneys

Abstract. In this note, we discuss the notion of symmetric self-duality of shaded planar algebras, which allows us to lift shadings on subfactor planar algebras to obtain $\mathbb{Z}/2\mathbb{Z}$-graded unitary fusion categories. This finishes the proof that there are unitary fusion categories with fusion graphs 4442 and 3333.

Planar algebras have proven to be useful in the construction [**Pet10**, **BMPS12**] and classification [**JMS14**, **AMP15**] of subfactors and fusion categories. In recent articles, we used planar algebras to construct subfactor planar algebras with principal graphs 4442, 3333, and 2221 [**MP15**],

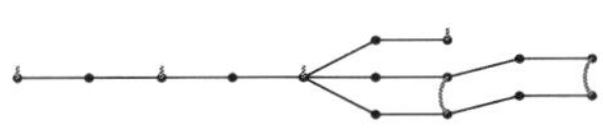 , 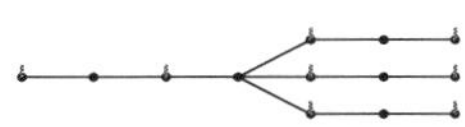 , and 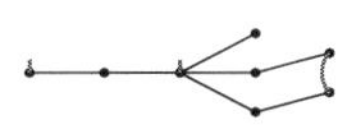 ,

and a new subfactor with principal graphs 22221 with interesting dual data [**LMP15**] (see Example 2.2 below). This 22221 subfactor turns out to be an example of a new parameterized family of unshaded subfactor planar algebras related to quantum subgroups [**Liu15**]. The 2221 subfactor was originally constructed by Izumi [**Izu01**], as was the 3333 subfactor [**Izu18**].

In [**CMS11**, Appendix], Ostrik constructed a $\mathbb{Z}/2$-graded unitary fusion category with fusion graph 2221. Thus upon constructing 4442 and 3333, we naturally wondered whether we could lift the shading on our subfactor planar algebras to get $\mathbb{Z}/2$-graded fusion categories with these fusion graphs.

We showed that these subfactor planar algebras are *symmetrically self-dual*, i.e., there is a planar algebra isomorphism Φ from $\mathcal{P}_\bullet = (\mathcal{P}_+, \mathcal{P}_-)$ to its dual $\overline{\mathcal{P}}_\bullet = (\mathcal{P}_-, \mathcal{P}_+)$ such that $\Phi_\mp \circ \Phi_\pm = \mathbf{1}_\pm$. We furthermore claimed that we can lift the shading on a symmetrically self-dual subfactor planar algebra to obtain an unshaded factor planar algebra [**BHP12**]. This would mean the associated tensor

2010 *Mathematics Subject Classification.* Primary 46L37, 18D10.

The first author was supported by a grant from Templeton Religion Trust and an AMS Simons travel grant.

The second author was supported by Discovery Projects 'Subfactors and symmetries' DP140100732 and 'Low dimensional categories' DP160103479, and a Future Fellowship 'Quantum symmetries' FT170100019 from the Australian Research Council.

The third author was supported by an AMS Simons travel grant and NSF DMS grants 1500387/1655912 and 1654159.

©2020 American Mathematical Society

category of projections is a $\mathbb{Z}/2$-graded unitary fusion category $\mathcal{C}$ whose even graded part is the even part of $\mathcal{P}_\bullet$ and whose odd graded part is the odd part of $\mathcal{P}_\bullet$.

In this note, we complete the proof of this claim to complete the construction of these categories.

THEOREM A. *Given a symmetrically self-dual shaded planar algebra* $(\mathbb{P}_\bullet, \Phi)$, *there is an unshaded planar algebra* $\mathbb{P}_\bullet$ *such that* $\mathbb{P}_\bullet$ *is obtained from* $\mathbb{P}_\bullet$ *by reshading (as in Definition* 1.4 *below).*

COROLLARY B. *There are* $\mathbb{Z}/2$*-graded unitary fusion categories with fusion graphs 4442 and 3333.*

1. Unshaded and shaded planar algebras. We refer the reader to [**Pet10, Jon12**] for the definition of a shaded subfactor planar algebra, and to [**BHP12, HP17**] for the definition of an unshaded factor planar algebra. As we need to shade unshaded planar algebras and lift the shading on shaded planar algebras, we introduce the following notation to help keep things as simple as possible.

NOTATION 1.1. We denote an unshaded planar algebra by the unshaded symbol $\mathbb{P}_\bullet$, and we denote a shaded planar algebra by the shaded symbol $\mathbb{P}_\bullet$. We write $\mathbb{U}$ for an unshaded planar tangle and $\mathbb{Z}$ for the unshaded partition function. Similarly, we write $\mathbb{S}$ for a shaded planar tangle and $\mathbb{Z}$ for the shaded partition function. We use the symbol $\mathcal{T}$ to denote a tangle which may be shaded or unshaded.

Given a shaded tangle $\mathbb{S}$, let $\mathbb{S}^{\mathrm{op}}$ be the shaded planar tangle obtained from $\mathbb{S}$ by reversing the shading. Note that forgetting the shading of $\mathbb{S}$ gives an unshaded tangle $\mathbb{S}$. In Definition 1.6 below, given an unshaded tangle $\mathbb{U}$, we define a special shaded tangle $\mathbb{U}$.

REMARK 1.2. We caution the reader that for an unshaded planar algebra $\mathbb{P}_\bullet$, the space $\mathbb{P}_n$ corresponds to boxes with n strands, whereas for a shaded planar algebra $\mathbb{P}_\bullet$, the space $\mathbb{P}_n$ corresponds to boxes with $2n$ strands.

DEFINITION 1.3. Given a shaded planar algebra $\mathbb{P}_\bullet$, its *dual planar algebra* $\overline{\mathbb{P}}_\bullet$ is the planar algebra obtained by reversing the shading. The planar algebra $\mathbb{P}_\bullet$ is *self-dual* if there is a shaded planar algebra isomorphism $\Phi : \mathbb{P}_\bullet \to \overline{\mathbb{P}}_\bullet$. Note that Φ can be written as

$$(\Phi_+, \Phi_-) : (\mathbb{P}_+, \mathbb{P}_-) \longrightarrow (\mathbb{P}_-, \mathbb{P}_+).$$

This means for a shaded planar tangle $\mathbb{S}$ with s input disks, $\Phi \circ \mathbb{S} = \mathbb{S}^{\mathrm{op}} \circ (\bigotimes^s \Phi)$.

If moreover $\Phi^2 = \mathbf{1}_{\mathbb{P}_\bullet}$, which is equivalent to $\Phi_\mp \circ \Phi_\pm = \mathbf{1}_{\mathbb{P}_\pm}$, then $\mathbb{P}_\bullet$ is called *symmetrically self-dual* [**MP15**].

One source of symmetrically self-dual shaded planar algebras is unshaded planar algebras.

DEFINITION 1.4. Let $\mathbb{P}_\bullet$ be an unshaded planar algebra such that $\mathbb{P}_{2n+1} = (0)$ for all $n \geq 0$. Define $\mathbb{P}_\bullet$ as follows. For all $n \geq 0$, let $\mathbb{P}_{n,\pm} = \mathbb{P}_{2n}$. Suppose that for a shaded planar tangle $\mathbb{S}$, the unshaded tangle $\mathbb{S}$ is obtained from forgetting the shading of $\mathbb{S}$. Notice that since $\mathbb{S}$ is shaded, $\mathbb{S}$ has an even number of boundary points on all disks. We define $\mathbb{Z}(\mathbb{S}) = \mathbb{Z}(\mathbb{S})$, i.e., the action of $\mathbb{S}$ is simply obtained by forgetting the shading.

Finally, note that switching the shading on $\mathbb{P}_\bullet$ is obviously a symmetric self-duality, since the action $\mathbb{Z}$ is independent of the shading, and merely depends on the unshaded action $\mathbb{Z}$.

Theorem A says that all symmetrically self-dual shaded planar algebras come from unshaded planar algebras. This theorem is both unsurprising and relatively tedious to prove. At this point a number of fusion categories have been discovered first in the form of shaded planar algebras (see [**MP15, LMP15**]). The motivation of this article is to complete the construction of these categories.

To prove Theorem A, we introduce the following notation.

NOTATION 1.5. Suppose $\mathcal{T}$ has t input disks. For $1 \leq i \leq t$, we let $D_i(\mathcal{T})$ denote the i-th input disk of $\mathcal{T}$, and we let $D_0(\mathcal{T})$ denote the output disk of $\mathcal{T}$. We write $k_i(\mathcal{T})$ to denote the number of boundary points of $D_i(\mathcal{T})$. Finally, if $\mathcal{T}$ is shaded, we write $\pm_i(\mathcal{T})$ to denote whether the distinguished interval of the disk $D_i(\mathcal{T})$ is in an unshaded or shaded region respectively.

DEFINITION 1.6. For an unshaded planar tangle $\mathbb{U}$ whose input and output disks all have an even number of boundary points, we define a shaded tangle $\operatorname{sh}(\mathbb{U}) = \mathbf{U}$ by checkerboard shading $\mathbb{U}$ using the rule that the region meeting the distinguished interval of the output disk is unshaded.

If $\mathbb{U}, \mathbb{V}$ are unshaded planar tangles such that $\mathbb{V}$ has $k_0(\mathbb{V}) = k_i(\mathbb{U})$ boundary points, then the shading map sh behaves as follows:

$$\operatorname{sh}(\mathbb{U} \circ_i \mathbb{V}) = \begin{cases} \mathbf{U} \circ_i \mathbf{V} & \text{if } \pm_i(\mathbf{U}) = + \\ \mathbf{U} \circ_i \mathbf{V}^{\mathrm{op}} & \text{if } \pm_i(\mathbf{U}) = - \end{cases}$$

where $\mp_i(\mathbf{U}) = - \pm_i (\mathbf{U})$.

We now define an unshaded planar algebra from a symmetrically self-dual planar algebra. The proof of Theorem A will then consist of showing the action $\mathbb{Z}$ is well-defined.

DEFINITION 1.7. Given a symmetrically self-dual shaded planar algebra $(\mathbf{P}_\bullet, \Phi)$, we define a $\mathbb{Z}/2$-graded unshaded planar algebra $\mathbb{P}_\bullet$ as follows. Let

$$\mathbb{P}_n = \begin{cases} \mathbf{P}_{n/2,+} & \text{if } n \text{ is even} \\ (0) & \text{if } n \text{ is odd.} \end{cases}$$

We now define the action $\mathbb{Z}$. First, given an unshaded tangle $\mathbb{U}$ so that $k_i(\mathbb{U}) \in 2\mathbb{Z}$ for all $0 \leq i \leq t$, define

$$\psi_i(\mathbf{U}) = \begin{cases} \mathbf{1}_{k_i(\mathbf{U})} & \text{if } \pm_i (\mathbf{U}) = + \\ \Phi_+ & \text{if } \pm_i (\mathbf{U}) = -. \end{cases}$$

We define the action of $\mathbb{U}$ on $\mathbb{P}_\bullet$ by

$$\mathbb{Z}(\mathbb{U}) = \mathbf{Z}(\mathbf{U}) \circ \left(\bigotimes_{i=1}^{t} \psi_i(\mathbf{U}) \right) : \bigotimes_{i=1}^{t} \mathbf{P}_{k_i,+} \to \mathbf{P}_{k_0,+}.$$

Note that this defines a map $\bigotimes_{i=1}^{t} \mathbb{P}_{2k_i} \to \mathbb{P}_{2k_0}$.

We must now show that $\mathbb{P}_\bullet$ is well-defined, i.e., that the action of a composite of tangles $\mathbb{Z}(\mathbb{U} \circ_i \mathbb{V})$ is equal to the composite of the actions of the tangles $\mathbb{Z}(\mathbb{U}) \circ_i \mathbb{Z}(\mathbb{V})$. We first give an illustrative example.

EXAMPLE 1.8. Let $\mathbb{F}$ be the unshaded 1-click rotation tangle, and let $\mathbf{F} : \mathbf{P}_{k,-} \to \mathbf{P}_{k,+}$ be the shaded 1-click rotation. Note that $\mathbb{Z}(\mathbb{F}) = \mathbf{Z}(\mathbf{F}) \circ \Phi_+$. Then

$$\begin{aligned}
\mathbb{Z}(\mathbb{F}) \circ \mathbb{Z}(\mathbb{F}) &= (\mathbf{Z}(\mathbf{F}) \circ \Phi_+) \circ (\mathbf{Z}(\mathbf{F}) \circ \Phi_+)\\
&= (\mathbf{Z}(\mathbf{F}) \circ \Phi_+) \circ \big(\Phi_- \circ \mathbf{Z}(\mathbf{F}^{\mathrm{op}})\big)\\
&= \mathbf{Z}(\mathbf{F}) \circ (\Phi_+ \circ \Phi_-) \circ \mathbf{Z}(\mathbf{F}^{\mathrm{op}})\\
&= \mathbf{Z}(\mathbf{F}) \circ \mathbf{Z}(\mathbf{F}^{\mathrm{op}})\\
&= \mathbf{Z}(\mathbf{F} \circ \mathbf{F}^{\mathrm{op}})\\
&= \mathbf{Z}(\mathrm{sh}(\mathbb{F} \circ \mathbb{F}))\\
&= \mathbb{Z}(\mathbb{F} \circ \mathbb{F}).
\end{aligned}$$

PROOF OF THEOREM A. Suppose $\mathbb{U}, \mathbb{V}$ are unshaded planar tangles with u, v input disks respectively, with $u \geq 1$, such that $\mathbb{V}$ has $k_0(\mathbb{V}) = k_i(\mathbb{U})$ boundary points. We show that $\mathbb{Z}(\mathbb{U} \circ_i \mathbb{V}) = \mathbb{Z}(\mathbb{U}) \circ_i \mathbb{Z}(\mathbb{V})$. We treat the two cases for $\pm_i(\mathbf{U})$ separately.

Case 1: Suppose that $\pm_i(\mathbf{U}) = +$, so $\psi_i(\mathbf{U}) = \mathbf{1}_{k_i(\mathbf{U})}$. Then we have

$$\begin{aligned}
\mathbb{Z}(\mathbb{U}) \circ_i \mathbb{Z}(\mathbb{V}) &= \left(\mathbf{Z}(\mathbf{U}) \circ \left(\bigotimes_{i=1}^{u} \psi_\ell(\mathbf{U})\right)\right) \circ_i \left(\mathbf{Z}(\mathbf{V}) \circ \left(\bigotimes_{j=1}^{v} \psi_j(\mathbf{V})\right)\right)\\
&= \mathbf{Z}(\mathbf{U} \circ_i \mathbf{V}) \circ \left(\bigotimes_{j=1}^{u+v-1} \psi_j(\mathbf{U} \circ_i \mathbf{V})\right)\\
&= \mathbb{Z}(\mathbb{U} \circ_i \mathbb{V}).
\end{aligned}$$

Case 2: Suppose that $\pm_i(\mathbf{U}) = -$, so $\psi_i(\mathbf{U}) = \Phi_+$. First, note that for all $1 \leq j \leq v$,

$$\Phi_{\pm_j(\mathbf{V})} \circ \psi_j(\mathbf{V}) = \begin{Bmatrix} \Phi_- \circ \Phi_+ = \mathbf{1}_{k_j,+} & \text{if } \pm_j(\mathbf{V}) = + \\ \Phi_+ \circ \mathbf{1}_{k_j,+} = \Phi_+ & \text{if } \pm_j(\mathbf{V}) = - \end{Bmatrix} = \psi_j(\mathbf{V}^{\mathrm{op}}).$$

Then we have that

$$\begin{aligned}
&\mathbb{Z}(\mathbb{U}) \circ_i \mathbb{Z}(\mathbb{V})\\
&\quad= \left(\mathbf{Z}(\mathbf{U}) \circ \left(\bigotimes_{\ell=1}^{u} \psi_\ell(\mathbf{U})\right)\right) \circ_i \left(\mathbf{Z}(\mathbf{V}) \circ \left(\bigotimes_{j=1}^{v} \psi_j(\mathbf{V})\right)\right)\\
&\quad= \left(\mathbf{Z}(\mathbf{U}) \circ \left(\bigotimes_{\ell\neq i} \psi_\ell(\mathbf{U})\right)\right) \circ_i \left(\Phi_+ \circ \mathbf{Z}(\mathbf{V}) \circ \left(\bigotimes_{j=1}^{v} \psi_j(\mathbf{V})\right)\right)\\
&\quad= \left(\mathbf{Z}(\mathbf{U}) \circ \left(\bigotimes_{\ell\neq i} \psi_\ell(\mathbf{U})\right)\right)\\
&\qquad\qquad \circ_i \left(\Phi_+ \circ \mathbf{Z}(\mathbf{V}) \circ \left(\bigotimes_{j=1}^{v} \Phi_{\mp_j(\mathbf{V})}\right) \circ \left(\bigotimes_{j=1}^{v} \Phi_{\pm_j(\mathbf{V})} \circ \psi_j(\mathbf{V})\right)\right)
\end{aligned}$$

$$
\begin{aligned}
&= \left(\mathbb{Z}(\mathbb{U}) \circ \left(\bigotimes_{\ell \neq i} \psi_\ell(\mathbb{U})\right)\right) \circ_i \left(\mathbb{Z}(\mathbb{V}^{\mathrm{op}}) \circ \left(\bigotimes_{j=1}^{v} \psi_j(\mathbb{V}^{\mathrm{op}})\right)\right)\\
&= \mathbb{Z}(\mathbb{U} \circ_i \mathbb{V}^{\mathrm{op}}) \circ \left(\bigotimes_{j=1}^{u+v-1} \psi_j(\mathbb{U} \circ_i \mathbb{V}^{\mathrm{op}})\right)\\
&= \mathbb{Z}(\mathbb{U} \circ_i \mathbb{V}). \qquad \square
\end{aligned}
$$

REMARK 1.9. Definition 1.4 and Theorem A show that there is a bijective correspondence between symmetric self-dualities on subfactor planar algebras and unshaded $\mathbb{Z}/2\mathbb{Z}$-graded factor planar algebras. This gives one way to construct examples of $\mathbb{Z}/2\mathbb{Z}$-graded extensions of the principal even half $\mathcal{C}_0$ of a finite depth subfactor planar algebra by its odd half $\mathcal{C}_1$.

The complete answer to this question was given by **[GJS15]**, using the techniques of **[ENO10]**. In particular, **[GJS15]** gives a cohomological condition under which any invertible self-dual $\mathcal{C}_0$-bimodule category $\mathcal{C}_1$ gives exactly 2 such extensions of the form $\mathcal{C}_0 \oplus \mathcal{C}_1$.

It is possible there might be extensions of the form $\mathcal{C}_0 \oplus \mathcal{C}_1$ which are not generated by symmetrically self-dual objects. However, we have no such examples at this time.

2. Examples.

EXAMPLE 2.1. In **[MP15]**, we showed that several spoke subfactors are symmetrically self-dual, in particular, the 4442, 3333 ($3^{\mathbb{Z}/2\mathbb{Z}\times\mathbb{Z}/2\mathbb{Z}}$), and 2221 subfactors with principal graphs

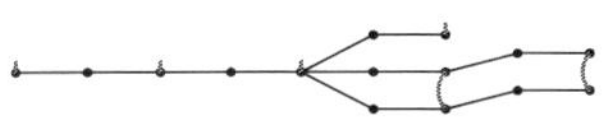 , 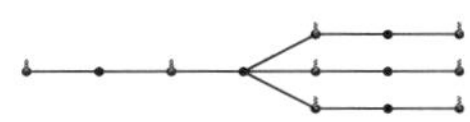 , and 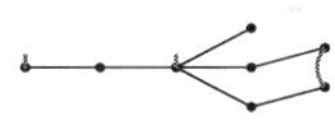 .

EXAMPLE 2.2. Many braided subfactor planar algebras coming from BMW algebras **[Wen88, BJL17]** are really unshaded. In **[LMP15]**, we introduced a notion of σ-braided subfactor planar algebras, and we showed our examples with principal graphs

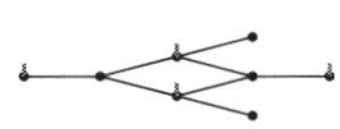 and 

are symmetrically self-dual. The former one is a special case of BMW planar algebras and $\sigma = \pm 1$. The latter one is new and $\sigma = \pm i$. Their $\mathbb{Z}_2$ fixed point planar algebras have principle graphs $2^{\mathbb{Z}_2\times\mathbb{Z}_2}1$ and $2^{\mathbb{Z}_4}1$ respectively (22221).

EXAMPLE 2.3. The $\sigma = \pm i$ subfactor in Example 2.2 turns out to be a special case of a new parameterized family of unshaded subfactor planar algebras **[Liu15]** related to quantum subgroups.

EXAMPLE 2.4. For a finite group G, the group subfactor $R \subset R \rtimes G$ is self-dual if and only if G is abelian. For G abelian, the group subfactor $R \subset R \rtimes G$ has many symmetric self-dualities, which we parametrize in §3 below via symmetric bicharacters. The resulting unitary fusion categories obtained via Theorem A are Tambara-Yamagami categories **[TY98]**.

Recall that given an abelian group A, a non-degenerate symmetric bicharacter $\chi : A \times A \to S^1$, and a sign $\pm$, the Tambara-Yamagami category $\mathcal{TY}(A, \chi, \pm)$ is a skeletal category with objects $a \in A$ and an object m satisfying the following fusion rules:

(1) $a \otimes b = ab$ for all $a, b \in A$
(2) $a \otimes m = m \otimes a = m$ for all $a \in A$, and
(3) $m \otimes m = \bigoplus_{a \in A} a$.

It was shown in [**TY98**] that the categories $\mathcal{TY}(A, \chi, \pm)$ and $\mathcal{TY}(A', \chi', \pm')$ are equivalent if and only if there is an isomorphism $A \to A'$ which sends $\chi \to \chi'$, and the signs are equal: $\pm = \pm'$. (See also [**ENO10**, Example 9.4].)

For example, Tambara-Yamagami showed in [**TY98**, Theorem 4.1] that there are only two inequivalent symmetric bicharacters on $\mathbb{Z}/2\mathbb{Z} \oplus \mathbb{Z}/2\mathbb{Z}$, which gives 4 inequivalent categories. Similarly, it is straightforward to show that there are only 2 inequivalent symmetric bicharacters on $\mathbb{Z}/4\mathbb{Z}$, which gives 4 inequivalent categories.

By [**GHR13**], all Tambara-Yamagami categories are unitary, since they are weakly group theoretical. However, to get a factor planar algebra, we need that m is symmetrically self-dual, i.e., m has Frobenius-Schur indicator 1 [**NS07**]. The Frobenius-Schur indicators for Tambara-Yamagami categories were completely worked out in [**Shi11**], where it was shown that $\nu_2(a) = \delta_{a^2,e}$ and $\nu_2(m) = \pm$, the sign in $\mathcal{TY}(A, \chi, \pm)$. Hence we must have $\pm = +$ to get a factor planar algebra.

EXAMPLE 2.5. The m-interval Jones-Wassermann subfactors for modular tensor categories are symmetrically self-dual for all $m \geq 1$ [**LX16**]. This result was inspired by the open problem to construct a conformal net whose representation category is a prescribed unitary modular tensor category. We refer the reader to [**LX16**] for further references related to Jones-Wassermann subfactors of conformal nets.

In particular, if we start with a unitary modular tensor category whose fusion ring is a finite abelian group A, then its m-interval Jones-Wassermann subfactor, for any $m \geq 2$, gives a symmetrically self-dual group subfactor (with group A), as well as a Tambara-Yamagami category as in Example 2.4. Moreover, the symmetric bicharacter χ of A is determined by the modular S matrix of the modular tensor category.

3. Group subfactors: symmetrical self-dualities and symmetric bicharacters. Given an abelian group A, we obtain a group subfactor planar algebra $\mathbb{P}_\bullet^A$. The minimal projections of $\mathbb{P}_{2,\pm}^A$ are indexed by the groups elements in A. These projections correspond to invertible objects in the category of projections [**MPS10, BHP12**], where the fusion corresponds to the coproduct, denoted $*$. We denote the minimal projections of $\mathbb{P}_{2,+}^A$ by P_g for $g \in A$ so that $dP_g * P_h = P_{gh}$, where $d = \sqrt{|A|}$.

The group subfactor planar algebra $\mathbb{P}_\bullet^A$ is an exchange relation planar algebra [**Lan02**]. By [**Liu16**, Theorem 2.26], $\Phi : \mathbb{P}_\bullet^A \to \overline{\mathbb{P}^A}_\bullet$ is a shaded planar algebra $*$-isomorphism of exchange relation planar algebras if and only if Φ preserves the structure of 2-boxes $\mathbb{P}_{2,\pm}^A$, which consists of the adjoint operator $(\cdot)^*$, the trace, the contragredient (rotation by π), the multiplication (stacking), and the coproduct $*$.

THEOREM 3.1. *Let A be a finite abelian group, and let $\mathfrak{F}_S : \mathbb{P}^A_{2,+} \to \mathbb{P}^A_{2,-}$ be the string Fourier transform (one-click rotation). If $\chi(g,h)$ is a non-degenerate bicharacter on A, then*

$$\Phi_-(\mathfrak{F}_S(P_g)) := \sum_{h\in A} \frac{1}{d}\chi(g,h)P_h \qquad\qquad \Phi_+ := \mathfrak{F}_S^{-1}\Phi_-\mathfrak{F}_S$$

extends uniquely to a shaded planar algebra $*$-isomorphism $\Phi : \mathbb{P}^A_\bullet \to \overline{\mathbb{P}^A}_\bullet$.

Conversely, if $\Phi : \mathbb{P}^A_\bullet \to \overline{\mathbb{P}^A}_\bullet$ is a shaded planar algebra $*$-isomorphism, then the coefficients $\chi(g,h)$ defined above give a bicharacter on A.

PROOF. We define $\Phi_\pm$ as in the statement and extend the map linearly to labelled tangles in the universal planar algebra generated by 2-boxes. By [**Liu16**, Theorem 2.26], it is enough to prove that Φ_- preserves the adjoint operator, the trace, the contragredient, the multiplication, and the coproduct on 2-boxes.

Step 1: the extension Φ_- preserves the adjoint operator:

$$\begin{aligned}(\Phi_-(\mathfrak{F}_S(P_g)))^* &= \sum_{h\in A} \frac{1}{d}\overline{\chi(g,h)}P_h \\ &= \sum_{h\in A} \frac{1}{d}\chi(g^{-1},h)P_h = \mathfrak{F}_S(P_{g^{-1}}) = \mathfrak{F}_S^{-1}(P_g) = (\mathfrak{F}_S(P_g))^*.\end{aligned}$$

Step 2: the extension Φ_- preserves the trace:

$$\mathrm{Tr}(\Phi_-(\mathfrak{F}_S(P_g))) = \sum_{h\in A} \frac{1}{d}\chi(g,h) = d\delta_{g=e} = \mathrm{Tr}(\mathfrak{F}_S(P_g)),$$

where $\delta_{g=e}$ is the Kronecker function, and e is the identity of the group A.

Step 3: the extension Φ_- preserves the contragredient:

$$\begin{aligned}\overline{\Phi_-(\mathfrak{F}_S(P_g))} &= \sum_{h\in A} \frac{1}{d}\chi(g,h)\overline{P_h} \\ &= \sum_{h\in A} \frac{1}{d}\chi(g,h)P_{h^{-1}} \\ &= \sum_{h\in A} \frac{1}{d}\chi(g,h^{-1})P_h \\ &= \sum_{h\in A} \frac{1}{d}\overline{\chi(g,h)}P_h \\ &= \sum_{h\in A} \frac{1}{d}\chi(g^{-1},h)P_h \\ &= \mathfrak{F}_S(P_{g^{-1}}) \\ &= \mathfrak{F}_S(\overline{P_g}).\end{aligned}$$

Step 4: the extension Φ_- preserves the multiplication:

$$\begin{aligned}\Phi_-(\mathfrak{F}_S(P_{g_1}))\Phi_-(\mathfrak{F}_S(P_{g_2})) &= \sum_{h\in A}\frac{1}{d^2}\chi(g_1,h)\chi(g_2,h)P_h\\ &= \sum_{h\in A}\frac{1}{d^2}\chi(g_1g_2,h)P_h\\ &= \frac{1}{d}\Phi_-(\mathfrak{F}_S(P_{g_1g_2}))\\ &= \Phi_-(\mathfrak{F}_S(P_{g_1}*P_{g_2}))\\ &= \Phi_-(\mathfrak{F}_S(P_{g_1})\mathfrak{F}_S(P_{g_2})).\end{aligned}$$

Step 5: the extension Φ_- preserves the coproduct:

$$\begin{aligned}\Phi_-(\mathfrak{F}_S(P_{g_1}))*\Phi_-(\mathfrak{F}_S(P_{g_2}) &= \sum_{h_1,h_2\in A}\frac{1}{d^2}\chi(g_1,h_1)\chi(g_2,h_2)P_{h_1}*P_{h_2}\\ &= \sum_{h_1,h\in A}\frac{1}{d^3}\chi(g_1,h_1)\chi(g_2,h_1^{-1}h)P_h\\ &= \sum_{h_1,h\in A}\frac{1}{d^3}\chi(g_1g_2^{-1},h_1)\chi(g_2,h)P_h\\ &= \sum_{h_1\in A}\frac{1}{d^2}\chi(g_1g_2^{-1},h_1)\mathfrak{F}_S(P_{g_2})\\ &= \delta_{g_1,g_2}\mathfrak{F}_S(P_{g_2})\\ &= \mathfrak{F}_S(P_{g_1}P_{g_2})\\ &= \mathfrak{F}_S(P_{g_1})*\mathfrak{F}_S(P_{g_2}).\end{aligned}$$

Therefore Φ extends uniquely to a shaded planar algebra $*$-isomorphism $\Phi:\mathbb{P}^A_\bullet\to\overline{\mathbb{P}^A}_\bullet$. Conversely, if $\Phi:\mathbb{P}^A_\bullet\to\overline{\mathbb{P}^A}_\bullet$ is a shaded planar algebra $*$-isomorphism, then by definition,

$$\chi(g,h)=d\,\mathrm{Tr}(\Phi_-(\mathfrak{F}_S(P_g))P_h).$$

In particular $\chi(e,h)=\mathrm{Tr}(P_h)=1$. By the computation in Step 3,

$$\Phi_-(\mathfrak{F}_S(P_{g_1}))\Phi_-(\mathfrak{F}_S(P_{g_2}))=\Phi_-(\mathfrak{F}_S(P_{g_1})\mathfrak{F}_S(P_{g_2}))$$

implies that $\chi(g_1g_2,h)=\chi(g_1,h)\chi(g_2,h)$. Thus $\chi(g,h)$ is a character with respect to g. On the other hand,

$$\chi(g,h)=d\,\mathrm{Tr}(\Phi_-(\mathfrak{F}_S(P_g))P_h)=d\,\mathrm{Tr}(\mathfrak{F}_S(P_g)\Phi_-^{-1}(P_h))$$

and in particular $\chi(g,e) = d\,\mathrm{Tr}(\mathfrak{F}_S(P_g)P_e) = 1$. Moreover,

$$\begin{aligned}
\chi(g,h_1h_2) &= d\,\mathrm{Tr}(\mathfrak{F}_S(P_g)\Phi_-^{-1}(P_{h_1h_2})) \\
&= d\,\mathrm{Tr}(P_g\mathfrak{F}_S^{-1}(\Phi_-^{-1}(P_{h_1h_2}))) \\
&= d^2\,\mathrm{Tr}(P_g\mathfrak{F}_S^{-1}(\Phi_-^{-1}(P_{h_1} * P_{h_2}))) \\
&= d^2\,\mathrm{Tr}(P_g\mathfrak{F}_S^{-1}((\Phi_-^{-1}(P_{h_1}) * \Phi^{-1}(P_{h_2}))) \\
&= d^2\,\mathrm{Tr}(P_g\mathfrak{F}_S^{-1}(\Phi_-^{-1}(P_{h_1}))\mathfrak{F}_S^{-1}(\Phi^{-1}(P_{h_2}))) \\
&= d^2\,\mathrm{Tr}(P_g\mathfrak{F}_S^{-1}(\Phi_-^{-1}(P_{h_1})))\,\mathrm{Tr}(P_g\mathfrak{F}_S^{-1}(\Phi^{-1}(P_{h_2}))) \\
&= \chi(g,h_1)\chi(g,h_2).
\end{aligned}$$

Therefore $\chi(g,h)$ is a bicharacter. □

THEOREM 3.2. *The bicharacter $\chi(g,h)$ in Theorem 3.1 is symmetric if and only if $\Phi^2 = 1$, namely $\mathscr{P}_\bullet^A$ is symmetrically self-dual.*

PROOF. Note that $\Phi_+\Phi_- = \mathfrak{F}_S^{-1}\Phi_-\mathfrak{F}_S\Phi_-$, so $\Phi^2 = 1$ if and only if $(\Phi_-\mathfrak{F}_S)^2 = \mathfrak{F}_S^2$.

If the bicharacter $\chi(g,h)$ is symmetric, then

$$(\Phi_-\mathfrak{F}_S)^2(P_g) = \sum_{h,k\in A} \frac{1}{d^2}\chi(h,k)\chi(g,h)P_k = \sum_{h,k\in A} \frac{1}{d^2}\chi(g{+}k,h)P_k = P_{g^{-1}} = \mathfrak{F}_S^2(P_g).$$

Thus $(\Phi_-\mathfrak{F}_S)^2 = \mathfrak{F}_S^2$ and $\Phi^2 = 1$.

On the other hand, if $\Phi^2 = 1$, then

$$\begin{aligned}
\chi(g,h) &= d\,\mathrm{Tr}(\Phi_-(\mathfrak{F}_S(P_g))P_h) \\
&= d\,\mathrm{Tr}(\mathfrak{F}_S^{-1}(\Phi_-(\mathfrak{F}_S(P_g)))\mathfrak{F}_S(P_h)) \\
&= d\,\mathrm{Tr}(\Phi_-^{-1}(P_g)\mathfrak{F}_S(P_h)) \\
&= d\,\mathrm{Tr}(P_g\Phi_-(\mathfrak{F}_S(P_h))) \\
&= d\,\mathrm{Tr}(\Phi_-(\mathfrak{F}_S(P_h))P_g) \\
&= \chi(h,g),
\end{aligned}$$

and the bicharacter is symmetric. □

For a finite abelian group A and a non-degenerate symmetric bicharacter $\chi : A\times A \to S^1$, the $\mathbb{Z}/2$-graded unitary fusion category corresponding to the unshaded planar algebra $\mathscr{P}_\bullet^A$ is the Tambara-Yamagami category. We refer the readers to **[LX16]** for a more general case, where A is replaced by a unitary modular tensor category and χ is replaced by the modular S matrix.

Acknowledgement. We would like to thank Noah Snyder for helpful conversations.

References

[AMP15] Narjess Afzaly, Scott Morrison, and David Penneys, *The classification of subfactors with index at most* $5\frac{1}{4}$, 2015. arXiv:1509.00038, to appear Mem. Amer. Math. Soc.

[BMPS12] Stephen Bigelow, Scott Morrison, Emily Peters, and Noah Snyder, *Constructing the extended Haagerup planar algebra*, Acta Math. **209** (2012), no. 1, 29–82. MR2979509, arXiv:0909.4099, DOI:10.1007/s11511-012-0081-7. MR2979509

[BJL17] Dietmar Bisch, Vaughan F. R. Jones, and Zhengwei Liu, *Singly generated planar algebras of small dimension, Part III*, Trans. Amer. Math. Soc. **369** (2017), no. 4, 2461–2476, DOI 10.1090/tran/6719. MR3592517

[BHP12] Arnaud Brothier, Michael Hartglass, and David Penneys, *Rigid* C^**-tensor categories of bimodules over interpolated free group factors*, J. Math. Phys. **53** (2012), no. 12, 123525, 43. MR3405915 DOI:10.1063/1.4769178 arXiv:1208.5505. MR3405915

[CMS11] Frank Calegari, Scott Morrison, and Noah Snyder, *Cyclotomic integers, fusion categories, and subfactors*, Comm. Math. Phys. **303** (2011), no. 3, 845–896. MR2786219 DOI:10.1007/s00220-010-1136-2 arXiv:1004.0665. MR2786219 (2012e:18013)

[ENO10] Pavel Etingof, Dmitri Nikshych, and Victor Ostrik, *Fusion categories and homotopy theory*, Quantum Topol. **1** (2010), no. 3, 209–273, DOI 10.4171/QT/6. With an appendix by Ehud Meir. MR2677836

[GHR13] César Galindo, Seung-Moon Hong, and Eric C. Rowell, *Generalized and quasi-localizations of braid group representations*, Int. Math. Res. Not. IMRN **3** (2013), 693–731, DOI 10.1093/imrn/rnr269. MR3021796

[GJS15] Pinhas Grossman, David Jordan, and Noah Snyder, *Cyclic extensions of fusion categories via the Brauer-Picard groupoid*, Quantum Topol. **6** (2015), no. 2, 313–331, DOI 10.4171/QT/64. MR3354332

[HP17] Michael Hartglass and David Penneys, C^**-algebras from planar algebras I: Canonical* C^**-algebras associated to a planar algebra*, Trans. Amer. Math. Soc. **369** (2017), no. 6, 3977–4019, DOI 10.1090/tran/6781. MR3624399

[Izu01] Masaki Izumi, *The structure of sectors associated with Longo-Rehren inclusions. II. Examples*, Rev. Math. Phys. **13** (2001), no. 5, 603–674, DOI 10.1142/S0129055X01000818. MR1832764

[Izu18] Masaki Izumi, *The classification of* 3^n *subfactors and related fusion categories*, Quantum Topol. **9** (2018), no. 3, 473–562, DOI 10.4171/QT/113. MR3827808

[Jon12] Vaughan F. R. Jones, *Quadratic tangles in planar algebras*, Duke Math. J. **161** (2012), no. 12, 2257–2295. MR2972458, arXiv:1007.1158, DOI:10.1215/00127094-1723608. MR2972458

[JMS14] Vaughan F. R. Jones, Scott Morrison, and Noah Snyder, *The classification of subfactors of index at most 5*, Bull. Amer. Math. Soc. (N.S.) **51** (2014), no. 2, 277–327, DOI 10.1090/S0273-0979-2013-01442-3. MR3166042

[Lan02] Zeph A. Landau, *Exchange relation planar algebras*, Proceedings of the Conference on Geometric and Combinatorial Group Theory, Part II (Haifa, 2000), 2002, pp. 183–214. MR1950890 DOI:10.1023/A:1021296230310. MR1950890

[Liu15] Zhengwei Liu, *Singly generated planar algebras of small dimension, part IV*, 2015. arXiv:1507.06030.

[Liu16] Zhengwei Liu, *Exchange relation planar algebras of small rank*, Trans. Amer. Math. Soc. **368** (2016), no. 12, 8303–8348, DOI 10.1090/tran/6582. MR3551573

[LMP15] Zhengwei Liu, Scott Morrison, and David Penneys, *1-Supertransitive Subfactors with Index at Most* $6\frac{1}{5}$, Comm. Math. Phys. **334** (2015), no. 2, 889–922. MR3306607, arXiv:1310.8566, DOI:10.1007/s00220-014-2160-4. MR3306607

[LX16] Zhengwei Liu and Feng Xu, *Jones-Wassermann subfactors for modular tensor categories*, 2016. arXiv:1612.08573.

[MP15] Scott Morrison and David Penneys, *Constructing spoke subfactors using the jellyfish algorithm*, Trans. Amer. Math. Soc. **367** (2015), no. 5, 3257–3298, DOI 10.1090/S0002-9947-2014-06109-6. MR3314808

[MPS10] Scott Morrison, Emily Peters, and Noah Snyder, *Skein theory for the* D_{2n} *planar algebras*, J. Pure Appl. Algebra **214** (2010), no. 2, 117–139, DOI 10.1016/j.jpaa.2009.04.010. MR2559686

[NS07] Siu-Hung Ng and Peter Schauenburg, *Higher Frobenius-Schur indicators for pivotal categories*, Hopf algebras and generalizations, Contemp. Math., vol. 441, Amer. Math. Soc., Providence, RI, 2007, pp. 63–90, DOI 10.1090/conm/441/08500. MR2381536

[Pet10] Emily Peters, *A planar algebra construction of the Haagerup subfactor*, Internat. J. Math. **21** (2010), no. 8, 987–1045, DOI 10.1142/S0129167X10006380. MR2679382

[Shi11] Kenichi Shimizu, *Frobenius-Schur indicators in Tambara-Yamagami categories*, J. Algebra **332** (2011), 543–564, DOI 10.1016/j.jalgebra.2011.02.002. MR2774703

[TY98] Daisuke Tambara and Shigeru Yamagami, *Tensor categories with fusion rules of self-duality for finite abelian groups*, J. Algebra **209** (1998), no. 2, 692–707, DOI 10.1006/jabr.1998.7558. MR1659954

[Wen88] Hans Wenzl, *Hecke algebras of type A_n and subfactors*, Invent. Math. **92** (1988), no. 2, 349–383. MR936086 DOI:10.1007/BF01404457.

Contemporary Mathematics
Volume **747**, 2020
https://doi.org/10.1090/conm/747/15039

Q-systems and compact W*-algebra objects

Corey Jones and David Penneys

Abstract. We show that given a rigid C*-tensor category, there is an equivalence of categories between normalized irreducible Q-systems, also known as connected unitary Frobenius algebra objects, and compact connected W*-algebra objects. Although this result could be proved as a corollary of our previous article on realizations of algebra objects and discrete subfactors, we prove it here directly via categorical methods without passing through subfactor theory.

1. Introduction

The standard invariant of a finite index II_1 subfactor was first defined as a λ-lattice [**Pop95**]. Since, it has been reinterpreted as a planar algebra [**Jon99**] and a Q-system, or unitary Frobenius algebra object, in a rigid C*-tensor category [**Lon89, Müg03, BKLR15**]. Using this third language, the standard invariant of a finite index irreducible II_1 subfactor is an irreducible Q-system, i.e., the underlying algebra object A is *connected*, meaning $\dim(\mathcal{C}(1_{\mathcal{C}}, A)) = 1$.

In [**JP17a**], we introduced the notion of a W*-algebra object in a rigid C*-tensor category. In [**JP17b**], we characterized an extremal irreducible discrete inclusion of factors $(N \subseteq M, E)$ where N is type II_1 and $E : M \to N$ is a faithful normal conditional expectation in terms of a connected W*-algebra object $\mathbf{A}$ in (an ind-completion of) a rigid C*-tensor category $\mathcal{C}$, and a fully faithful embedding $\mathcal{C} \hookrightarrow \mathsf{Bim}_{\mathsf{bf}}^{\mathsf{sp}}(N)$, the spherical/extremal bifinite bimodules over N. (Such an embedding always exists for $N = L\mathbb{F}_\infty$ by [**PS03, BHP12**].) This means that we may also view the standard invariant of a finite index irreducible II_1 subfactor as a *compact* connected W*-algebra object, i.e., one which actually lies not only in the ind-completion of $\mathcal{C}$, but also in $\mathcal{C}^\natural$, the image of the Yoneda embedding of $\mathcal{C}$ into $\operatorname{ind}(\mathcal{C})$.[1] Thus passing through our theorem [**JP17b**, Thm. A and §7.2], we get an equivalence between irreducible Q-systems and compact connected W*-algebra objects.

2010 *Mathematics Subject Classification.* Primary 46L37, 18D10.

Corey Jones was supported by Discovery Projects 'Subfactors and symmetries' DP140100732 and 'Low dimensional categories' DP160103479 from the Australian Research Council.

David Penneys was supported by NSF grants DMS-1500387/1655912 and 1654159.

This work was supported by EPSRC grant no EP/K032208/1.

[1]Warning: $\mathcal{C}^\natural$ is identical to $\mathcal{C}$ as a linear tensor category, but it has no dagger structure!

©2020 American Mathematical Society

The purpose of the present article is to provide a direct proof of the equivalence between irreducible Q-systems and compact connected W*-algebra objects in an arbitrary rigid C*-tensor category without passing through subfactor theory. In doing so, we observe that the correct notion of morphisms between Q-systems is that of involutive algebra morphism from [**Vic11**, Def. 3.4 and 3.9]. We refer the reader to Section 2.4 below for the definition.

Our main theorem is:

THEOREM A. *Let $\mathcal{C}$ be a rigid* C*-*tensor category.*[2] *There is an equivalence of categories*

$$\left\{\begin{array}{l}\text{Compact connected W*-algebra}\\ \text{objects } \mathbf{A} \in \mathcal{C}^{\natural} \text{ with unital } *\text{-}\\ \text{algebra morphisms}\end{array}\right\} \cong \left\{\begin{array}{l}\text{Normalized irreducible Q-}\\ \text{systems in } \mathcal{C} \text{ with involutive}\\ \text{algebra morphisms}\end{array}\right\}.$$

In this article, we model $\operatorname{ind}(\mathcal{C})$ as the category of linear functors $\mathcal{C}^{\mathrm{op}} \to \mathsf{Vec}$, denoted by $\mathsf{Vec}(\mathcal{C})$. While this category has no dagger structure, it does have an *involutive structure* (see Section 2.1) which behaves like complex conjugation on complex vector spaces. We also use the category $\mathsf{Hilb}(\mathcal{C})$ of linear dagger functors $\mathcal{C}^{\mathrm{op}} \to \mathsf{Hilb}$, which is equivalent to the unitary ind-category of $\mathcal{C}$ defined in [**NY16**]. Now $\mathsf{Hilb}(\mathcal{C})$ has both an involutive structure similar to complex conjugation and a dagger structure given by taking adjoints of bounded natural transformations. Indeed, these involutions are compatible, and we call $\mathsf{Hilb}(\mathcal{C})$ a *bi-involutive* W*-category. The relationship between $\mathsf{Vec}(\mathcal{C})$ and $\mathsf{Hilb}(\mathcal{C})$ is analogous to the relationship between vector spaces and Hilbert spaces.

The Yoneda embedding is a fully faithful monoidal functor $\mathcal{C} \hookrightarrow \mathsf{Vec}(\mathcal{C})$ given by $c \mapsto \mathcal{C}(-, c)$ which is compatible with the involutive structures (complex conjugation functors) on both categories. As alluded to above, the image of $\mathcal{C}$ in $\mathsf{Vec}(\mathcal{C})$ denoted $\mathcal{C}^{\natural}$ no longer has an adjoint on the level of morphisms! In comparison, we have a Yoneda embedding $\mathcal{C} \hookrightarrow \mathsf{Hilb}(\mathcal{C})$, whose image is a rigid C*-tensor category which is dagger equivalent to $\mathcal{C}$. This is explained in detail in Section 2.2.

The proof of Theorem A proceeds as follows. In Section 3, starting with an irreducible Q-system $A \in \mathcal{C}$, we define a compact connected W*-algebra object $\mathbf{A} = \mathcal{C}(-, A) \in \mathcal{C}^{\natural} \subset \mathsf{Vec}(\mathcal{C})$ by taking its image under the Yoneda embedding. The difficult direction is going the other way; it is easy to get a vector space from a Hilbert space by forgetting the adjoint, but it is more difficult (and usually not possible) to canonically obtain a Hilbert space from a vector space! In particular, if we want to produce Q-system (unitary Frobenius algebra object) in $\mathcal{C}$ starting with an algebra $\mathbf{A} \in \mathcal{C}^{\natural} \subset \mathsf{Vec}(\mathcal{C})$, we need to endow the vector spaces $\mathbf{A}(a)$ with inner products such that $\mathbf{A}$ becomes a dagger functor, i.e., we need to produce an object in $\mathsf{Hilb}(\mathcal{C})$. Luckily, there is a canonical way to do this for a compact connected W*-algebra object $\mathbf{A} \in \mathsf{Vec}(\mathcal{C})$, which yields a canonical compact Hilbert space object $\mathbf{L}^2\mathbf{A} \in \mathcal{C} \subset \mathsf{Hilb}(\mathcal{C})$ (see Section 2.6 for more details).

In Section 4, starting with $\mathbf{A}$, we embed the compact Hilbert space object $\mathbf{L}^2\mathbf{A} \in \mathsf{Hilb}(\mathcal{C})$ into the (non-irreducible) Q-system $\mathbf{L}^2\mathbf{A} \otimes \overline{\mathbf{L}^2\mathbf{A}}$, which gives a canonical projector $p \in \operatorname{End}_{\mathcal{C}}(H \otimes \overline{H})$ where $H \in \mathcal{C}$ is such that $\mathbf{L}^2\mathbf{A}(a) = \mathcal{C}(a, H)$ for all $a \in \mathcal{C}$. This is similar to the strategy of [**Vic11**, Lem. 3.21], where it is shown that an involutive subalgebra of a unitary Frobenius algebra is again Frobenius.

[2] In this article, we assume our rigid C*-tensor category $\mathcal{C}$ has simple unit object, is additive, and is idempotent complete. By [**NT13**, Cor. 2.2.8], such $\mathcal{C}$ are semi-simple.

Conceptually, this is analogous to the fact that a unital $*$-subalgebra of a finite dimensional von Neumann algebra is again a von Neumann algebra. We then prove the equivalence of categories in Sections 5.1 and 5.2.

While our proof is conceptual from an operator algebraic perspective, focusing completely on the algebra objects, it is important to note that there is a second conceptual approach passing through proper module categories. Taking this approach, this result should be viewed as an extension of [**NY17**, Appendix A]. One could expand the proof of [**NY17**, Thm. A.1] to prove an equivalence between the category of irreducible Q-systems and the category of proper cyclic $\mathcal{C}$-module W*-categories with simple basepoint. One would then apply our equivalence between W*-algebra objects and cyclic $\mathcal{C}$-module W*-categories [**JP17a**, Thm. 3.24] to get another proof of Theorem A.

Acknowledgements. The authors would like to thank Marcel Bischoff, André Henriques, and Jamie Vicary for helpful discussions.

The authors would like to thank the Isaac Newton Institute for Mathematical Sciences, Cambridge, for support and hospitality during the programme Operator Algebras: Subfactors and their Applications where work on this paper was undertaken.

2. Background

We refer the reader to [**HP17**, §2.1-2.2] and [**JP17a**, §2.1-2.3] for background on rigid C*-tensor categories and their graphical calculus.

2.1. Involutive and bi-involutive tensor categories. Suppose $\mathcal{C}$ is a tensor category. An *involutive* structure on $\mathcal{C}$ consists of an anti-linear anti-monoidal equivalence $\overline{\cdot} : \mathcal{C} \to \mathcal{C}$ together with a monoidal natural isomorphism $\varphi : \mathrm{id} \Rightarrow \overline{\overline{\cdot}}$. For $a, b \in \mathcal{C}$, there are associative natural isomorphisms $\nu_{a,b} : \overline{a} \otimes \overline{b} \to \overline{b \otimes a}$ and a distinguished isomorphism $r : 1 \to \overline{1}$. The natural isomorphisms ν, φ, r satisfy various coherence relations; see [**JP17a**, §2.2] for more details. If $\mathcal{C}$ is a dagger category, we call an involutive structure *bi-involutive* if $\overline{\cdot}$ is a dagger functor, and the coherence natural isomorphisms ν, φ, r are all unitary. Whenever possible, we suppress all structure isomorphisms, including associators, unitors, and ν, φ, r.

A rigid C*-tensor category $\mathcal{C}$ has a canonical bi-involutive structure; see [**HP17**, §2.2] or [**JP17a**, §2.3] for the precise definitions. Applying both the conjugate $\overline{\cdot}$ and the adjoint $*$ to morphisms in $\mathcal{C}$ gives a contravariant dual functor $(\cdot)^\vee : \mathcal{C} \to \mathcal{C}$:

$$\mathcal{C}(a,b) \ni \psi \mapsto \psi^\vee := \overline{\psi}^* = \overline{\psi^*} = \quad \text{(1)}$$

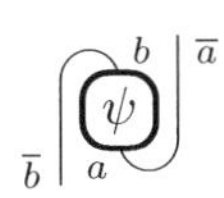

$$\in \mathcal{C}(\overline{b}, \overline{a}).$$

The equation (1) implies that we actually have *three* commuting involutions on morphisms: $(\cdot)^*, \overline{\cdot}, (\cdot)^\vee$. Here, $(\cdot)^*, \overline{\cdot}$ are anti-linear and $(\cdot)^\vee$ is linear, and combining any two in either order gives the third.

2.2. Tensor categories associated to $\mathcal{C}$. Fix a rigid C*-tensor category $\mathcal{C}$, and let $\mathrm{Irr}(\mathcal{C})$ be a set of representatives for the isomorphism classes of simple objects of $\mathcal{C}$.

DEFINITION 2.1 ([**LR97**, §2]). We say that morphisms $R_c : 1_\mathcal{C} \to \overline{c} \otimes c$ and $S_c : 1_\mathcal{C} \to c \otimes \overline{c}$ are *solutions to the conjugate equations* if they satisfy the zig-zag axioms. They are called *standard solutions* if they additionally satisfy the balancing condition

$$R_c^* \circ (\mathrm{id}_{\overline{c}} \otimes f) \circ R_c = S_c^* \circ (f \otimes \mathrm{id}_{\overline{c}}) \circ S_c \qquad \text{for all } f \in \mathrm{End}_{\mathcal{C}}(c).$$

DEFINITION 2.2. We rapidly recall the tensor categories associated to $\mathcal{C}$ which arise in the study of $*$-algebra objects. We refer the reader to [**JP17a, JP17b**] for more details on these categories.

- $\mathcal{C}^\natural$ is the rigid involutive tensor category obtained from $\mathcal{C}$ by forgetting the adjoint $*$.
- $\mathsf{Vec}(\mathcal{C})$ is the tensor category of linear functors $(\mathcal{C}^\natural)^{\mathrm{op}} \to \mathsf{Vec}$ with linear natural transformations and the Day convolution tensor product. Note that $\mathsf{Vec}(\mathcal{C})$ is equivalent to $\mathrm{ind}(\mathcal{C}^\natural)$.
- $\mathsf{Hilb}(\mathcal{C})$ is the W*-tensor category of linear dagger tensor functors $\mathcal{C}^{\mathrm{op}} \to \mathsf{Hilb}$ with bounded linear natural transformations and the Day convolution tensor product. Note that $\mathsf{Hilb}(\mathcal{C})$ is equivalent to the Neshveyev-Yamashita unitary ind category of $\mathcal{C}$ [**NY16**].

NOTATION 2.3. Typically, we use bold letters for functors, like $\mathbf{F} \in \mathsf{Vec}(\mathcal{C})$ and $\mathbf{H} \in \mathsf{Hilb}(\mathcal{C})$, and we will use regular letters for objects in $\mathcal{C}^\natural$ and $\mathcal{C}$.

Recall that the Yoneda embedding $a \mapsto \mathbf{a} := \mathcal{C}^\natural(\cdot, a)$ gives an embedding of involutive tensor categories $\mathcal{C}^\natural \hookrightarrow \mathsf{Vec}(\mathcal{C})$ and similarly, we have an embedding of bi-involutive tensor categories $\mathcal{C} \hookrightarrow \mathsf{Hilb}(\mathcal{C})$ also denoted $b \mapsto \mathbf{b} := \mathcal{C}(\cdot, b)$. (We explicitly distinguish between the two when confusion can arise.) The morphism spaces of $\mathcal{C}$ come equipped with inner products given by

(2) $\langle f|g\rangle_{\mathcal{C}(a,b)} :=$ 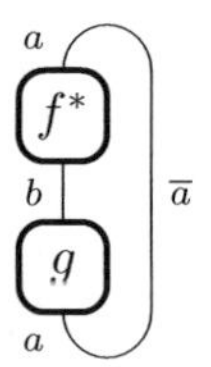

.

It is straightforward to verify that $\mathbf{b} : \mathcal{C}^{\mathrm{op}} \to \mathsf{Hilb}$ where each $\mathcal{C}(a, b)$ has the inner product as above gives a fully faithful linear dagger functor dagger $\mathcal{C}^{\mathrm{op}} \to \mathsf{Hilb}$. (Indeed, the proof that $\mathcal{C}(\cdot, \psi^*) = \mathcal{C}(\cdot, \psi)^*$ for morphisms ψ in $\mathcal{C}$ follows as in [**JP17a**, Rem. 2.28].)

An object $\mathbf{V} \in \mathsf{Vec}(\mathcal{C})$ is called *compact* if $\mathbf{V}$ is in the essential image of the Yoneda embedding $\mathcal{C}^\natural \hookrightarrow \mathsf{Vec}(\mathcal{C})$, i.e., there is a corresponding $V \in \mathcal{C}^\natural$ and a natural isomorphism $\mathbf{V} \cong \mathcal{C}^\natural(\cdot, V)$. Recall that given two such representing objects $a, b \in \mathcal{C}^\natural$ and natural isomorphisms $\alpha : \mathbf{a} \Rightarrow \mathbf{V}$ and $\beta : \mathbf{b} \Rightarrow \mathbf{V}$, there is a unique isomorphism $\psi \in \mathcal{C}^\natural(a, b)$ such that $\beta^{-1} \circ \alpha = \mathcal{C}^\natural(\cdot, \psi) \in \mathrm{Hom}_{\mathsf{Vec}(\mathcal{C})}(\mathbf{a}, \mathbf{b})$. In the sequel, we will immediate replace a compact $\mathbf{V} \in \mathsf{Vec}(\mathcal{C})$ with $\mathcal{C}^\natural(\cdot, V)$ for some $V \in \mathcal{C}^\natural$.

An object $\mathbf{H} \in \mathsf{Hilb}(\mathcal{C})$ is called *compact* if it is in the essential image of the Yoneda embedding $\mathcal{C} \hookrightarrow \mathsf{Hilb}(\mathcal{C})$, i.e., there is a corresponding $H \in \mathcal{C}$ and a *unitary* natural isomorphism $\mathbf{H} \cong \mathcal{C}(\cdot, H)$. Observe that if we have two representing objects $a, b \in \mathcal{C}$ for a compact $\mathbf{H} \in \mathsf{Hilb}(\mathcal{C})$ together with two unitary natural isomorphisms $\alpha : \mathbf{a} \Rightarrow \mathbf{H}$ and $\beta : \mathbf{b} \Rightarrow \mathbf{H}$, then the unique isomorphism $\beta^{-1} \circ \alpha \in \mathrm{Hom}_{\mathsf{Hilb}_{\mathcal{C}}}(\mathbf{a}, \mathbf{b})$ is necessarily unitary. Now as the Yoneda embedding $\mathcal{C} \hookrightarrow \mathsf{Hilb}(\mathcal{C})$ is a fully faithful dagger functor, we see that the unique isomorphism $\psi \in \mathcal{C}(a, b)$ such that $\beta^{-1} \circ \alpha =$

$\mathcal{C}(\cdot,\psi) \in \operatorname{Hom}_{\mathsf{Hilb}(\mathcal{C})}(\mathbf{a},\mathbf{b})$ is unitary. Indeed,

$$\mathcal{C}(\cdot,\psi^*\circ\psi) = \mathcal{C}(\cdot,\psi^*)\circ\mathcal{C}(\cdot,\psi) = \mathcal{C}(\cdot,\psi)^*\circ\mathcal{C}(\cdot,\psi) = \mathrm{id}_{\mathbf{a}} \qquad \Longrightarrow \qquad \psi^*\circ\psi = \mathrm{id}_a,$$

and similarly, $\psi\circ\psi^* = \mathrm{id}_b$. Again, we immediately replace a compact $\mathbf{H}\in\mathsf{Hilb}(\mathcal{C})$ with $\mathcal{C}(\cdot,H)$.

The next proposition helps us identify compact objects in $\mathsf{Hilb}(\mathcal{C})$.

PROPOSITION 2.4. *Suppose* $\mathbf{H}\in\mathsf{Hilb}(\mathcal{C})$ *such that each* $\mathbf{H}(a)$ *is finite dimensional, and there are only finitely many* $c\in\operatorname{Irr}(\mathcal{C})$ *such that* $\mathbf{H}(c)\neq 0$. *Then* $\mathbf{H}$ *is compact.*

PROOF. We must find an $H\in\mathcal{C}$ and a unitary natural isomorphism $\theta:\mathbf{H}\Rightarrow\mathcal{C}(\cdot,H)$. First, define $H := \bigoplus_{c\in\operatorname{Irr}(\mathcal{C})} c^{\oplus\dim\mathbf{H}(c)}\in\mathcal{C}$ which exists by additivity of $\mathcal{C}$. Now for $a\in\mathcal{C}$, we have a unitary isomorphism natural in $a\in\mathcal{C}$:

$$\bigoplus_{c\in\operatorname{Irr}(\mathcal{C})} \mathcal{C}(a,c)\otimes_c\mathcal{C}(c,H)\cong\mathcal{C}(a,H)$$

given by $\psi\otimes\phi\mapsto\phi\circ\psi$. Here, the inner product on $\mathcal{C}(a,c)\otimes_c\mathcal{C}(c,H)$ is given by

$$\langle\psi_1\otimes\phi_1,\psi_2\otimes\phi_2\rangle := d_c^{-1}\langle\psi_1,\psi_2\rangle_{\mathcal{C}(a,c)}\langle\phi_1,\phi_2\rangle_{\mathcal{C}(c,H)},$$

where the purpose of this factor of d_c^{-1} is to make the above map unitary.

We now construct a unitary isomorphism natural in $a\in\mathcal{C}$

$$\bigoplus_{c\in\operatorname{Irr}(\mathcal{C})} \mathcal{C}(a,c)\otimes_c\mathbf{H}(c)\cong\mathbf{H}(a)$$

given by $\mathcal{C}(a,c)\otimes_c\mathbf{H}(c)\ni\psi\otimes\xi\mapsto\mathbf{H}(\psi)[\xi]\in\mathbf{H}(a)$. Again, the inner product on each summand is balanced by a factor of d_c^{-1} for unitarity (see also [**JP17a**, (13)]).

Now for each $c\in\operatorname{Irr}(\mathcal{C})$, since $\dim(\mathbf{H}(c)) = \dim(\mathcal{C}(c,H))$, we may choose an arbitrary unitary isomorphism $\mathbf{H}(c)\cong\mathcal{C}(c,H)$. Finally, $\mathbf{H}\cong\mathcal{C}(\cdot,H)$ via the composite unitary isomorphism

$$\mathbf{H}(a)\cong\bigoplus_{c\in\operatorname{Irr}(\mathcal{C})}\mathcal{C}(a,c)\otimes_c\mathcal{C}(c,H)\cong\bigoplus_{c\in\operatorname{Irr}(\mathcal{C})}\mathcal{C}(a,c)\otimes_c\mathbf{H}(c)\cong\mathbf{H}(a)$$

which is natural in $a\in\mathcal{C}$ by construction. □

2.3. C* Frobenius algebra objects. We now introduce C* Frobenius algebra objects in rigid C*-tensor categories following [**BKLR15**, §3.1]. Of particular importance will be Q-systems, which are unitary Frobenius algebra objects discussed in Section 2.4. Some other general references on Frobenius algebras (and Q-systems) in rigid C*-tensor categories include [**Lon89**, **LR97**, **Müg03**, **Vic11**].

DEFINITION 2.5. An *algebra object* in a tensor category $\mathcal{T}$ is a triple (A,m,i) where $A\in\mathcal{T}$ and $m\in\mathcal{T}(A\otimes A,A)$ and $i\in\mathcal{T}(1_{\mathcal{T}},A)$ are morphisms satisfying the following relations:

- (associativity) $m\circ(m\otimes\mathrm{id}_A) = m\circ(\mathrm{id}_A\otimes m)$.
- (unitality) $m\circ(i\otimes\mathrm{id}_A) = \mathrm{id}_A = m\circ(\mathrm{id}_A\otimes i)$.

A coalgebra object is defined similarly by reversing the arrows.

An algebra object is called *connected* if $\dim(\mathcal{T}(1_{\mathcal{T}},A)) = 1$.

We make heavy use of the graphical calculus for morphisms in tensor categories, in which m is denoted by a trivalent vertex, and i is denoted by a univalent vertex. We now specialize to algebras in a rigid C*-tensor category $\mathcal{C}$.

DEFINITION 2.6. A C* *Frobenius algebra object* is an algebra object $(A, m, i) \in \mathcal{C}$ (which automatically implies (A, m^*, i^*) is a coalgebra object) such that the following condition holds:

- (Frobenius) $m^* \circ m = (\mathrm{id}_A \otimes m) \circ (m^* \otimes \mathrm{id}_A) = (m \otimes \mathrm{id}_A) \circ (\mathrm{id}_A \otimes m^*)$.

A C* Frobenius algebra is called *separable* [**Müg03**, Def. 3.13] or *special* [**FRS02**, (6)], [**BKLR15**, Def. 3.2] if $m \circ m^* = \lambda \,\mathrm{id}_A$ for some constant $\lambda \in \mathbb{C}$. In this case, since $\mathcal{C}$ is C*, we automatically have $\lambda > 0$. Since $1_\mathcal{C}$ is simple and $\mathcal{C}$ is C*, we automatically have $i^* \circ i = \lambda' \,\mathrm{id}_{1_\mathcal{C}}$ for some non-zero constant $\lambda' > 0$. We call a C* Frobenius algebra *normalized* if $\lambda' = 1$.[3]

EXAMPLE 2.7. Suppose $c \in \mathcal{C}$, and consider $c \otimes \overline{c}$ with multiplication $m := \mathrm{id}_c \otimes \mathrm{ev}_c \otimes \mathrm{id}_{\overline{c}}$ and unit $i := \mathrm{coev}_c$. Then $(c \otimes \overline{c}, m, i)$ is a separable C* Frobenius algebra, connected if and only if c is simple. Moreover, it is well known that $c \otimes \overline{c}$ is a real/symmetrically self-dual object.

LEMMA 2.8. *Suppose (A, m, i) is an algebra object in $\mathcal{C}$. The Frobenius condition above is equivalent to the following two conditions:*

- *(Self-duality) There exist maps $\widetilde{\mathrm{ev}}_A : A \otimes A \to 1_\mathcal{C}$ and $\widetilde{\mathrm{coev}}_A : 1_\mathcal{C} \to A \otimes A$ which satisfy the zig-zag axioms, and*
- *(Rotational invariance) $m^* = (\mathrm{id}_A \otimes m) \circ (\widetilde{\mathrm{coev}}_A \otimes \mathrm{id}_A) = (m \otimes \mathrm{id}_A) \circ (\mathrm{id}_A \otimes \widetilde{\mathrm{coev}}_A)$.*

PROOF. One passes from the Frobenius condition to the above two conditions by setting $\widetilde{\mathrm{ev}}_A = i^* \circ m$ and $\widetilde{\mathrm{coev}}_A = m^* \circ i$ and using the Frobenius relation. One passes from the two above conditions to the Frobenius condition by trading m^* for m using the rotation relation and using associativity. □

2.4. Unitary Frobenius algebras/Q-systems and involutive morphisms.

DEFINITION 2.9. Given two algebra objects $(A, m_A, i_A), (B, m_B, i_B)$ in a tensor category $\mathcal{T}$, a morphism $\theta \in \mathcal{T}(A, B)$ is called an *algebra morphism* if $m_B \circ (\theta \otimes \theta) = \theta \circ m_A$ and $i_B \circ \theta = i_A$.

Given a C* Frobenius algebra $A \in \mathcal{C}$, following [**Vic11**, Def. 3.4],[4] we can define two canonical isomorphisms $\sigma^L_A, \sigma^R_A \in \mathcal{C}(A, \overline{A})$ given by

Notice that the cup in the diagram on the left is coev_A, and the cup in the diagram on the right is ev_A^*. By [**Vic11**, Lem. 3.5(13)], $\overline{\sigma^L_A} \circ \sigma^L_A = \varphi_A$ and $\overline{\sigma^R_A} \circ \sigma^R_A = \varphi_A$, and thus both are involutions.

LEMMA 2.10. *The following are equivalent for a separable* C* *Frobenius algebra object $(A, m, i) \in \mathcal{C}$.*

(1) *$R := i_A^* \circ m_A$ and $S := i_A^* \circ m_A$ are standard solutions to the conjugate equations,*

[3]Note that our definition of normalized Q-system agrees with [**NY16**]. Other normalizations include $\lambda = 1$ as in [**FRS02**, (6)] or $\lambda = \lambda'$ as in [**Müg03**, Def. 3.13] or [**BKLR15**, Def. 3.2].

[4]To translate between our conventions for bi-involutive categories and those in [**Vic11**], our $(\cdot)^*$ is his $(\cdot)^\dagger$, our $\overline{\cdot}$ is his $(\cdot)_*$, and our $(\cdot)^\vee$ is his $(\cdot)^*$.

(2) $i^* \circ m \circ m^* \circ i = d_A \operatorname{id}_{1_{\mathcal{C}}}$,
(3) σ_A^L *is unitary, and*
(4) $\sigma_A^R = \sigma_A^L$.

PROOF. (3) if and only if (4) follows by [**Vic11**, Lem. 3.11].[5] (1) implies (2) is immediate, since (2) is satisfied by standard solutions. (2) implies (1) follows from [**LR97**, §6].[6] To show (1) implies (3), suppose $R := i_A^* \circ m_A$ and $S := i_A^* \circ m_A$ are standard solutions to the conjugate equations. Then so are $R' = \operatorname{ev}_A^*$ and $S' = \operatorname{coev}_A$, and we must have $\sigma_A^L = (R \otimes \operatorname{id}_A) \circ (\operatorname{id}_A \otimes \operatorname{coev}_A)$ is unitary by [**NT13**, Prop. 2.2.13]. Finally, to show (3) implies (2), if $\sigma_A := \sigma_A^L$ is unitary, then it is easy to calculate that for $R = S = i_A^* \circ m_A$,

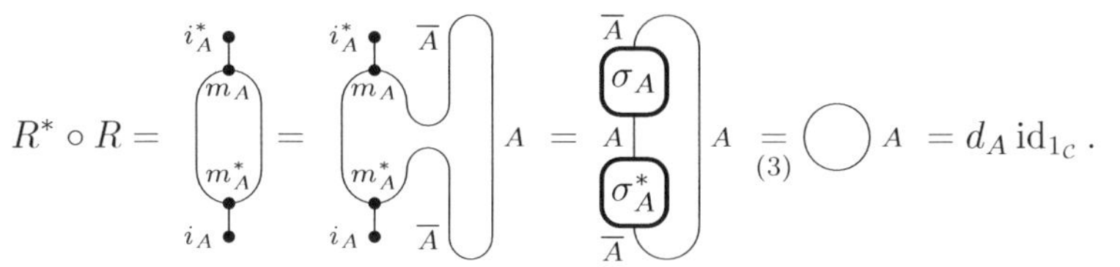

□

In the event that the conditions in Lemma 2.10 hold, we write $\sigma_A := \sigma_A^L = \sigma_A^R$.

DEFINITION 2.11. A *unitary Frobenius algebra* or a *Q-system* is a separable C* Frobenius algebra such that the equivalent conditions in Lemma 2.10 hold [**BKLR15**, Def. 3.2 and 3.8]. A Q-system is called *irreducible* if the underlying algebra object $A \in \mathcal{C}$ is connected. A Q-system is called *normalized* if its underlying C* Frobenius algebra is normalized.

REMARK 2.12. By condition (1) in Lemma 2.10, the underlying object $A \in \mathcal{C}$ of a Q-system is *real* a.k.a. *symmetrically self-dual* [**Müg03**, Rem. 5.6(3)]. Note that a Q-system is exactly a unitary †-Frobenius monoid in the sense of [**Vic11**, Def. 3.9].

COROLLARY 2.13. *A connected* C* *Frobenius algebra is automatically an irreducible Q-system.*

PROOF. Let (A, m, i) be a connected C* Frobenius algebra. By [**BKLR15**, Lem. 3.3], A is separable. Moreover, $R = S = m^* \circ i$ are automatically standard solutions by [**LR97**, §6].[7] □

EXAMPLE 2.14. The separable C* Frobenius algebra $(c \otimes \overline{c}, m, i)$ from Example 2.7 is a Q-system since $R^* \circ R = d_c^2$. However, one must replace the multiplication and unit by $m = d_c^{1/2}(\operatorname{id}_c \otimes \operatorname{ev}_c \otimes \operatorname{id}_{\overline{c}})$ and $i = d_c^{-1/2} \operatorname{coev}_c$ to get a normalized Q-system. As before, the Q-system is irreducible if and only if c is simple.

The next definition is adapted from [**Vic11**, Def. 2.21] and gives the correct notion of morphism between irreducible Q-systems for our purposes.

DEFINITION 2.15. Suppose $(A, m_A, i_A), (B, m_B, i_B)$ are algebra objects in $\mathcal{C}$. An algebra morphism $\theta \in \mathcal{C}(A, B)$ is called *involutive* if $\sigma_B \circ \theta = \overline{\theta} \circ \sigma_A$.

[5]The condition $\sigma_A^L = \sigma_A^R$ also appears in the non-unitary setting in [**FRS02**] near displayed equation (7) therein.

[6]We warn the reader that the proof in [**LR97**, §6] for a separable C* Frobenius algebra involves constructing a subfactor, and thus the Jones index is the minimal index [**Lon89**].

[7]This was first observed by [**Müg03**, Rem. 5.6(3)]. Note that the proof in [**LR97**, §6] relies on subfactor theory. The more general result [**NY17**, Thm. 2.9] does not pass through subfactor theory and implies R, S are standard solutions.

The recent article [**NY17**] introduces another notion of isomorphism between (unitary) C*-Frobenius algebras. The say an algebra isomorphism $\theta \in \mathcal{C}(A, B)$ is a C*-Frobenius algebra isomorphism if

$$m_A \circ [\mathrm{id}_A \otimes (\theta^* \circ \theta)] = \theta^* \circ \theta \circ m_A. \tag{3}$$

We now prove that involutive algebra morphisms satisfy (3).

LEMMA 2.16. *Every involutive algebra morphism $\theta \in \mathcal{C}(A, B)$ satisfies* (3).

PROOF. First, using that θ is involutive and (1), we have

(4) 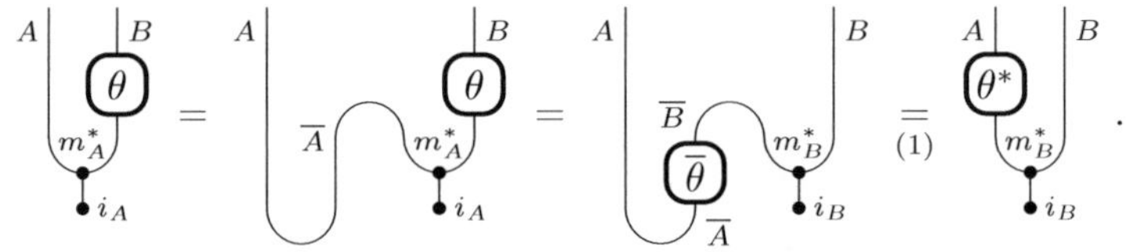

Now using the adjoint of (4), together with Lemma 2.8, and that $\theta^* \in \mathcal{C}(B, A)$ is a coalgebra morphism, we have

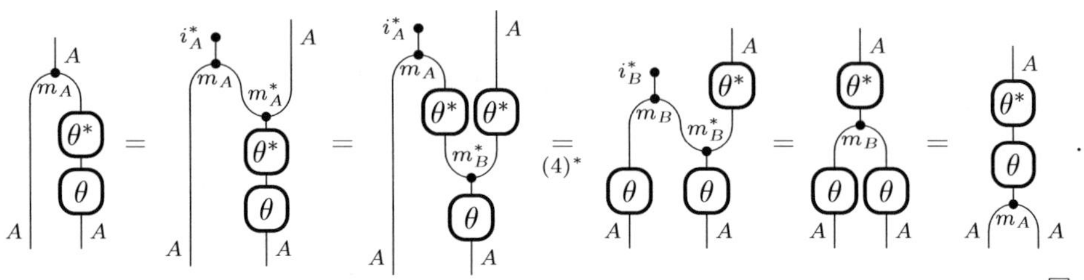

□

2.5. Compact connected W*-algebra objects. An algebra object $(\mathbf{A}, m, i) \in \mathsf{Vec}(\mathcal{C})$ may be viewed as a lax monoidal functor $(\mathbf{A}, \mu, i) : (\mathcal{C}^\natural)^{\mathrm{op}} \to \mathsf{Vec}$ by [**JP17a**, Prop. 3.3]. If $\mathbf{A} \in \mathsf{Vec}(\mathcal{C})$ is a compact algebra object, the corresponding object $A \in \mathcal{C}^\natural$ satisfying $\mathbf{A}(a) = \mathcal{C}^\natural(a, A)$ for $a \in \mathcal{C}$ is easily seen to be an algebra object.

DEFINITION 2.17. A $*$-structure on an algebra object $\mathbf{A} \in \mathsf{Vec}(\mathcal{C})$ is a collection of conjugate linear natural isomorphisms $j_a : \mathbf{A}(a) \to \mathbf{A}(\overline{a})$ satisfying the following axioms:

- (conjugate naturality) for all $\psi \in \mathcal{C}(b, a)$ and $f \in \mathbf{A}(a)$, $j_b(f \circ \psi) = j_a(f) \circ \overline{\psi}$.
- (involutive) for all $f \in \mathbf{A}(a)$, $j_{\overline{a}}(j_a(f)) = f$.
- (unital) $j_{1_\mathcal{C}}(i) = i$.
- (monoidal) for all $f \in \mathbf{A}(a)$ and $g \in \mathbf{A}(b)$, $j_{a\otimes b}(m \circ (f \otimes g)) = m \circ (j_b(g) \otimes j_a(f))$.

EXAMPLE 2.18. Suppose $c \in \mathcal{C}$, and consider the normalized Q-system $(c \otimes \overline{c}, m, i) \in \mathcal{C}$ from Example 2.14. By the Yoneda embedding, $\mathbf{c} \otimes \overline{\mathbf{c}} : (\mathcal{C}^\natural)^{\mathrm{op}} \to \mathsf{Vec}$ is given by $(\mathbf{c} \otimes \overline{\mathbf{c}})(a) = \mathcal{C}^\natural(a, c \otimes \overline{c})$. The $*$-algebra structure on $\mathbf{c} \otimes \overline{\mathbf{c}}$ is given by

$$\begin{aligned}
\mu_{a,b} : (\mathbf{c}\otimes\overline{\mathbf{c}})(a)\otimes(\mathbf{c}\otimes\overline{\mathbf{c}})(b) &\to (\mathbf{c} \otimes \overline{\mathbf{c}})(a \otimes b) && f \otimes g \mapsto d_c^{1/2}(\mathrm{id}_c \otimes \mathrm{ev}_c \otimes \mathrm{id}_{\overline{c}}) \circ (f \otimes g)\\
i \in (\mathbf{c} \otimes \overline{\mathbf{c}})(1_\mathcal{C}) &= \mathcal{C}^\natural(1_\mathcal{C}, c \otimes \overline{c}) && i = d_c^{-1/2}\,\mathrm{coev}_c\\
j_a : (\mathbf{c} \otimes \overline{\mathbf{c}})(a) &\to (\mathbf{c} \otimes \overline{\mathbf{c}})(\overline{a}) && f \mapsto \overline{f}
\end{aligned}$$

(this definition of j suppresses the isomorphism between the real object $c \otimes \overline{c}$ and its conjugate). Then $\mathbf{c} \otimes \overline{\mathbf{c}}$ is a compact $*$-algebra object in $\mathsf{Vec}(\mathcal{C})$ which is connected if and only if c is simple. The corresponding object in $\mathcal{C}^\natural$ is obviously $c \otimes \overline{c}$.

It was shown in [**JP17a**, Thm. 3.20] that $*$-structures on algebra objects in $\mathsf{Vec}(\mathcal{C})$ are equivalent to dagger structures on the cyclic $\mathcal{C}$-module category $(\mathcal{M}_{\mathbf{A}}, \mathbf{A})$ whose objects are the $\mathbf{c} \otimes \mathbf{A}$ for $c \in \mathcal{C}$ and whose morphisms are right $\mathbf{A}$-module morphisms. Indeed, when $\mathbf{A} \in \mathsf{Vec}(\mathcal{C})$ is compact, we have

$$\mathcal{M}_{\mathbf{A}}(\mathbf{a} \otimes \mathbf{A}, \mathbf{b} \otimes \mathbf{A}) \cong \mathcal{C}(a, b \otimes A)$$

with composition given for $f \in \mathcal{C}(a, b \otimes A)$ and $g \in \mathcal{C}(b, c \otimes A)$ given by

$$g \circ_{\mathcal{M}_{\mathbf{A}}} f := (\mathrm{id}_c \otimes m) \circ_{\mathcal{C}} (g \otimes \mathrm{id}_A) \circ f.$$

The dagger structure on $\mathcal{M}_{\mathbf{A}}$ induced by j is given on $f \in \mathcal{C}(a, b \otimes A)$ by

$$f^* := j_{\overline{b} \otimes a}\Big((\mathrm{ev}_b \otimes \mathrm{id}_A) \circ f \Big) \circ (\mathrm{coev}_a \otimes \mathrm{id}_{\overline{b}}).$$

Conversely, given a cyclic $\mathcal{C}$-module dagger category $(\mathcal{M}, m)$, we recover the lax monoidal functor $\mathbf{A} : (\mathcal{C}^\natural)^{\mathrm{op}} \to \mathsf{Vec}$ by $\mathbf{A}(a) = \mathcal{M}(a \otimes m, m)$, where μ and i are given by

(5) $$\mu_{a,b}(f \otimes g) := f \circ (\mathrm{id}_a \otimes g) \qquad \text{and} \qquad i := \mathrm{id}_m \in \mathbf{A}(1) = \mathrm{End}_{\mathcal{M}}(m),$$

and the $*$-structure j is given on $f \in \mathbf{A}(a)$ by

(6) $$j_a(f) := (\mathrm{ev}_a \otimes \mathrm{id}_m) \circ (\mathrm{id}_{\overline{a}} \otimes f^*).$$

We see that $\mathbf{A}$ is compact if and only if $\mathcal{M}(a \otimes m, m)$ is always finite dimensional, and non-zero for only finitely many $a \in \mathrm{Irr}(\mathcal{C})$ (this property of $\mathcal{M}$ is called being *proper* in [**NY17**] or being *cofinite* in [**AC15**]).

Just as being a C*-algebra is a property of a complex $*$-algebra and not extra structure, being a C*-category is a property of a dagger category and not extra structure. We refer the reader to [**JP17a**, §2.1] for more details.

Definition 2.19. A $*$-algebra object $(\mathbf{A}, m, i, j) \in \mathsf{Vec}(\mathcal{C})$ is a C*/W*-algebra object if the cyclic $\mathcal{C}$-module dagger category $\mathcal{M}_{\mathbf{A}}$ is a C*/W*-category. If $\mathbf{A}$ is locally finite, meaning $\mathbf{A}(a)$ is finite dimensional for all $a \in \mathcal{C}$, then all morphisms spaces of $\mathcal{M}_{\mathbf{A}}$ are finite dimensional. In this case, $\mathcal{M}_{\mathbf{A}}$ is a C*-category if and only if $\mathcal{M}_{\mathbf{A}}$ is a W*-category.

Example 2.20. Given a Hilbert space object $\mathbf{H} \in \mathsf{Hilb}(\mathcal{C})$, we define $\mathbf{B}(\mathbf{H}) \in \mathsf{Vec}(\mathcal{C})$ to be the W*-algebra object corresponding to the cyclic $\mathcal{C}$-module W*-subcategory of $\mathsf{Hilb}(\mathcal{C})$ whose objects are of the form $\mathbf{c} \otimes \mathbf{H}$. Thus $\mathbf{B}(\mathbf{H})(a) = \mathrm{Hom}_{\mathsf{Hilb}(\mathcal{C})}(\mathbf{a} \otimes \mathbf{H}, \mathbf{H})$ for all $a \in \mathcal{C}$, where μ, i, and j are given by the formulas in (5) and (6).

Example 2.21. The $*$-algebra object $\mathbf{c} \otimes \overline{\mathbf{c}} = \mathcal{C}^\natural(\cdot, c \otimes \overline{c}) \in \mathsf{Vec}(\mathcal{C})$ from Example 2.18 is a W*-algebra object. To see this, we construct a $*$-algebra natural isomorphism $\pi : \mathbf{B}(\mathbf{c}) \Rightarrow \mathbf{c} \otimes \overline{\mathbf{c}}$, where the instance of $\mathbf{c}$ inside $\mathbf{B}(\mathbf{c})$ is the linear dagger functor $\mathbf{c} = \mathcal{C}(\cdot, c) : \mathcal{C}^{\mathrm{op}} \to \mathsf{Hilb}$, which is a Hilbert space object under the Yoneda embedding. For $a \in \mathcal{C}$ and $f \in \mathbf{B}(\mathbf{c})(a) = \mathcal{C}(a \otimes c, c)$, we define

$\pi_a(f) := d_c^{-1/2}$ 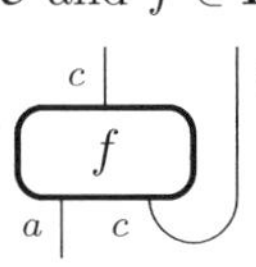 $\in \mathcal{C}(a, c \otimes \overline{c}) = (\mathbf{c} \otimes \overline{\mathbf{c}})(a).$

The reader can check π is a $*$-algebra natural isomorphism. (It is similar to, but easier than, the proof of Theorem 5.1 below.) Finally, $\mathbf{B}(\mathbf{c})$ is the W*-algebra object corresponding to the cyclic $\mathcal{C}$-module W*-category $(\mathcal{C}, c)$, and thus $\mathbf{c} \otimes \overline{\mathbf{c}}$ is W*.

2.6. Inner products on hom spaces of $\mathcal{C}$. Recall that each hom space $\mathcal{C}(a,b)$ comes equipped with the usual trace inner product (2). For all $a, b \in \mathcal{C}$, we pick orthonormal bases $\mathrm{ONB}(a,b)$ for $\mathcal{C}(a,b)$ under the inner product (2).

For each $c \in \mathrm{Irr}(\mathcal{C})$ and $a \in \mathcal{C}$, we let $\mathrm{Isom}(c,a) = \{\sqrt{d_c} f | f \in \mathrm{ONB}(c,a)\}$, which is a maximal set of isometries in $\mathcal{C}(c,a)$ with mutually orthogonal ranges. Note that $\mathrm{Isom}(c,a)$ is an orthonormal basis for $\mathcal{C}(c,a)$ with the modified inner product given by

$$\langle f|g\rangle_{\mathrm{Isom}} \, \mathrm{id}_c = f^* \circ g. \tag{7}$$

DEFINITION 2.22. The 1-*strand rotation* or *Fourier transform* $\mathcal{F}$ on $\mathcal{C}(a, b\otimes c)$ for $a, b, c \in \mathcal{C}$ is given by

$\mathcal{C}(c, a\otimes b) \ni f \mapsto$ 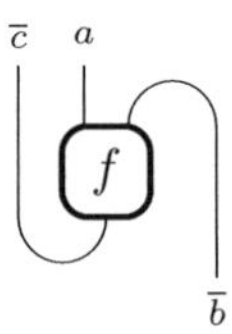

$\in \mathcal{C}(\overline{b}, \overline{c}\otimes a).$

Note that if $a, b, c \in \mathrm{Irr}(\mathcal{C})$ and $f \in \mathrm{Isom}(c, a\otimes b)$, then a weighting is required to produce another isometry:

$$\mathrm{Isom}(c, a\otimes b) \ni f \mapsto \left(\frac{d_b}{d_c}\right)^{1/2} \mathcal{F}(f)$$

produces an isometry in $\mathcal{C}(\overline{b}, \overline{c}\otimes a)$, and thus

$$\left\{ \left(\frac{d_b}{d_c}\right)^{1/2} \mathcal{F}(f) \middle| f \in \mathrm{Isom}(c, a\otimes b) \right\} \subset \mathcal{C}(\overline{b}, \overline{c}\otimes a)$$

is a new orthonormal basis under inner product (7).

Suppose now that $\mathbf{A} \in \mathsf{Vec}(\mathcal{C})$ is a W*-algebra object. We have two $\mathbf{A}(1_{\mathcal{C}})$-valued inner products on each vector space $\mathbf{A}(a)$ given by

(8) $\langle f|g\rangle_a :=$ 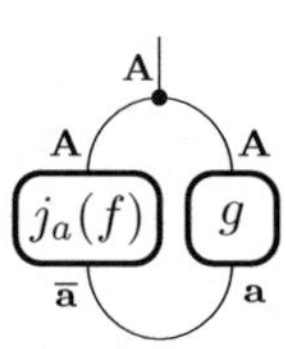

and ${}_a\langle g, f\rangle :=$ 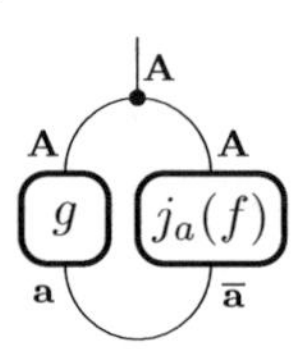

Recall from [**JP17a**] that a normal state on $\mathbf{A}$ is a normal state on the W*-algebra $\mathbf{A}(1_{\mathcal{C}})$. Given a normal state ϕ on $\mathbf{A}$, we can define two $\mathbb{C}$-valued inner products on each vector space $\mathbf{A}(a)$ by composing the $\mathbf{A}(1_{\mathcal{C}})$-valued inner products in (8) above with ϕ. The state ϕ is called *tracial* if $\phi(\langle f|g\rangle_a) = \phi({}_a\langle g, f\rangle)$ for all $f, g \in \mathbf{A}(a)$ for all $a \in \mathcal{C}$.

When $\mathbf{A}$ is connected, by [**JP17b**, Prop. 2.8], $\mathbf{A}(a)$ is finite dimensional for each $a \in \mathbf{A}$. Moreover, there is a unique normal state on $\mathbf{A}(1_{\mathcal{C}}) = \mathbb{C}$ given by $i_{\mathbf{A}} \mapsto 1_{\mathbb{C}}$. We thus say $\mathbf{A}$ is *tracial* if $\langle f|g\rangle_a = {}_a\langle g, f\rangle$ for all $f, g \in \mathbf{A}(a)$.

Identifying $\mathbf{A}(1_{\mathcal{C}}) = \mathbb{C}$, we get two finite dimensional Hilbert space structures on each $\mathbf{A}(a)$ using the right and left inner products $\langle f|g\rangle_a$ and ${}_a\langle g, f\rangle$ respectively. By [**JP17a**, Thm. 4.18 and §4.5], defining $\mathbf{L}^2\mathbf{A} : \mathcal{C}^{\mathrm{op}} \to \mathsf{Hilb}$ by $\mathbf{L}^2\mathbf{A}(a) = \mathbf{A}(a)$ with the right inner product actually yields a Hilbert space object in $\mathsf{Hilb}(\mathcal{C})$, where $\mathbf{L}^2\mathbf{A}(\psi) = \mathbf{A}(\psi)$ for morphisms $\psi \in \mathcal{C}$. Indeed, using the $*$-structure j of $\mathbf{A}$, we have $\mathbf{A}(\overline{\psi}) \circ j_b = j_a \circ \mathbf{A}(\psi)$ for all $\psi \in \mathcal{C}(a, b)$. For $f \in \mathbf{A}(a)$, $g \in \mathbf{A}(b)$, and $\psi \in \mathcal{C}(a, b)$, we see that $\langle \mathbf{L}^2\mathbf{A}(\psi^*)[f]|g\rangle_b = \langle f|\mathbf{L}^2\mathbf{A}(\psi)[g]\rangle_a$ by calculating

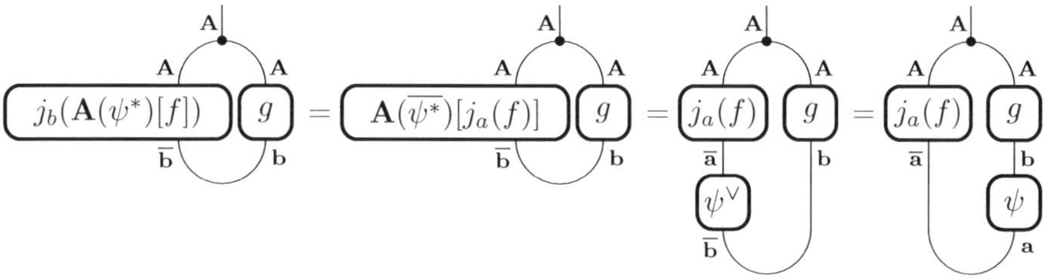

We showed in [**JP17b**, Prop. 2.6] that every compact connected W*-algebra object is tracial. Thus if $\mathbf{A} = \mathcal{C}^\natural(-, A)$ for $A \in \mathcal{C}^\natural$, the right inner product equals the left inner product on $\mathbf{A}(a) = \mathcal{C}^\natural(a, A)$:

(9) $\langle f|g\rangle_a :=$ 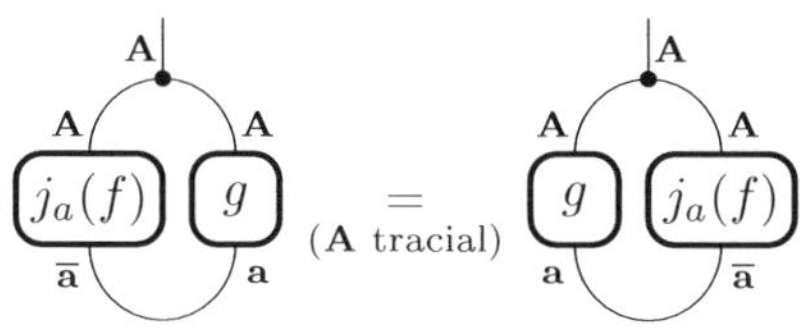 $= {}_a\langle g, f\rangle$

Then $\mathbf{L}^2\mathbf{A} \in \mathsf{Hilb}(\mathcal{C})$ with inner product (9) is compact by Proposition 2.4, and we get an object in $\mathcal{C}$ up to canonical unitary isomorphism.

The GNS representation of $\mathbf{A}$ gives a canonical faithful $*$-algebra natural transformation $\lambda : \mathbf{A} \Rightarrow \mathbf{B}(\mathbf{L}^2\mathbf{A})$ [**JP17a**, Eq. (20)]. Since $\mathbf{A}$ is connected, the inner product (9) satisfies

(10) $\langle f|g\rangle_a \,\mathrm{id}_{\mathbf{L}^2\mathbf{A}} =$ 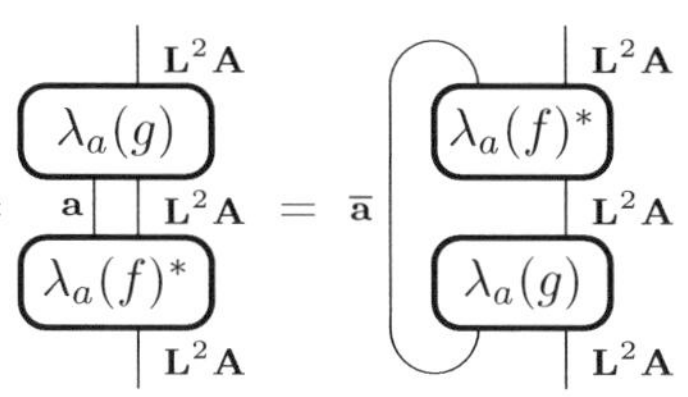

Finally, given a $*$-algebra natural transformation $\theta : \mathbf{A} \Rightarrow \mathbf{B}$ of connected W*-algebra objects, notice that $\mathbf{L}^2\mathbf{A}(a) \ni f \mapsto \theta(f) \in \mathbf{L}^2\mathbf{B}(a)$ gives a canonical isometry $\mathbf{L}^2\theta : \mathbf{L}^2\mathbf{A} \Rightarrow \mathbf{L}^2\mathbf{B}$ since

(11) $\langle \theta_a(f)|\theta_a(g)\rangle_a^{\mathbf{B}}\, i_{\mathbf{B}}$

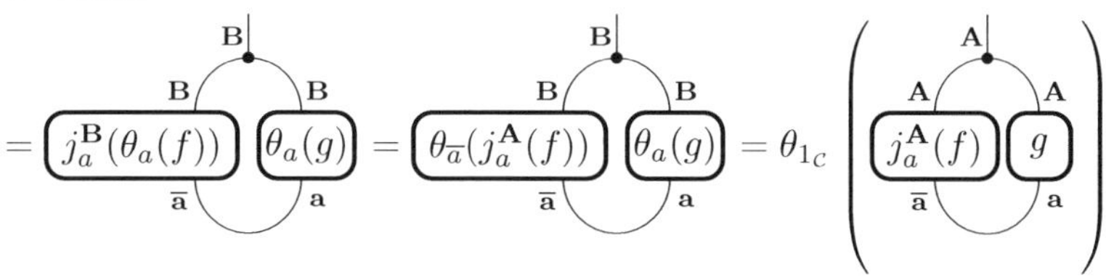

$$= \theta_{1_{\mathcal{C}}}(\langle f|g\rangle_a^{\mathbf{A}} i_{\mathbf{A}}) = \langle f|g\rangle_a^{\mathbf{A}} \theta_{1_{\mathcal{C}}}(i_{\mathbf{A}}) = \langle f|g\rangle_a^{\mathbf{A}} i_{\mathbf{B}}$$

Noticing that the previous argument was independent of the W*-algebra objects being compact, we conclude the following.

COROLLARY 2.23. *Any $*$-algebra natural transformation $\theta : \mathbf{A} \Rightarrow \mathbf{B}$ between connected* W*-*algebra objects in* $\mathsf{Vec}(\mathcal{C})$ *is injective and induces an isometry* $\mathbf{L}^2\theta : \mathbf{L}^2\mathbf{A} \Rightarrow \mathbf{L}^2\mathbf{B}$ *with respect to the right (respectively left) inner products.*

3. From Q-systems to W*-algebra objects

In this section, we define a functor $\mathbf{W}^*$ from normalized irreducible Q-systems to compact connected W*-algebra objects.

3.1. Objects.

Suppose we have a normalized irreducible Q-system (A, m, i) in $\mathcal{C}$. We now define a compact connected W*-algebra object $\mathbf{W}^*(A) := (\mathbf{A}, \mu, i, j) \in \mathcal{C}^\natural$.

DEFINITION 3.1. For $a \in \mathcal{C}^\natural$, we define $\mathbf{A}(a) = \mathcal{C}^\natural(a, A)$. We define the laxitor $\mu_{a,b} : \mathbf{A}(a) \otimes \mathbf{A}(b) \to \mathbf{A}(a \otimes b)$ to be the map $f \otimes g \mapsto m \circ (f \otimes g)$ and the same unit map $i \in \mathbf{A}(1_\mathcal{C}) = \mathcal{C}(1_\mathcal{C}, A)$. It is straightforward to verify $(\mathbf{A}, \mu, i)$ is a lax monoidal functor $(\mathcal{C}^\natural)^{\mathrm{op}} \to \mathsf{Vec}$. Since $A \in \mathcal{C}$, we have $\mathbf{A}$ is compact, and $\mathbf{A}(1_\mathcal{C})$ is clearly one dimensional, and thus $\mathbf{A}$ is connected.

We define a $*$-structure j on $\mathbf{A}$ by

$$\mathcal{C}^\natural(a, A) = \mathbf{A}(a) \ni f \xmapsto{j_a} \sigma_A^{-1} \circ \overline{f} \in \mathcal{C}^\natural(\overline{a}, A) = \mathbf{A}(\overline{a}),$$

where $\sigma_A \in \mathcal{C}(A, \overline{A})$ is the unitary isomorphism from Lemma 2.10. Using that $\overline{\sigma_A} \circ \sigma_A = \varphi_A$ by [**Vic11**, Lem. 3.5(13)], it is a worthwhile exercise to verify $(j_a)_{a \in \mathcal{C}}$ satisfies the axioms of a $*$-structure on $\mathbf{A}$ from Definition 2.17.

PROPOSITION 3.2. *The compact connected $*$-algebra object* $(\mathbf{A}, \mu, i, j) \in \mathsf{Vec}(\mathcal{C})$ *is* W*.

PROOF. Consider the cyclic $\mathcal{C}$-module dagger category $(\mathcal{M}_A, A)$ of 'free' right A-modules in $\mathcal{C}$ with basepoint $A \in \mathcal{C}$ and dagger structure given by the adjoint in $\mathcal{C}$. That is, objects of $\mathcal{M}_A$ are exactly the $c \otimes A$ for $c \in \mathcal{C}$, and the morphisms are right A-module morphisms. By Lemma 2.8, for every $g \in \mathcal{M}_A(a \otimes A, b \otimes A)$, $g^* \in \mathcal{M}_A(b \otimes A, a \otimes A)$.

Thus $\mathcal{M}_A$ is a dagger sub-category of $\mathcal{C}$ which is closed under finite direct sums. This means that $\mathcal{M}_A$ is automatically W* by the finite dimensional bicommutant theorem [**Jon15**, Thm. 3.2.1] which says a unital $*$-subalgebra of a finite dimensional von Neumann algebra is again a von Neumann algebra. Indeed, for all $c \in \mathcal{C}$,

$$\mathcal{M}_A(c \otimes A, c \otimes A) \subseteq \mathcal{C}(c \otimes A, c \otimes A)$$

is a unital $*$-subalgebra of a finite dimensional von Neumann algebra. This is sufficient to show $\mathcal{M}_A$ is W* by Roberts' 2×2 trick [**GLR85**] (see also [**JP17a**, §2.1 and Prop. 3.26]).

We claim that $(\mathbf{A}, \mu, i, j)$ is $*$-isomorphic to the W*-algebra object $\mathbf{B}$ corresponding to the cyclic $\mathcal{C}$-module W*-category $(\mathcal{M}_A, A)$ under [**JP17a**, Thm. 3.20]. The unit, multiplication, and $*$-structure of $\mathbf{A}$ was given in Definition 3.1. The corresponding structures for $\mathbf{B}$ are given by the formulas in (5) and (6).

We now construct a unital $*$-algebra natural isomorphism $\theta : \mathbf{A} \Rightarrow \mathbf{B}$. For $a \in \mathcal{C}$, we define $\theta_a : \mathbf{A}(a) = \mathcal{C}(a, A) \to \mathbf{B}(a) = \mathcal{M}(a \otimes A, A)$ by

(12) $\mathcal{C}(a, A) \ni f \mapsto$ 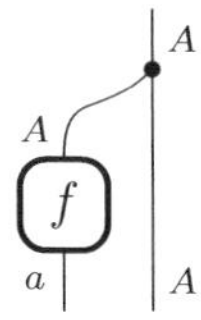 $\in \mathcal{M}_A(a \otimes A, A).$

The inverse is given by precomposing with $\mathrm{id}_a \otimes i_A$. These formulas are similar to [**KO02**, Fig. 4] in the context of the free module functor [**KO02, BN11, HPT16**].

Now a straightforward computation verifies that θ is a unital $*$-algebra morphism. We show that θ intertwines the $*$-structures, and we leave the other calculations to the reader. By [**JP17a**, Thm. 3.20] (see (6)), the $*$-structure on the algebra corresponding to $(\mathcal{M}_A, A)$ is given by

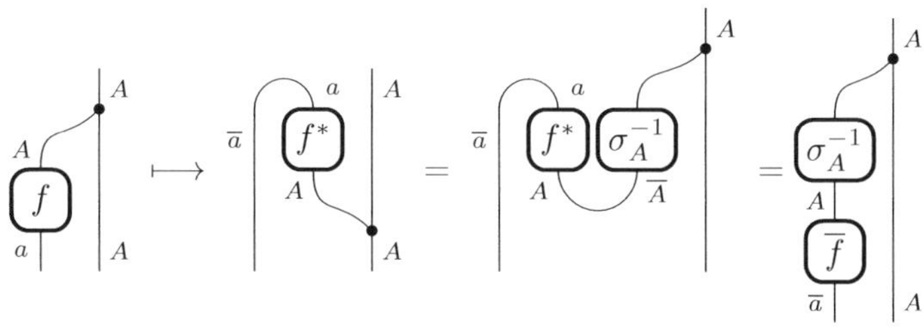

which is exactly the image of $j_a(f)$ under the isomorphism (12). □

Recall that we define $\mathbf{L}^2\mathbf{A} \in \mathsf{Hilb}(\mathcal{C})$ by $\mathbf{L}^2\mathbf{A}(a) = \mathbf{A}(a)$ with inner product (9). Now $\mathbf{L}^2\mathbf{A} \in \mathsf{Hilb}(\mathcal{C})$ is compact by Proposition 2.4, and thus there is an $H \in \mathcal{C}$ such that $\mathbf{L}^2\mathbf{A}(a) = \mathcal{C}(a, H)$ for all $a \in \mathcal{C}$.

LEMMA 3.3. *The object $H \in \mathcal{C}$ is canonically unitarily isomorphic to A.*

PROOF. By fully faithfulness of the Yoneda embedding $\mathcal{C} \hookrightarrow \mathsf{Hilb}(\mathcal{C})$, it suffices to show that for all $a \in \mathcal{C}$, the inner product (2) on $\mathcal{C}(a, A)$ and the inner product (9) on

$$\mathcal{C}(a, H) = \mathbf{H}(a) = \mathbf{L}^2\mathbf{A}(a) = \mathbf{A}(a) = \mathcal{C}^\natural(a, A)$$

agree. Indeed, for $f, g \in \mathcal{C}(a, A)$, by Lemma 2.10 and Remark 2.12 together with (1) and (9), we have

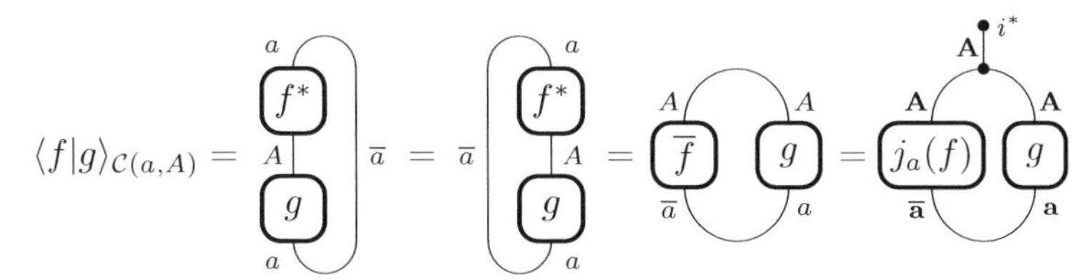

$$= \langle f|g\rangle_a(i^* \circ i) = \langle f|g\rangle_a$$

where $i^* \circ i = \mathrm{id}_{1_\mathcal{C}}$ as in Definition 2.6 since A is normalized. Here, the 5th equality above follows from (9), where the factor of i in the right inner product is suppressed. □

REMARK 3.4. Notice that the cyclic $\mathcal{C}$-module W*-category $(\mathcal{M}_A, A)$ constructed in the proof of Proposition 3.2 naturally includes into the full cyclic $\mathcal{C}$-module W*-subcategory of $\mathcal{C}$ generated by A with objects of the form $c \otimes A$ for

$c \in \mathcal{C}$ together with all morphisms. By Lemma 3.3, this latter category is dagger equivalent to the cyclic $\mathcal{C}$-module W*-subcategory of $\mathsf{Hilb}(\mathcal{C})$ generated by $\mathbf{L}^2\mathbf{A}$, which corresponds to the W*-algebra object $\mathbf{B}(\mathbf{L}^2\mathbf{A})$ by [**JP17a**, Ex. 3.35]. Thus (12) provides an explicit description of the canonical GNS faithful $*$-algebra natural transformation $\lambda : \mathbf{A} \Rightarrow \mathbf{B}(\mathbf{L}^2\mathbf{A})$.

3.2. Morphisms. Suppose $\theta \in \mathcal{C}(A, B)$ is an involutive algebra morphism as in Section 2.4. We now define a $*$-algebra natural transformation $\mathbf{W}^*(\theta) : \mathbf{W}^*(A) \Rightarrow \mathbf{W}^*(B)$. We use the notation $\mathbf{W}^*(A) = (\mathbf{A}, \mu^{\mathbf{A}}, i_{\mathbf{A}}, j^{\mathbf{A}})$ and recall $\mathbf{A}(a) = \mathcal{C}^\natural(a, A)$. We have similar notation for $\mathbf{W}^*(B)$.

DEFINITION 3.5. Given an involutive algebra morphism $\theta \in \mathcal{C}(A, B)$, we define $\mathbf{W}^*(\theta) : \mathbf{A} \Rightarrow \mathbf{B}$ by defining the a-component $\mathbf{W}^*(\theta)_a$ for $a \in \mathcal{C}$ by

$$\mathcal{C}^\natural(a, A) = \mathbf{A}(a) \ni f \mapsto \theta \circ f \in \mathbf{B}(a) = \mathcal{C}^\natural(a, B).$$

Note that $\mathbf{W}^*(\theta)$ is automatically a natural transformation. That θ is an algebra morphism implies $\mathbf{W}^*(\theta)$ is an algebra natural transformation, and involutivity of θ implies $\mathbf{W}^*(\theta)$ is a $*$-algebra natural transformation. Indeed, for $f \in \mathbf{A}(a)$,

$$\begin{aligned} j^{\mathbf{B}}_a(\mathbf{W}^*(\theta)_a(f)) &= j^{\mathbf{B}}_a(\theta \circ f) = \sigma_B^{-1} \circ \overline{\theta \circ f} = \sigma_B^{-1} \circ \overline{\theta} \circ \overline{f} \\ &= \theta \circ \sigma_A^{-1} \circ \overline{f} = \theta \circ j^{\mathbf{A}}_a(f) = \mathbf{W}^*(\theta)_{\overline{a}}(j^{\mathbf{A}}_a(f)). \end{aligned}$$

PROPOSITION 3.6. $\mathbf{W}^*$ *is a functor.*

PROOF. We must prove that $\mathbf{W}^*$ preserves identities and composites. If $\theta = \mathrm{id}_A \in \mathcal{C}(A, A)$, it is obvious that $\mathbf{W}^*(\theta) = \mathrm{id}_{\mathbf{A}}$. Moreover, if we have $\theta_1 \in \mathcal{C}(A, B)$ and $\theta_2 \in \mathcal{C}(B, C)$, then for all $a \in \mathcal{C}$ and $f \in \mathbf{A}(a)$, we have

$$\mathbf{W}^*(\theta_1 \circ \theta_2)_a(f) = \theta_1 \circ \theta_2 \circ f = (\mathbf{W}^*(\theta_1) \circ \mathbf{W}^*(\theta_2))_a(f).$$

We are finished. □

4. From W*-algebra objects to Q-systems

4.1. Objects. Suppose $\mathbf{A} \in \mathsf{Vec}(\mathcal{C})$ is a compact connected W*-algebra object, and define $\mathbf{H} := \mathbf{L}^2\mathbf{A} \in \mathsf{Hilb}(\mathcal{C})$, the Hilbert space object defined by $\mathbf{H}(a) = \mathbf{L}^2\mathbf{A}(a) = \mathbf{A}(a)$ with inner product (9). By Proposition 2.4, $\mathbf{H}$ is compact, and we let $H \in \mathcal{C}$ be the corresponding object satisfying $\mathbf{H}(a) = \mathcal{C}(a, H)$ for all $a \in \mathcal{C}$. Recall $\mathbf{A}$ has a canonical state corresponding to the state on $\mathbf{A}(1_{\mathcal{C}}) \cong \mathbb{C}$ via $i_A \mapsto 1_{\mathbb{C}}$. The GNS representation [**JP17a**, Eq. (20)] gives a canonical faithful $*$-algebra natural transformation $\lambda : \mathbf{A} \Rightarrow \mathbf{B}(\mathbf{L}^2\mathbf{A})$.

Now since $H \in \mathcal{C}$, we get a canonical compact W*-algebra object $\mathbf{H} \otimes \overline{\mathbf{H}} : (\mathcal{C}^\natural)^{\mathrm{op}} \to \mathsf{Vec}$ by $(\mathbf{H} \otimes \overline{\mathbf{H}})(a) = \mathcal{C}(a, H \otimes \overline{H})$ as in Examples 2.18 and 2.21, which is isomorphic to the compact W*-algebra object $\mathbf{B}(\mathbf{H}) \in \mathsf{Vec}(\mathcal{C})$.

DEFINITION 4.1. Define $\pi : \mathbf{A} \Rightarrow \mathbf{H} \otimes \overline{\mathbf{H}}$ by

$\mathbf{A}(a) \ni f \overset{\pi_a}{\longmapsto} d_H^{-1/2}$ 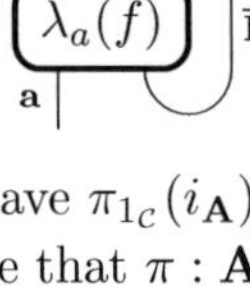 $\in \mathrm{Hom}_{\mathsf{Hilb}(\mathcal{C})}(\mathbf{a}, \mathbf{H} \otimes \overline{\mathbf{H}}) \cong \mathcal{C}(a, H \otimes \overline{H}) = (\mathbf{H} \otimes \overline{\mathbf{H}})(a).$

In particular, we have $\pi_{1_{\mathcal{C}}}(i_{\mathbf{A}}) = d_H^{-1/2} \operatorname{coev}_H$ since λ is unital.

It is easy to see that $\pi : \mathbf{A} \Rightarrow \mathbf{H} \otimes \overline{\mathbf{H}}$ is the faithful $*$-algebra natural transformation corresponding to $\lambda : \mathbf{A} \Rightarrow \mathbf{B}(\mathbf{H})$ under the $*$-algebra natural isomorphism $\mathbf{B}(\mathbf{H}) \cong \mathbf{H} \otimes \overline{\mathbf{H}}$.

NOTATION 4.2. For each $a \in \operatorname{Irr}(\mathcal{C})$, suppose we have $f, f' \in \mathbf{A}(a)$, and define $\alpha = \pi_a(f)$ and $\alpha' = \pi_a(f')$. We then have the following identities from (10):

(13)

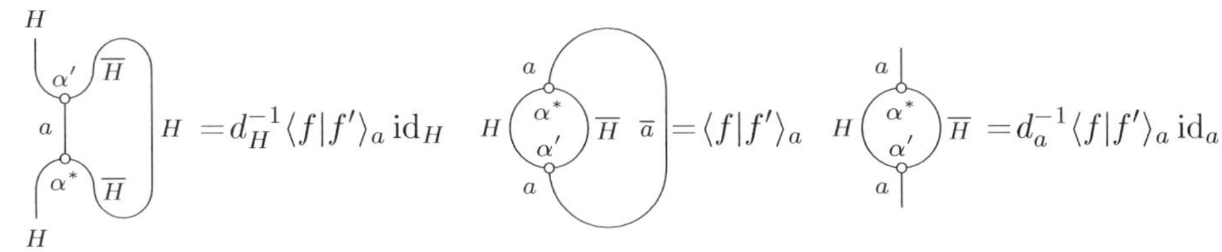

For each $a \in \operatorname{Irr}(\mathcal{C})$, pick an orthonormal basis $\operatorname{ONB}(\mathbf{A}(a))$ of $\mathbf{A}(a)$ under the inner product (9), and define

$$\mathcal{B}_a^{\mathbf{A}} = \left\{ d_a^{1/2}\pi_a(f) \middle| f \in \operatorname{ONB}(\mathbf{A}(a)) \right\}.$$

We will suppress the superscript and merely write $\mathcal{B}_a$ when no confusion can arise. By convention, we pick $\operatorname{ONB}(\mathbf{A}(1_{\mathcal{C}})) = \{i_{\mathbf{A}}\}$, and thus $\mathcal{B}_{1_{\mathcal{C}}} = \{d_H^{-1/2}\operatorname{coev}_H\}$. This normalization means that for $\alpha, \alpha' \in \mathcal{B}_a$, we have

(14) 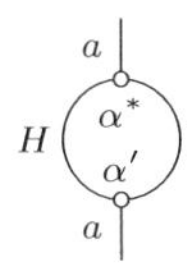 $= \delta_{\alpha=\alpha'} \operatorname{id}_a .$

Thus $\mathcal{B}_a$ is an orthonormal basis for $\pi_a(\mathbf{A}(a))$ under the new inner product (14).

DEFINITION 4.3. We define the following distinguished element of $\mathcal{C}(H\otimes\overline{H}, H\otimes\overline{H})$:

$$p = \sum_{a\in\operatorname{Irr}(\mathcal{C})}\sum_{\alpha\in\mathcal{B}_a}$$ 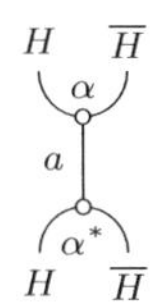 .

Note that the element p is independent of the choices of orthonormal bases $\mathcal{B}_a$ for $a \in \operatorname{Irr}(\mathcal{C})$ under the inner product (14).

LEMMA 4.4. *The element p is a symmetrically self-dual orthogonal projector. That is, $p^2 = p^* = p$, and $(p \otimes \operatorname{id}_{H\otimes\overline{H}}) \circ \operatorname{coev}_{H\otimes\overline{H}} = (\operatorname{id}_{H\otimes\overline{H}} \otimes p) \circ \operatorname{coev}_{H\otimes\overline{H}}$.*

PROOF. It is obvious that $p^* = p$, and it follows readily from (14) that $p^2 = p$. To verify the final condition, for each $a \in \operatorname{Irr}(\mathcal{C})$, there is a unique $a' \in \operatorname{Irr}(\mathcal{C})$ with $a' \cong \overline{a}$. We pick a unitary isomorphism $\gamma_a \in \mathcal{C}(a', \overline{a})$, and we see that

$(p\otimes\operatorname{id}_{H\otimes\overline{H}})\circ\operatorname{coev}_{H\otimes\overline{H}} = \sum_{a\in\operatorname{Irr}(\mathcal{C})}\sum_{\alpha\in\mathcal{B}_a}$ 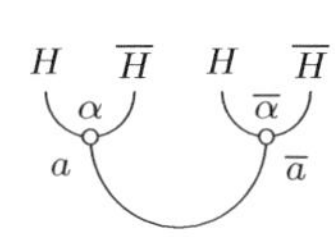 $= \sum_{a\in\operatorname{Irr}(\mathcal{C})}\sum_{\alpha\in\mathcal{B}_a}$ 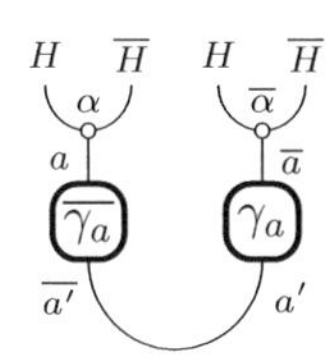

Now since $\{\overline{\alpha} \circ \gamma_a | \alpha \in \mathcal{B}_a\}$ is an orthonormal basis for $\pi_{a'}(\mathbf{A}(a'))$ under the inner product (14), and since p is independent of the choice of orthonormal bases $\mathcal{B}_a$, we see that the right hand side above is equal to

$$\sum_{a \in \operatorname{Irr}(\mathcal{C})} \sum_{\alpha \in \mathcal{B}_a}$$ 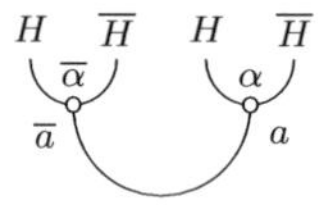

$$= (\mathrm{id}_{H \otimes \overline{H}} \otimes p) \circ \mathrm{coev}_{H \otimes \overline{H}}. \qquad \square$$

We now define important structure constants which are essentially certain 6j symbols.

DEFINITION 4.5. For $a, b, c \in \operatorname{Irr}(\mathcal{C})$, $\alpha \in \mathcal{B}_a$, $\beta \in \mathcal{B}_b$, $\gamma \in \mathcal{B}_c$, and $\delta \in \mathcal{C}(c, a \otimes b)$, we define the *tetrahedral structure constant*

$$\Delta\left(\begin{array}{ccc} a & b & c \\ \alpha & \beta & \gamma \end{array} \middle| \delta \right) =$$ 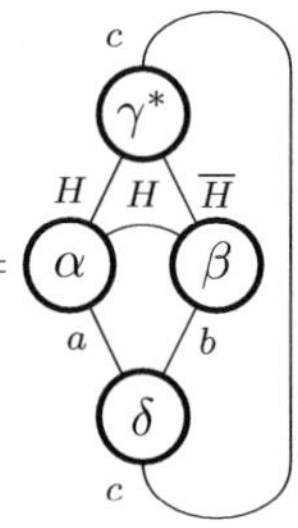

.

Note that these structure constants have a $\mathbb{Z}/3\mathbb{Z}$ symmetry:
(15)
$$\Delta\left(\begin{array}{ccc} a & b & c \\ \alpha & \beta & \gamma \end{array} \middle| \delta \right) = \Delta\left(\begin{array}{ccc} \overline{c} & a & \overline{b} \\ \overline{\gamma} & \alpha & \overline{\beta} \end{array} \middle| \mathcal{F}(\delta) \right) = \Delta\left(\begin{array}{ccc} b & \overline{c} & \overline{a} \\ \beta & \overline{\gamma} & \overline{\alpha} \end{array} \middle| \mathcal{F}^2(\delta) \right).$$

EXAMPLE 4.6. Using the tetrahedral structure constants, we get the following two important identities. Both are proved by expanding the left hand sides out using the respective inner products.

(1) For all $a, b, c \in \operatorname{Irr}(\mathcal{C})$ and $\alpha \in \mathcal{B}_a$ and $\beta \in \mathcal{B}_b$, and $\delta \in \operatorname{Isom}(c, a \otimes b)$, expanding using the inner product (14) on $\mathcal{B}_c$ yields

(16) 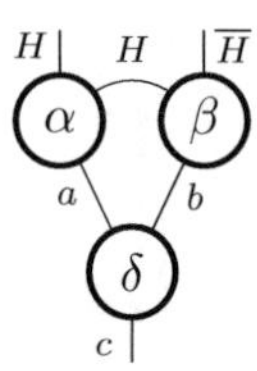

$$= d_c^{-1} \sum_{\gamma \in \mathcal{B}_c} \Delta\left(\begin{array}{ccc} a & b & c \\ \alpha & \beta & \gamma \end{array} \middle| \delta \right)$$ 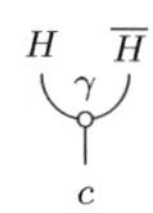

.

(2) For all $a, b, c \in \operatorname{Irr}(\mathcal{C})$, $\alpha \in \mathcal{B}_a$, $\beta \in \mathcal{B}_b$, and $\gamma \in \mathcal{B}_c$, expanding using the inner product on $\operatorname{Isom}(c, a \otimes b)$ yields

(17) 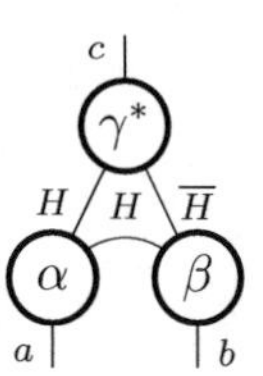

$$= d_c^{-1} \sum_{\delta \in \operatorname{Isom}(c, a \otimes b)} \Delta\left(\begin{array}{ccc} a & b & c \\ \alpha & \beta & \gamma \end{array} \middle| \delta \right)$$ 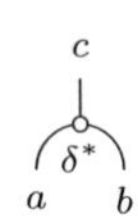

.

LEMMA 4.7. *The following two morphisms in* $\mathcal{C}(H \otimes \overline{H} \otimes H \otimes \overline{H}, H \otimes \overline{H})$ *are equal:*

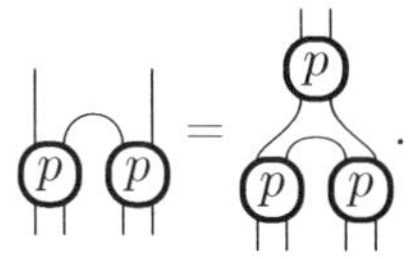

PROOF. Using (16) and (17) and the fusion relation

 $= \sum_{c \in \mathrm{Irr}(\mathcal{C})} \sum_{\delta \in \mathrm{Isom}(c, a\otimes b)}$ 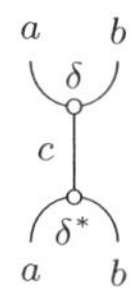 ,

it is straightforward to show that both morphisms are equal to

$$\sum_{a,b,c\in \mathrm{Irr}(\mathcal{C})} d_c^{-1} \sum_{\substack{\alpha\in\mathcal{B}_a\\ \beta\in\mathcal{B}_b\\ \gamma\in\mathcal{B}_c}} \sum_{\delta\in\mathrm{Isom}(c,a\otimes b)} \Delta\left(\begin{array}{ccc|c} a & b & c & \delta \\ \alpha & \beta & \gamma & \end{array}\right) \tag{18}$$

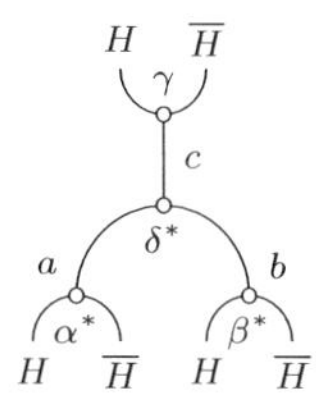

$$= \sum_{a,b,c\in \mathrm{Irr}(\mathcal{C})} \sum_{\substack{\alpha\in\mathcal{B}_a\\ \beta\in\mathcal{B}_b\\ \gamma\in\mathcal{B}_c}} \sum_{\delta\in\mathrm{ONB}(c,a\otimes b)} \Delta\left(\begin{array}{ccc|c} a & b & c & \delta \\ \alpha & \beta & \gamma & \end{array}\right)$$

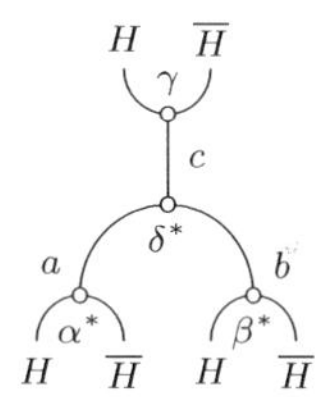

as $\mathrm{Isom}(c, a \otimes b) = \sqrt{d_c}\,\mathrm{ONB}(c, a \otimes b)$. □

REMARK 4.8. Notice that the the morphism (18) is invariant under cyclic rotation, since the sum is independent of the choice of $\mathrm{ONB}(c, a \otimes b)$, and the 1-click rotation $\mathcal{F}$ produces an orthonormal basis of $\mathcal{C}(\overline{b}, \overline{c} \otimes a)$ under the standard inner product (2) which caps off all strings. Finally, we use the $\mathbb{Z}/3\mathbb{Z}$ symmetry of the tetrahedral structure constants (15).

COROLLARY 4.9. *The morphism* 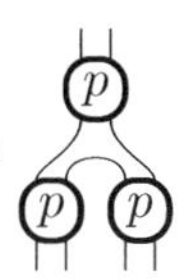 *remains unchanged by removing any single* p.

PROOF. Immediate from Lemmas 4.4 and 4.7 using 2-strand rotations. □

LEMMA 4.10. *Capping off* p *produces* id_H*, i.e.,*  $= \mathrm{id}_H$.

PROOF. Using the definition of $\mathcal{B}_a$ and applying (13), we have

$$= \sum_{a\in\mathrm{Irr}(\mathcal{C})}\sum_{\alpha\in\mathcal{B}_a}$$
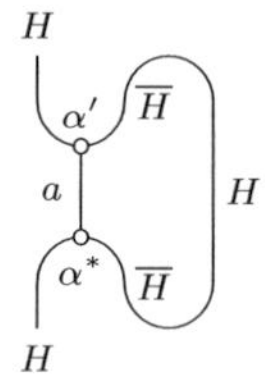
$$= \sum_{a\in\mathrm{Irr}(\mathcal{C})} \dim(\mathbf{A}(a))\frac{d_a}{d_H}\,\mathrm{id}_H = \mathrm{id}_H\,. \qquad \square$$

A version of the following proposition appears in [**BKLR15**].

PROPOSITION 4.11. *The object* $\mathrm{im}(p) \subset H \otimes \overline{H}$ *with multiplication and unit morphisms*

$$m = d_H^{1/2}$$
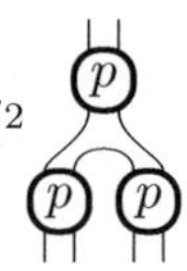
$$i = d_H^{-1/2}$$
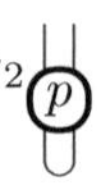

is a normalized irreducible Q-system in $\mathcal{C}$.

PROOF. A straightforward calculation using Corollary 4.9 and Lemma 4.10 shows that $(\mathrm{im}(p), m, i)$ is a separable C* Frobenius algebra. For example, removing the middle copy of p from $m^* \circ m$, applying isotopy, and re-introcuing a copy of p yields the Frobenius relation. (See also the argument in [**LP15**, Lem. 5.8].) Separability follows from removing a middle copy of p from $m \circ m^*$ and applying Lemma 4.10, which shows $m \circ m^* = d_H\,\mathrm{id}_H$. By definition of p we have $i^* \circ i = \mathrm{id}_{1_\mathcal{C}}$, and thus $\mathrm{im}(p)$ is a normalized Q-system. It is obvious that $\mathrm{im}(p)$ is connected as $\dim(\mathcal{B}_{1_\mathcal{C}}) = 1$, and thus the Q-system is irreducible. $\square$

LEMMA 4.12. *For all* $a \in \mathcal{C}$ *and* $g \in \mathbf{A}(a)$, $p \circ \pi_a(g) = \pi_a(g)$. *In particular, the unit for the Q-system in Proposition* 4.11 *is given by* $i = d_H^{-1/2}(p \circ \mathrm{coev}_H) = d_H^{-1/2}\,\mathrm{coev}_H$.

PROOF. For $a \in \mathcal{C}$ and $g \in \mathbf{A}(a)$, we have

(19) $$p \circ \pi_a(g) = d_H^{-1/2} \sum_{c\in\mathrm{Irr}(\mathcal{C})}\sum_{\alpha\in\mathcal{B}_c}$$
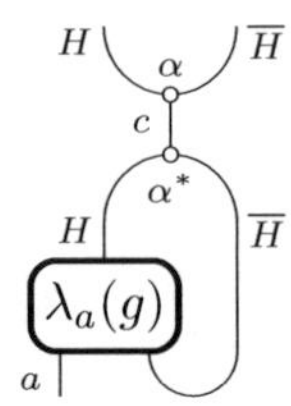

$$= d_H^{-3/2} \sum_{c\in\mathrm{Irr}(\mathcal{C})}\ \sum_{f\in\mathrm{ONB}(\mathbf{A}(c))} d_c$$
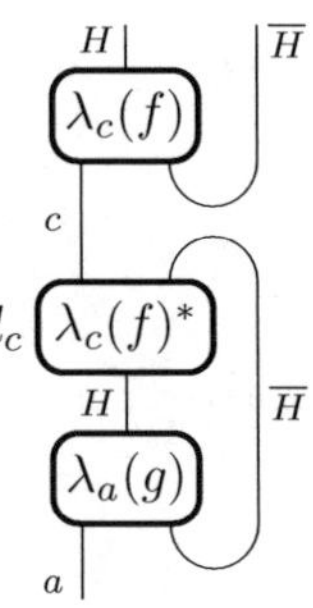

where $\mathrm{ONB}(\mathbf{A}(c))$ is with respect to inner product (9). For $c \in \mathrm{Irr}(\mathcal{C})$ and $f \in \mathrm{ONB}(\mathbf{A}(c))$, we take inner products against an orthonormal basis $\mathrm{Isom}(c, a)$ for inner product (7) on $\mathcal{C}(c, a)$ to see

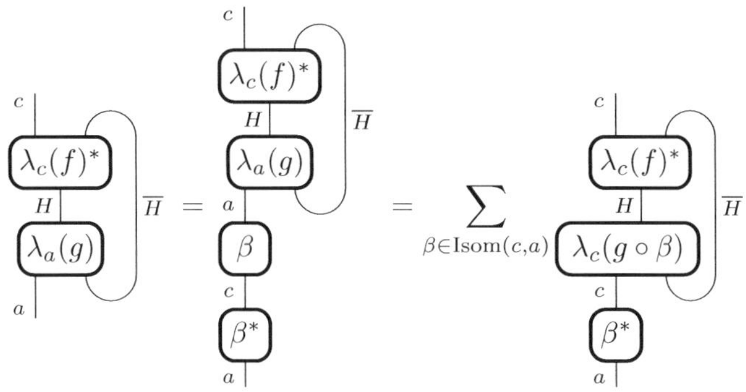

$$= \frac{d_H}{d_c} \sum_{\beta \in \mathrm{Isom}(c,a)} \langle f | g \circ \beta \rangle_c \beta^*$$

by (13). Plugging back in to (19), we obtain

$$(20) \qquad p \circ \pi_a(g) = d_H^{-1/2} \sum_{c \in \mathrm{Irr}(\mathcal{C})} \sum_{f \in \mathrm{ONB}(\mathbf{A}(c))} \sum_{\beta \in \mathrm{Isom}(c,a)} \langle f | g \circ \beta \rangle_c$$ 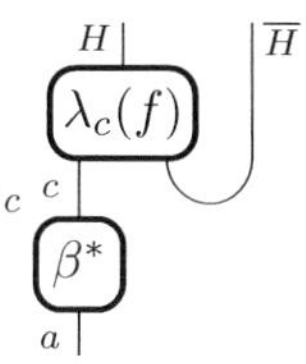 .

Now it is straightforward to verify that for all $f \in \mathbf{A}(c)$ and $\beta \in \mathcal{C}(c, a)$, $\langle f | g \circ \beta \rangle_c = \langle f \circ \beta^* | g \rangle_a$. Moreover,

$$\{ f \circ \beta^* | f \in \mathrm{ONB}(\mathbf{A}(c)),\ \beta \in \mathrm{Isom}(c, a),\ c \in \mathrm{Irr}(\mathcal{C}) \}$$

is an orthonormal basis for $\mathbf{A}(a)$ under inner product (9). Thus (20) simplifies to

$$p \circ \pi_a(g) = d_H^{-1/2} \sum_{h \in \mathrm{ONB}(\mathbf{A}(a))} \langle h | g \rangle_a$$ 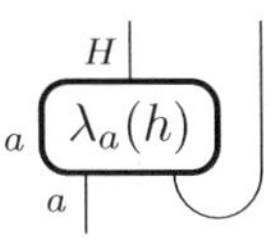 $= d_H^{-1/2}$ 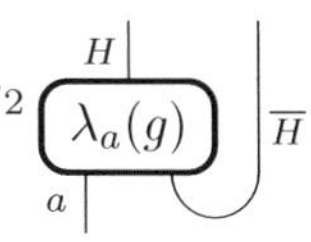 $= \pi_a(g).$

□

4.2. Morphisms. In the previous section, to each compact connected W*-algebra object $\mathbf{A} \in \mathsf{Vec}(\mathcal{C})$, we associated a canonical projection $p_{\mathbf{A}} \in \mathrm{End}_{\mathcal{C}}(H_{\mathbf{A}} \otimes \overline{H_{\mathbf{A}}})$ where $H_{\mathbf{A}} \in \mathcal{C}$ is the object representing the compact Hilbert space object $\mathbf{L}^2\mathbf{A}$, i.e., $\mathbf{L}^2\mathbf{A} = \mathcal{C}(a, H_{\mathbf{A}})$. We use the notation $Q(\mathbf{A}) = \mathrm{im}(p_{\mathbf{A}})$ for the subobject of $H_{\mathbf{A}} \otimes \overline{H_{\mathbf{A}}}$.

Suppose now we have a $*$-algebra natural transformation $\theta : \mathbf{A} \Rightarrow \mathbf{B}$ between compact connected W*-algebra objects.

NOTATION 4.13. Recall that for $a \in \mathcal{C}$, we defined

$$\mathcal{B}_a^{\mathbf{A}} = \left\{ d_a^{1/2} \pi_a^{\mathbf{A}}(f) \middle| f \in \mathrm{ONB}(\mathbf{A}(a)) \right\}$$

where $\mathrm{ONB}(\mathbf{A}(a))$ is an orthonormal basis for $\mathbf{A}(a)$ under the inner product (9). By a slight abuse of notation, for $\alpha = d_a^{1/2} \pi_a^{\mathbf{A}}(f) \in \mathcal{B}_a^{\mathbf{A}}$, we write $\theta(\alpha)$ for $d_a^{1/2} \pi_a^{\mathbf{B}}(\theta(f)) \in \mathcal{B}_a^{\mathbf{B}}$. Moreover, by (11), we may choose the orthonormal basis $\mathrm{ONB}(\mathbf{B}(a))$ so that it contains $\theta(\text{ONB}(\mathbf{A}(a)))$ as a subset. This means that $\theta(\mathcal{B}_a^{\mathbf{A}}) \subset \mathcal{B}_a^{\mathbf{B}}$ as orthonormal bases under inner product (14).

DEFINITION 4.14. We define the following morphisms in $\mathcal{C}$, which are clearly independent of the choice of orthonormal bases $\mathcal{B}_a^{\mathbf{A}}$ for $a \in \text{Irr}(\mathcal{C})$:

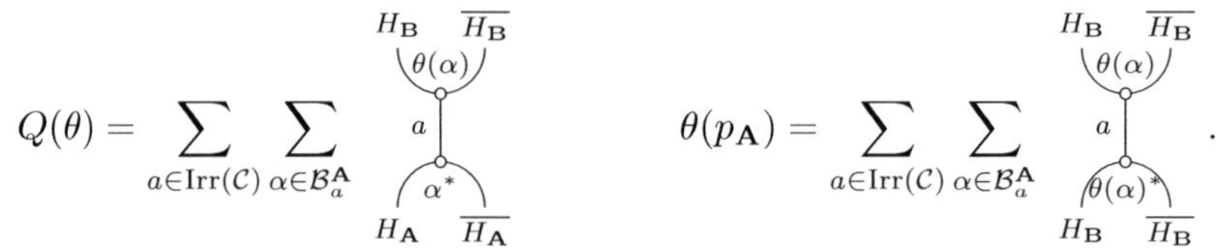

It is straightforward to verify using using (11) together with (13) and (14) that $\theta(p_{\mathbf{A}})$ is an orthogonal projection such that $\theta(p_{\mathbf{A}}) \le p_{\mathbf{B}}$, and $Q(\theta)$ is a partial isometry satisfying $Q(\theta)Q(\theta)^* = \theta(p_{\mathbf{A}})$ and $Q(\theta)^*Q(\theta) = p_{\mathbf{A}}$.

We have the following key lemma.

LEMMA 4.15. *For all $a \in \mathcal{C}$ and $f \in \mathbf{A}(a)$,*

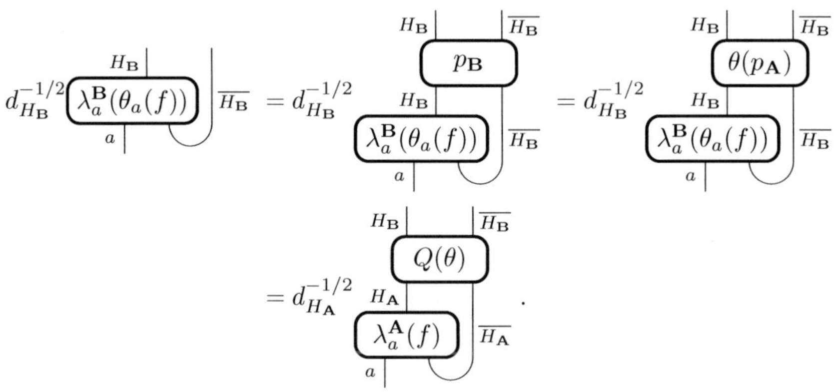

In particular, $\pi_a^{\mathbf{B}}(\theta(f)) = Q(\theta) \circ \pi_a^{\mathbf{A}}(f)$.

PROOF. The first equality follows from Lemma 4.12. The second equality follows from the fact that $\pi_a^{\mathbf{B}}(\theta_a(f))$ is orthogonal to $\mathcal{B}_a^{\mathbf{B}} \setminus \theta(\mathcal{B}_a^{\mathbf{A}})$. We omit the proof of the final equality, which is similar to the proof of Lemma 4.12 using (11) together with (13) and (14). $\square$

COROLLARY 4.16. *The morphism $Q(\theta) \in \mathcal{C}(Q(\mathbf{A}), Q(\mathbf{B}))$ is an involutive algebra morphism.*

PROOF. It is easy to see that $Q(\theta)$ is unital, since $\theta_{1_\mathcal{C}}(i_{\mathbf{A}}) = i_{\mathbf{B}}$. Thus by Lemma 4.15,

$$d_{H_{\mathbf{A}}}^{-1/2} Q(\theta) \circ \text{coev}_{H_{\mathbf{A}}} = Q(\theta) \circ \pi_{1_\mathcal{C}}^{\mathbf{A}}(i_{\mathbf{A}}) = \pi_{1_\mathcal{C}}^{\mathbf{B}}(\theta(i_{\mathbf{A}})) = \pi_{1_\mathcal{C}}^{\mathbf{B}}(i_{\mathbf{B}}) = d_{H_{\mathbf{B}}}^{-1/2} \text{coev}_{H_{\mathbf{B}}}.$$

To show $Q(\theta)$ is multiplicative, we first use multiplicativity of $\theta : \mathbf{A} \Rightarrow \mathbf{B}$ and $\pi^{\mathbf{B}} : \mathbf{B} \Rightarrow \mathbf{L^2B} \otimes \overline{\mathbf{L^2B}}$ and the relationship between the inner products (11), (13), and (14) to prove an identity similar to (17). For all $a, b, c \in \text{Irr}(\mathcal{C})$, $\alpha \in \mathcal{B}_a$, $\beta \in \mathcal{B}_b$, and $\gamma \in \mathcal{B}_c$, expanding using the inner product on $\mathrm{Isom}(c, a \otimes b)$ yields

(21)

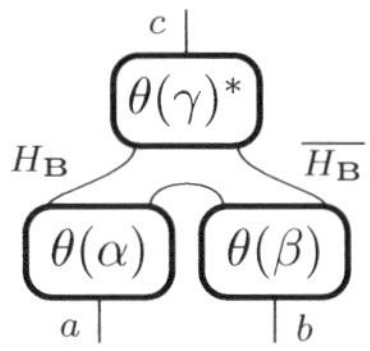

$$= \frac{d_{H_\mathbf{A}}^{1/2}}{d_c d_{H_\mathbf{B}}^{1/2}} \sum_{\delta \in \mathrm{Isom}(c, a\otimes b)} \Delta\left(\begin{array}{ccc|c} a & b & c & \delta \\ \alpha & \beta & \gamma & \end{array}\right)$$

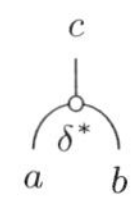

$$\underset{(17)}{=} \frac{d_{H_\mathbf{A}}^{1/2}}{d_{H_\mathbf{B}}^{1/2}}$$

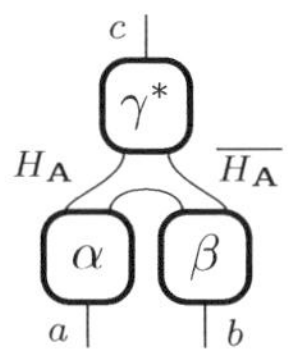

.

Now, similar to the proof of Lemma 4.7, we have

$m_{Q(\mathbf{B})} \circ (Q(\theta) \otimes Q(\theta))$

$:= d_{H_\mathbf{B}}^{1/2}$ 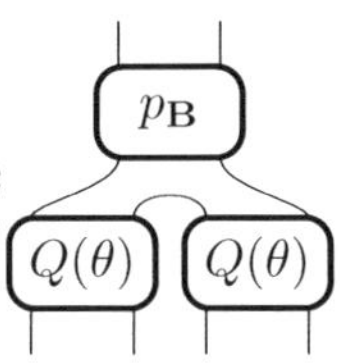 $= d_{H_\mathbf{B}}^{1/2}$ 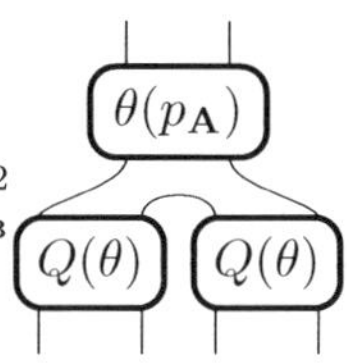 (Lem. 4.7)

$$= d_{H_\mathbf{A}}^{1/2} \sum_{a,b,c \in \mathrm{Irr}(\mathcal{C})} d_c^{-1} \sum_{\substack{\alpha \in \mathcal{B}_a \\ \beta \in \mathcal{B}_b \\ \gamma \in \mathcal{B}_c}} \sum_{\delta \in \mathrm{Isom}(c, a\otimes b)} \Delta\left(\begin{array}{ccc|c} a & b & c & \delta \\ \alpha & \beta & \gamma & \end{array}\right)$$

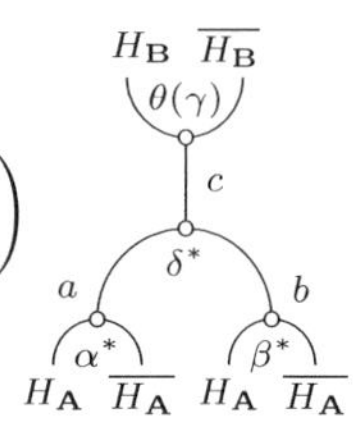

(by (21))

$= d_{H_\mathbf{A}}^{1/2}$ 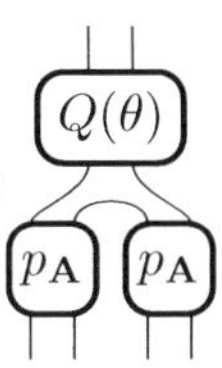 $=: Q(\theta) \circ m_{Q(\mathbf{A})}$ (by (17)).

Finally, $\sigma_{Q(\mathbf{A})} = p_\mathbf{A}$ and $\sigma_{Q(\mathbf{B})} = p_\mathbf{B}$, so invoutivity of $Q(\theta)$ is equivalent to $Q(\theta) = \overline{Q(\theta)}$, which is easily seen to hold by independence of the choice of $\mathcal{B}_a^\mathbf{A}$ as in the proof of Lemma 4.4. (This suppresses the isomorphism between real objects of the form $c \otimes \overline{c} \in \mathcal{C}$ and their conjugates as in Example 2.18.) $\square$

PROPOSITION 4.17. *Q is a functor.*

PROOF. It remains to show that Q preserves identities and composites. First, if $\theta = \mathrm{id}_\mathbf{A}$, then $Q(\theta) = p_\mathbf{A} = \mathrm{id}_{Q(\mathbf{A})}$. Second, if $\theta_1 : \mathbf{A} \Rightarrow \mathbf{B}$ and $\theta_2 : \mathbf{B} \Rightarrow \mathbf{C}$, then for each $a \in \mathcal{C}$, we choose $\mathrm{ONB}(\mathcal{C}(a))$ so as to include $\theta_2(\mathrm{ONB}(\mathbf{B}(a)))$ just as we chose $\mathrm{ONB}(\mathbf{B}(a))$ to include $\theta_1(\mathrm{ONB}(\mathbf{A}(a))$. Now for all $\alpha \in \mathcal{B}_a^\mathbf{A}$, $\theta_1(\alpha)$ is orthogonal to $\mathcal{B}_a^{\mathbf{B}} \setminus \theta_1(\mathcal{B}_a^{\mathbf{A}})$ under inner product (14), and thus

$$Q(\theta_2)Q(\theta_1) = \sum_{a,b\in\operatorname{Irr}(\mathcal{C})} \sum_{\substack{\alpha\in\mathcal{B}_a^{\mathbf{A}} \\ \beta\in\mathcal{B}_b^{\mathbf{B}}}}$$

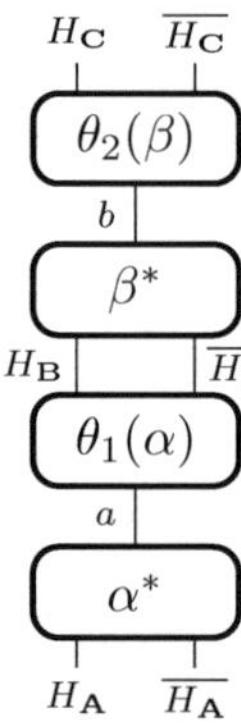

$$= \sum_{a\in\operatorname{Irr}(\mathcal{C})} \sum_{\alpha\in\mathcal{B}_a^{\mathbf{A}}}$$

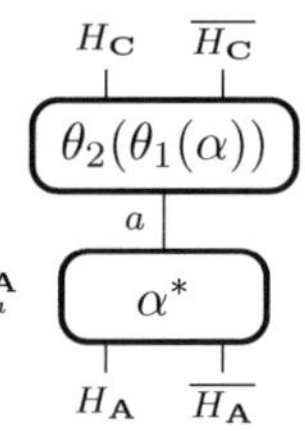

$$= Q(\theta_2 \circ \theta_1). \qquad \square$$

5. Equivalence of Q-systems and W*-algebra objects

We now prove the the functors $\mathbf{W}^*$ and Q constructed in Sections 3 and 4 respectively witness an equivalence of categories.

5.1. From W*-algebras to Q-systems and back. We begin by building a natural isomorphism $\eta : \mathrm{id} \Rightarrow \mathbf{W}^* \circ Q$.

Suppose we have a compact connected W*-algebra object $(\mathbf{A}, \mu^{\mathbf{A}}, i_{\mathbf{A}}, j^{\mathbf{A}}) \in \mathsf{Vec}(\mathcal{C})$ together with its canonical GNS faithful $*$-algebra natural transformation $\lambda : \mathbf{A} \Rightarrow \mathbf{B}(\mathbf{L}^2\mathbf{A})$. Recall that $Q(\mathbf{A}) = \operatorname{im}(p_{\mathbf{A}}) \subset H_{\mathbf{A}} \otimes \overline{H_{\mathbf{A}}} \in \mathcal{C}$ is an irreducible Q-system by Proposition 4.11. Consider the corresponding compact connected W*-algebra object $\mathbf{W}^*Q(\mathbf{A}) \in \mathsf{Vec}(\mathcal{C})$ as in Proposition 3.2, and notice that for all $a \in \mathcal{C}$, $\mathbf{W}^*Q(\mathbf{A})(a) = \mathcal{C}(a, Q(\mathbf{A}))$ with multiplication and unit induced from $Q(\mathbf{A})$, and $*$-structure given by conjugation, which is induced from the adjoint on the cyclic $\mathcal{C}$-module W*-category $(\mathcal{C}, B)$.

THEOREM 5.1. *The GNS representation $\lambda : \mathbf{A} \Rightarrow \mathbf{B}(\mathbf{L}^2\mathbf{A})$ induces a canonical $*$-algebra natural isomorphism $\eta^{\mathbf{A}} : \mathbf{A} \Rightarrow \mathbf{W}^*Q(\mathbf{A})$.*

PROOF. Using the shorthand $\mathbf{H}_{\mathbf{A}} = \mathbf{L}^2\mathbf{A}$, let $\pi^{\mathbf{A}} : \mathbf{A} \Rightarrow \mathbf{H}_{\mathbf{A}} \otimes \overline{\mathbf{H}_{\mathbf{A}}}$ be the induced faithful $*$-algebra natural transformation as in Definition 4.1. Note that $\pi_a^{\mathbf{A}}(\mathbf{A}(a))$ is exactly equal to

$$\mathbf{W}^*Q(a) = \mathcal{C}(a, Q(\mathbf{A})) = \mathcal{C}(a, \operatorname{im}(p_{\mathbf{A}})) \subset \mathcal{C}(a, H_{\mathbf{A}} \otimes \overline{H_{\mathbf{A}}}).$$

We define $\eta^{\mathbf{A}} : \mathbf{A} \Rightarrow \mathbf{W}^*Q(\mathbf{A})$ by $\eta_a^{\mathbf{A}}(f) = p_{\mathbf{A}} \circ \pi_a^{\mathbf{A}}(f) \in \mathcal{C}(a, \operatorname{im}(p_{\mathbf{A}}))$, i.e., $\eta^{\mathbf{A}}$ is just $\pi^{\mathbf{A}}$ considered as a morphism $\mathbf{A} \Rightarrow \mathbf{W}^*Q(\mathbf{A})$ rather than $\mathbf{A} \Rightarrow \mathbf{H}_{\mathbf{A}} \otimes \overline{\mathbf{H}_{\mathbf{A}}}$.

To show that $\eta^{\mathbf{A}} : \mathbf{A} \Rightarrow \mathbf{W}^*Q(\mathbf{A})$ is a $*$-algebra natural isomorphism, it remains to show that $\eta^{\mathbf{A}}$ intertwines the unit, multiplication, and $*$-structures of $\mathbf{A}$ and $\mathbf{W}^*Q(\mathbf{A})$. As in Lemma 4.12,

$$\pi_{1_{\mathcal{C}}}^{\mathbf{A}}(i_{\mathbf{A}}) = d_{H_{\mathbf{A}}}^{-1/2}$$

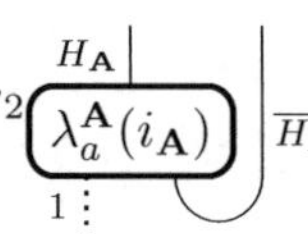

$$= d_{H_{\mathbf{A}}}^{-1/2} \operatorname{coev}_{H_{\mathbf{A}}} = i_{\mathbf{W}^*Q(\mathbf{A})}.$$

If $f \in \mathbf{A}(a)$ and $g \in \mathbf{A}(b)$, then we see that

$$\mu_{a,b}^{\mathbf{W}^*Q(\mathbf{A})}(\pi_a^{\mathbf{A}}(f) \otimes \pi_b^{\mathbf{A}}(g)) = \frac{d_{H_{\mathbf{A}}}^{1/2}}{d_{H_{\mathbf{A}}}}$$

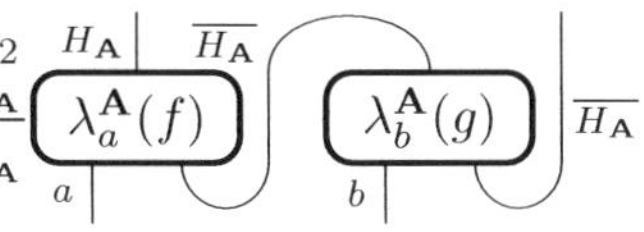

$$= d_{H_{\mathbf{A}}}^{-1/2}$$

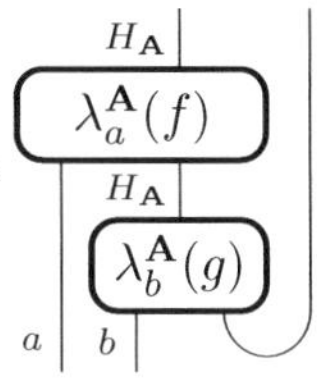

$$= d_{H_{\mathbf{A}}}^{-1/2}$$

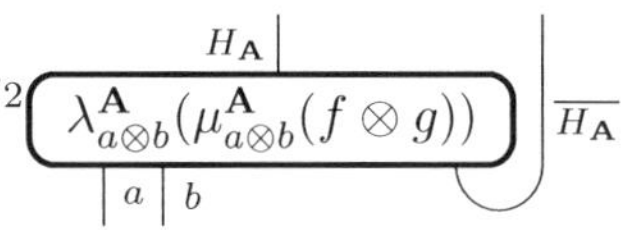

$$= \pi_{a\otimes b}^{\mathbf{A}}(\mu_{a,b}^{\mathbf{A}}(f \otimes g)).$$

Finally, if $f \in \mathbf{A}(a)$, using (1), together with the fact that λ preserves the $*$-structure, and the correspondence between the $*$-structure of $\mathbf{A}$ and the dagger structure of $\mathcal{M}_{\mathbf{A}}$ from [**JP17a**, Thm. 3.20], we have that

$$j_a^{\mathbf{W}^*Q(A)}(\pi_a^{\mathbf{A}}(f)) = d_{H_{\mathbf{A}}}^{-1/2}$$

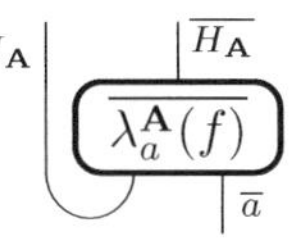

$$= d_{H_{\mathbf{A}}}^{-1/2}$$

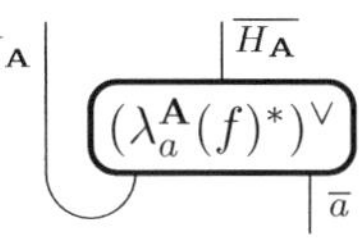

$$= d_{H_{\mathbf{A}}}^{-1/2}$$

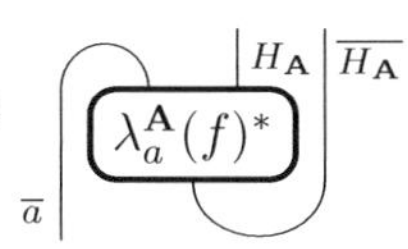

$$= d_{H_{\mathbf{A}}}^{-1/2}$$

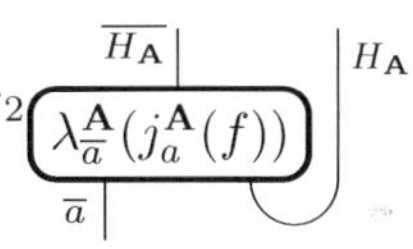

$$= \pi_{\overline{a}}^{\mathbf{A}}(j_a^{\mathbf{A}}(f)).$$

This completes the proof. □

PROPOSITION 5.2. *The $*$-algebra isomorphism $\eta^{\mathbf{A}} : \mathbf{A} \Rightarrow \mathbf{W}^*Q(\mathbf{A})$ is natural in $\mathbf{A}$.*

PROOF. Suppose $\theta : \mathbf{A} \Rightarrow \mathbf{B}$ is a natural $*$-algebra transformation. We must prove that $\mathbf{W}^*Q(\theta) \circ \eta^{\mathbf{A}} = \eta^{\mathbf{B}} \circ \theta$. We calculate that for $a \in \mathcal{C}$ and $f \in \mathbf{A}(a)$, we have

$$(\mathbf{W}^*Q(\theta)_a \circ \eta_a^{\mathbf{A}})(f) = d_{H_{\mathbf{A}}}^{-1/2}$$

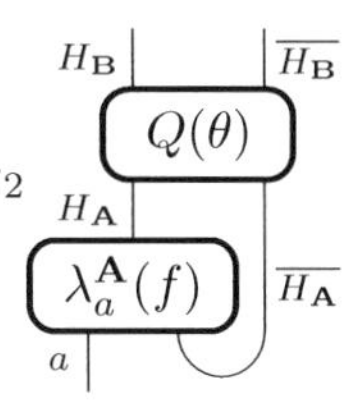

and

$$(\eta_a^{\mathbf{B}} \circ \theta_a)(f) = d_{H_{\mathbf{B}}}^{-1/2}$$

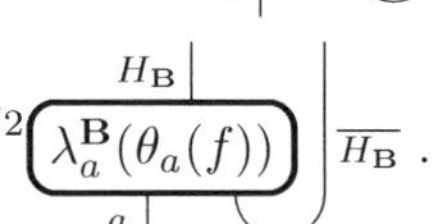

.

The equality of the above morphisms is exactly the identity $\pi_a^{\mathbf{B}}(\theta(f)) = Q(\theta) \circ \pi_a^{\mathbf{A}}(f)$ which was established in Lemma 4.15. □

5.2. From Q-systems to W*-algebras and back. We now build a natural isomorphism $\zeta : \mathrm{id} \Rightarrow Q\mathbf{W}^*$.

Starting with a normalized irreducible Q-system $(A, m_A, i_A) \in \mathcal{C}$, define the compact connected W*-algebra object $\mathbf{A} \in \mathsf{Vec}(\mathcal{C})$ by $\mathbf{A}(a) = \mathcal{C}^\natural(a, A)$ as in Proposition 3.2. By Lemma 3.3, the compact Hilbert space object $\mathbf{L}^2\mathbf{A}$ is given by $\mathbf{L}^2\mathbf{A}(a) = \mathcal{C}(a, A)$, and by Lemma 4.12, the canonical GNS faithful $*$-algebra natural transformation $\lambda : \mathbf{A} \Rightarrow \mathbf{B}(\mathbf{L}^2\mathbf{A})$ is given by (12). By Proposition 4.11, $Q\mathbf{W}^*(A) = \mathrm{im}(p) \subset A \otimes \overline{A} \in \mathcal{C}$ is a normalized irreducible Q-system.

THEOREM 5.3. *Under the GNS representation $\lambda : \mathbf{A} \Rightarrow \mathbf{B}(\mathbf{L}^2\mathbf{A})$, the identity map $\mathrm{id}_A \in \mathcal{C}(A, A)$ induces a canonical unitary involutive algebra isomorphism $\zeta_A \in \mathcal{C}(A, Q\mathbf{W}^*(A))$.*

PROOF. Recall that $\mathcal{C}(A, Q\mathbf{W}^*(A)) = \mathcal{C}(A, \mathrm{im}(p)) \subset \mathcal{C}(A, A \otimes \overline{A})$, and that $\mathcal{C}(A, \mathrm{im}(p))$ is exactly the image of $\pi_A : \mathbf{A}(A) \to (\mathbf{L}^2\mathbf{A} \otimes \overline{\mathbf{L}^2\mathbf{A}})(A) = \mathcal{C}(A, A \otimes \overline{A})$ by Lemma 4.12. By Remark 3.4,

$\lambda_A(\mathrm{id}_A) =$ 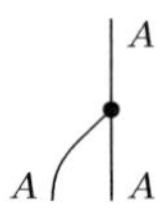 $\in \mathcal{C}(A \otimes A, A) \quad \Longrightarrow \quad \pi_A(\mathrm{id}_A) = d_A^{-1/2}$ 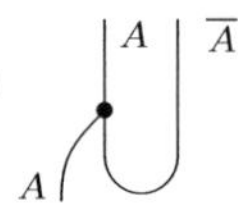 $=: \zeta_A$

Denoting $\zeta_A := \pi_A(\mathrm{id}_A)$, by Lemma 4.12, $\zeta_A = p \circ \zeta_A \in \mathcal{C}(A, \mathrm{im}(p)) = \mathcal{C}(A, Q\mathbf{W}^*(A))$.

We claim ζ_A is an involutive unitary isomorphism of Q-systems. First, since (A, m_A, i_A) is a normalized Q-system, $m_A \circ m_A^* = d_A \,\mathrm{id}_A$. Thus $\zeta_A^* \circ \zeta_A = \mathrm{id}_A$ by construction. Now $\zeta_A = p \circ \zeta_A$ by Lemma 4.12, and it is easy to see that $\zeta_A \circ \zeta_A^* = p = \mathrm{id}_{Q\mathbf{W}^*(A)}$. Indeed, $q := \zeta_A \circ \zeta_A^*$ is a subprojection of p such that $\dim(\mathrm{im}(q)) = \dim(A) = \dim(\mathrm{im}(p))$, which means $q = p$. Thus ζ_A is a unitary isomorphism.

We now show that ζ_A intertwines the unit and multiplications of A and B. First, by Lemma 4.12,

$\zeta_A \circ i_A = d_A^{-1/2}$ 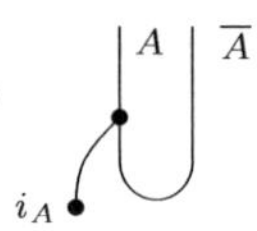 $= d_A^{-1/2} \mathrm{coev}_A = d_A^{-1/2}(p \circ \mathrm{coev}_A) = i_{Q\mathbf{W}^*(A)}.$

Second, using Corollary 4.9, $p \circ \zeta_A = \zeta_A$ by Lemma 4.12, and the Frobenius relation,

$m_{Q\mathbf{W}^*(A)} \circ (\zeta_A \otimes \zeta_A) = \dfrac{d_A^{1/2}}{d_A}$ 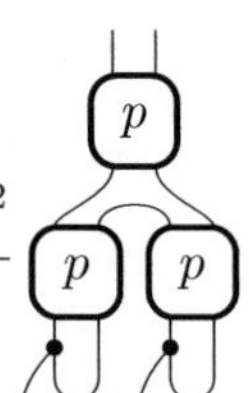 $= d_A^{-1/2}$ 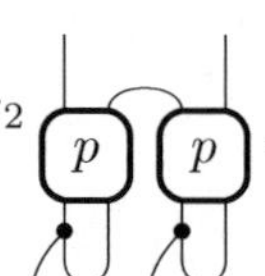 $= d_A^{-1/2}$ 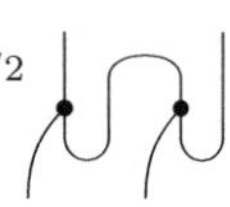

$= d_A^{-1/2}$ 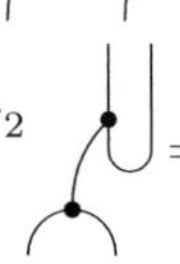 $= \zeta_A \circ m_A.$

Finally, we show ζ_A is involutive, i.e., $\sigma_{\mathrm{im}(p)} \circ \zeta_A = \overline{\zeta_A} \circ \sigma_A$. First, note that by Lemmas 4.4 and 4.7, $\sigma_{\mathrm{im}(p)} = p$ on the nose. Using (1) and Lemma 2.8 we compute

$$\overline{\zeta_A} \circ \sigma_A = (\zeta_A^*)^\vee \circ \sigma_A = d_A^{-1/2}$$
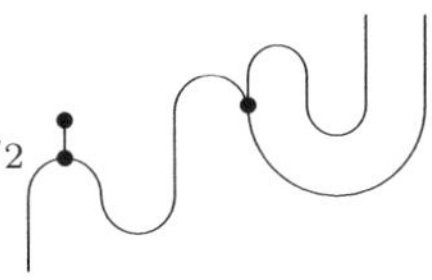
$$= d_A^{-1/2}$$
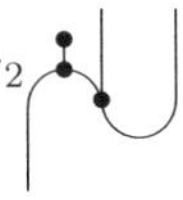
$$= d_A^{-1/2}$$
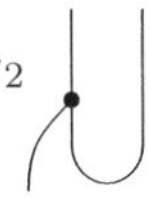
$$= \zeta_A.$$

Since $\sigma_{\mathrm{im}(p)} \circ \zeta_A = p \circ \zeta_A = \zeta_A$ by Lemma 4.12, we are finished. □

PROPOSITION 5.4. *The involutive algebra isomorphism $\zeta_A \in \mathcal{C}(A, Q\mathbf{W}^*(A))$ is natural in A.*

PROOF. Suppose $\theta \in \mathcal{C}(A, B)$ is an involutive algebra morphism. We must prove that $Q\mathbf{W}^*(\theta) \circ \zeta_A = \zeta_B \circ \theta$. Expanding the definition of $\theta(\alpha)$ for $\alpha \in \mathcal{B}_a^{\mathbf{W}^*(A)}$, one calculates that

$$Q\mathbf{W}^*(\theta) = d_A^{-1/2} d_B^{-1/2} \sum_{c \in \mathrm{Irr}(\mathcal{C})} \sum_{f \in \mathrm{ONB}(\mathbf{A}(a))} d_c$$
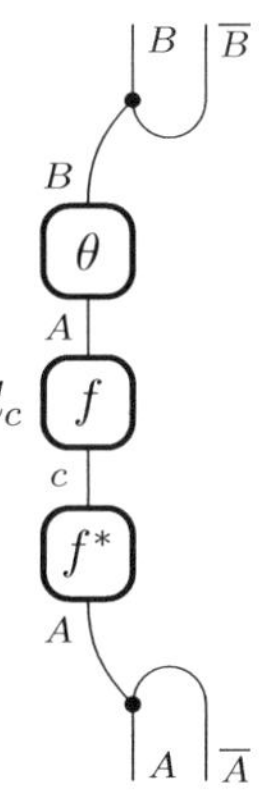

$$= d_A^{-1/2} d_B^{-1/2}$$
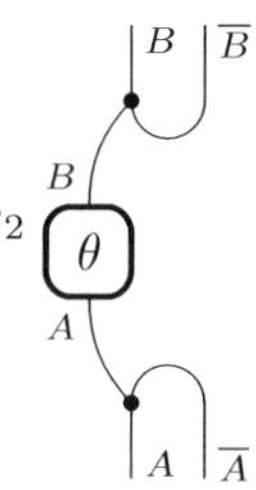

,

where in the second equality, we used that $\sqrt{d_c}\,\mathrm{ONB}(\mathbf{A}(c)) = \sqrt{d_c}\,\mathrm{ONB}(c, A) = \mathrm{Isom}(c, A)$. Finally, since A is normalized, we obtain

$$Q\mathbf{W}^*(\theta) \circ \zeta_A = d_B^{-1/2}$$
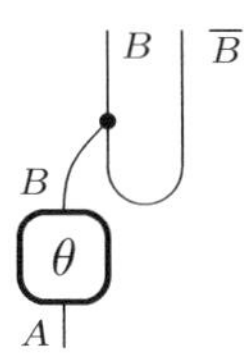

$$= \zeta_B \circ \theta.$$
□

References

[AC15] Yuki Arano and Kenny De Commer, *Torsion-freeness for fusion rings and tensor* C*-*categories*, J. Noncommut. Geom. **13** (2019), no. 1, 35–58, DOI 10.4171/JNCG/322. MR3941472

[BHP12] Arnaud Brothier, Michael Hartglass, and David Penneys, *Rigid C^*-tensor categories of bimodules over interpolated free group factors*, J. Math. Phys. **53** (2012), no. 12, 123525, 43, DOI 10.1063/1.4769178. MR3405915

[BKLR15] Marcel Bischoff, Yasuyuki Kawahigashi, Roberto Longo, and Karl-Henning Rehren, *Tensor categories and endomorphisms of von Neumann algebras—with applications to quantum field theory*, SpringerBriefs in Mathematical Physics, vol. 3, Springer, Cham, 2015. MR3308880

[BN11] Alain Bruguières and Sonia Natale, *Exact sequences of tensor categories*, Int. Math. Res. Not. IMRN **24** (2011), 5644–5705, DOI 10.1093/imrn/rnq294. MR2863377

[FRS02] Jürgen Fuchs, Ingo Runkel, and Christoph Schweigert, *Conformal correlation functions, Frobenius algebras and triangulations*, Nuclear Phys. B **624** (2002), no. 3, 452–468, DOI 10.1016/S0550-3213(01)00638-1. MR1882479

[GLR85] P. Ghez, R. Lima, and J. E. Roberts, *W^*-categories*, Pacific J. Math. **120** (1985), no. 1, 79–109. MR808930

[HP17] André Henriques and David Penneys, *Bicommutant categories from fusion categories*, Selecta Math. (N.S.) **23** (2017), no. 3, 1669–1708, DOI 10.1007/s00029-016-0251-0. MR3663592

[HPT16] André Henriques, David Penneys, and James Tener, *Categorified trace for module tensor categories over braided tensor categories*, Doc. Math. **21** (2016), 1089–1149. MR3578212

[Jon99] Vaughan F. R. Jones, *Planar algebras I*, 1999, arXiv:math.QA/9909027.

[Jon15] Vaughan F. R. Jones, *Von Neumann algebras*, 2015, https://math.vanderbilt.edu/jonesvf/VONNEUMANNALGEBRAS2015/VonNeumann2015.pdf.

[JP17a] Corey Jones and David Penneys, *Operator algebras in rigid C^*-tensor categories*, Comm. Math. Phys. **355** (2017), no. 3, 1121–1188, DOI 10.1007/s00220-017-2964-0. MR3687214

[JP17b] Corey Jones and David Penneys, *Realizations of algebra objects and discrete subfactors*, Adv. Math. **350** (2019), 588–661, DOI 10.1016/j.aim.2019.04.039. MR3948170

[KO02] Alexander Kirillov Jr. and Viktor Ostrik, *On a q-analogue of the McKay correspondence and the ADE classification of $\mathfrak{sl}_2$ conformal field theories*, Adv. Math. **171** (2002), no. 2, 183–227, DOI 10.1006/aima.2002.2072. MR1936496

[Lon89] Roberto Longo, *Index of subfactors and statistics of quantum fields. I*, Comm. Math. Phys. **126** (1989), no. 2, 217–247. MR1027496

[LP15] Zhengwei Liu and David Penneys, *The generator conjecture for 3^G subfactor planar algebras*, Proceedings of the 2014 Maui and 2015 Qinhuangdao conferences in honour of Vaughan F. R. Jones' 60th birthday, Proc. Centre Math. Appl. Austral. Nat. Univ., vol. 46, Austral. Nat. Univ., Canberra, 2017, pp. 344–366. MR3635674

[LR97] R. Longo and J. E. Roberts, *A theory of dimension*, K-Theory **11** (1997), no. 2, 103–159, DOI 10.1023/A:1007714415067. MR1444286

[Müg03] Michael Müger, *From subfactors to categories and topology. I. Frobenius algebras in and Morita equivalence of tensor categories*, J. Pure Appl. Algebra **180** (2003), no. 1-2, 81–157, DOI 10.1016/S0022-4049(02)00247-5. MR1966524

[NT13] Sergey Neshveyev and Lars Tuset, *Compact quantum groups and their representation categories*, Cours Spécialisés [Specialized Courses], vol. 20, Société Mathématique de France, Paris, 2013. MR3204665

[NY16] Sergey Neshveyev and Makoto Yamashita, *Drinfeld center and representation theory for monoidal categories*, Comm. Math. Phys. **345** (2016), no. 1, 385–434, DOI 10.1007/s00220-016-2642-7. MR3509018

[NY17] Sergey Neshveyev and Makoto Yamashita, *Categorically Morita equivalent compact quantum groups*, Doc. Math. **23** (2018), 2165–2216. MR3933035

[Pop95] Sorin Popa, *An axiomatization of the lattice of higher relative commutants of a subfactor*, Invent. Math. **120** (1995), no. 3, 427–445, DOI 10.1007/BF01241137. MR1334479

[PS03] Sorin Popa and Dimitri Shlyakhtenko, *Universal properties of $L(\mathbf{F}_\infty)$ in subfactor theory*, Acta Math. **191** (2003), no. 2, 225–257, DOI 10.1007/BF02392965. MR2051399

[Vic11] Jamie Vicary, *Categorical formulation of finite-dimensional quantum algebras*, Comm. Math. Phys. **304** (2011), no. 3, 765–796, DOI 10.1007/s00220-010-1138-0. MR2794547

Contemporary Mathematics
Volume **747**, 2020
https://doi.org/10.1090/conm/747/15040

Dimension as a quantum statistic and the classification of metaplectic categories

Paul Bruillard, Paul Gustafson, Julia Yael Plavnik, and Eric C. Rowell

ABSTRACT. We discuss several useful interpretations of the categorical dimension of objects in a braided fusion category, as well as some conjectures demonstrating the value of quantum dimension as a quantum statistic for detecting certain behaviors of anyons in topological phases of matter. From this discussion we find that objects in braided fusion categories with integral *squared* dimension have distinctive properties. A large and interesting class of non-integral modular categories such that every simple object has integral squared-dimensions are the metaplectic categories that have the same fusion rules as $SO(N)_2$ for some N. We describe and complete their classification and enumeration, by recognizing them as $\mathbb{Z}_2$-gaugings of cyclic modular categories (i.e. metric groups). We prove that any modular category of dimension $2^k m$ with m square-free and $k \leq 4$, satisfying some additional assumptions, is a metaplectic category. This illustrates anew that dimension can, in some circumstances, determine a surprising amount of the category's structure.

1. Introduction

The goal of this article is twofold: 1) to survey various notions of dimension for fusion categories and some properties determined (at least conjecturally) by dimension and 2) to classify of metaplectic categories of dimension $16k$ and apply it to the classification of categories with global dimension $16m$ with m square-free.

Dimensions of simple objects in fusion categories are one of the most ubiquitous invariants we encounter. Algebraically, a dimension function on a fusion category $\mathcal{C}$ is a generalization of a linear character for a finite group: it is an assignment of a complex number to each object $X \in \mathcal{C}$ that obeys the fusion rules. As the character table of a finite group G contains a significant amount of information about G itself (e.g. whether G is abelian, simple, or perfect), it is natural to ask: What information is contained in the dimensions of simple objects in a category? That is, how much of the structure, and which properties are determined by dimensions? The two dimension functions that one is most often interested in are the categorical and Frobenius-Perron (FP) dimensions. Fortunately, in the physically-relevant unitary setting the FP and categorical dimension coincide. For our purposes it is the FP

2010 *Mathematics Subject Classification.* Primary 18D10; Secondary 17B37.
PNNL Information Release: PNNL-SA-120943.
The research in this paper was, in part, conducted under the Laboratory Directed Research and Development Program at PNNL, a multi-program national laboratory operated by Battelle for the U.S. Department of Energy.

©2020 American Mathematical Society

dimension that is most useful, so we will focus on this particular function. In what follows we will simply refer to the FP dimension as the dimension.

Non-degenerate ribbon fusion (i.e. modular) categories model certain topological phases of matter [**38**], where the simple objects label anyons, and the isomorphism classes correspond to the anyon types. Here the word anyon is meant to convey that they are generalizations of bosons and fermions, which are (quasi)-particles with Bose/Fermi exchange statistics. In this interpretation the quantum dimensions of simple objects correspond asymptotically to the dimensions of state spaces of n identical anyons in a disk with boundary labeled by the trivial object (vacuum anyon). One may also ask: What properties of an anyon are determined by their (label's) dimension? As we view anyon systems through the lens of category theory, this leads to mathematical questions and conjectures. There are three important and seemingly unrelated properties of an anyon that are (at least conjecturally) controlled by the dimension: abelianness, localizability and universality. This motivates our perspective that dimension is the central quantum statistic for anyonic systems. We have the following (see Section 3):

CONJECTURE. *An anyon x is braiding universal if and only if $\dim(x)^2 \in \mathbb{Z}$, if and only if x is localizable.*

A coarser invariant of a fusion category is the global (FP) dimension: the sum of the squares of the dimension of the simple objects. Still, some properties of a modular category are determined by the global dimension. It is well-known (see [**8**, Cor. 3.14]) that there are finitely many fusion categories with a given fixed global dimension. In some cases one can go a bit further to give a complete classification of modular categories of a given global dimension in terms of other well-studied categories. We illustrate this principle by classifying certain modular categories of dimension $16m$ with m square-free in terms of metaplectic modular categories.

A critical ingredient of this classification is the classification of metaplectic modular categories, which we finish in this paper, section 4.1. Combining with results of [**1, 9**] we have the following

THEOREM. *Let $\mathcal{C}$ be a metaplectic modular category of dimension $4N$. Let $N = 2^a p_1^{a_1} \cdots p_s^{a_s}$ be the prime factorization (where $a \geq 0$ and $a_i > 0$). Then*

(a) *$\mathcal{C}$ is obtained from a pointed cyclic modular category $\mathcal{C}(\mathbb{Z}_N, q)$ by a $\mathbb{Z}_2$-gauging.*
(b) *If $a \leq 1$ then there are exactly 2^{s+1+a} inequivalent metaplectic modular categories of dimension $4N$, and if $a > 1$ there are $3(2^{s+2})$.*

As we mentioned above, for strictly weakly integral modular categories the structure can sometimes be determined by the dimension. Of course this is far from true for integral modular categories. There are many inequivalent modular categories of dimension 2^{2n}: simply take the twisted double of a finite group of order 2^n. On the other hand, the dimension of a non-integral category $\mathcal{C}$ can be factored over the Dedekind domain $\mathbb{Z}[\zeta_k]$ where k is the order of the T-matrix (see [**8**]) and it makes sense to look at such prime factorizations. For example:

QUESTION 1.1. *How much of the structure of a non-integral modular category $\mathcal{C}$ can be determined from the primes dividing the ideal $\langle \dim \mathcal{C} \rangle$ (in an appropriate $\mathbb{Z}[\zeta_k]$)?*

As metaplectic modular categories are ubiquitous among weakly integral modular categories as well as anyon models [**29**], an important open problem is to prove Conjectures 3.6 and 3.4 for metaplectic anyons. Some partial results are found in [**41**].

We structure the paper as follows: Section 2 contains the basic definitions that we use in subsequent sections, including various notions of dimension. Section 3 surveys properties of unitary braided fusion categories that are (conjecturally, in some cases) determined by dimension. Section 4 completes the classification and enumeration of metaplectic modular categories and Section 5 applies this classification to classify certain modular categories of dimension $16m$ with m square-free.

2. Preliminaries and Dimension Functions

We outline the axioms for the categories we are interested in, referring the reader to [**2, 18**] for further details.

A modular tensor category is a nondegenerate ribbon fusion category defined over a subfield $k \subset \mathbb{C}$. We unravel these adjectives with the following definitions.

(1) A **monoidal category** is a category with a tensor product $\otimes$ and an identity object $\mathbf{1}$ satisfying axioms that guarantee that the tensor product is associative (at least up to isomorphism) and that

$$\mathbf{1} \otimes X \cong X \otimes \mathbf{1} \cong X$$

for any object X.

(2) A monoidal category has left **duality** if there is a dual object X^* for each object X and morphisms

$$coev_X : \mathbf{1} \to X \otimes X^*, \qquad ev_X : X^* \otimes X \to \mathbf{1}$$

satisfying

$$\begin{aligned}(\mathbf{1}_X \otimes ev_X)(coev_X \otimes \mathbf{1}_X) &= \mathbf{1}_X,\\ (ev_X \otimes \mathbf{1}_{X^*})(\mathbf{1}_{X*} \otimes coev_X) &= \mathbf{1}_{X^*}.\end{aligned}$$

Right duality is defined similarly. The duality allows us to define (left) duals of morphisms too: for any
$f \in \mathrm{Hom}(X, Y)$ we define $f^* \in \mathrm{Hom}(Y^*, X^*)$ by:

$$f^* = (ev_Y \otimes \mathbf{1}_{X^*})(\mathbf{1}_{Y^*} \otimes f \otimes \mathbf{1}_{X^*})(\mathbf{1}_{Y^*} \otimes coev_X).$$

A monoidal category that has left and right duality is called **rigid**.

(3) A **braiding** in a monoidal category is a natural family of isomorphisms

$$c_{X,Y} : X \otimes Y \to Y \otimes X$$

satisfying

$$\begin{aligned}c_{X,Y\otimes Z} &= (\mathbf{1}_Y \otimes c_{X,Z})(c_{X,Y} \otimes \mathbf{1}_Z),\\ c_{X\otimes Y,Z} &= (c_{X,Z} \otimes \mathbf{1}_Y)(\mathbf{1}_X \otimes c_{Y,Z}).\end{aligned}$$

(4) A **twist** in a braided monoidal category is a natural family of isomorphisms

$$\theta_X : X \to X$$

satisfying:

$$\theta_{X\otimes Y} = c_{Y,X}c_{X,Y}(\theta_X \otimes \theta_Y).$$

(5) In the presence of a braiding, a twist and duality these structures are compatible if

$$\theta_{X^*} = (\theta_X)^*.$$

A braided monoidal category with a twist and a compatible duality is a **ribbon** category.

(6) An abelian k-linear category is **semisimple** if it has the property that every object X is isomorphic to a finite direct sum of *simple* objects–that is, objects X_i with $\mathrm{End}(X_i) \cong k$ satisfying the conclusion of Schur's Lemma:

$$\mathrm{Hom}(X_i, X_j) = 0 \quad \text{for } i \neq j.$$

(7) In a ribbon abelian k-linear category one may define a k-linear **trace** of endomorphisms. Let $f \in \mathrm{End}(X)$ for some object X. Set:

$$\mathrm{Tr}(f) = ev_X c_{X,X^*}(\theta_X f \otimes \mathbf{1}_{X^*}) coev_X.$$

Note that this is an element of $\mathrm{End}(\mathbf{1}) \cong k$, and so may be identified with a scalar. Similarly, since $\theta_X \in \mathrm{End}(X)$ for any object X, θ_{X_i} is a scalar map (as X_i is simple). We denote this scalar by θ_i.

(8) A semisimple rigid monoidal category is called a **fusion category** (of **rank** r) if it has finitely many isomorphism classes of simple objects enumerated as $\{X_0 = \mathbf{1}, X_1, \ldots, X_{r-1}\}$. In particular, the monoidal unit $\mathbf{1}$ is simple. It is known [**17**, Prop. 4.8.1] that in a fusion category left and right dualities coincide.

(9) A **premodular category** is a ribbon fusion category. A premodular category is **modular** if its so-called *S-matrix* with entries

$$S_{i,j} := \mathrm{Tr}(c_{X_j,X_i} \circ c_{X_i,X_j})$$

is invertible. Observe that S is a symmetric matrix.

(10) A **Hermitian** ribbon fusion category $\mathcal{C}$ is equipped with an additive involutive operation $\dagger$ on morphisms, $\dagger : \mathrm{Hom}(X, Y) \to \mathrm{Hom}(Y, X)$, that is compatible with $\otimes$, composition, braiding ($c^{\dagger}_{X,Y} = c^{-1}_{X,Y}$), twists and rigidity morphisms [**47**]. In particular $(f, g) := \mathrm{Tr}_{\mathcal{C}}(fg^{\dagger})$ is a non-degenerate Hermitian form on $\mathrm{Hom}(X, Y)$. If, in addition, $\dagger$ acts on $\mathbb{C}$ by complex conjugation and (f, g) is a positive definite form then we say $\mathcal{C}$ is **unitary**.

REMARK 2.1. The term "modular" comes from the following fact: if we set $T = (\delta_{i,j}\theta_i)_{ij}$ then the map:

$$\begin{pmatrix} 0 & -1 \\ 1 & 0 \end{pmatrix} \to S, \begin{pmatrix} 1 & 1 \\ 0 & 1 \end{pmatrix} \to T$$

defines a projective representation of the *modular group* $SL(2, \mathbb{Z})$. In fact, by renormalizing S and T one gets an honest representation of $SL(2, \mathbb{Z})$.

2.1. Dimension functions. Let $\mathcal{C}$ be a fusion category of rank r with $X_0 = \mathbf{1}, X_1, \ldots, X_{r-1}$ a collection of representatives of the distinct isomorphism classes of simple objects. The Grothendieck semi-ring $K_0(\mathcal{C})$ of $\mathcal{C}$ encodes the fusion rules of $\mathcal{C}$: it is the based $\mathbb{Z}_+$-ring with basis the simple isomorphism classes of objects and the operations induced by the direct sum $\oplus$ and the tensor product $\otimes$ [**39**]. A *dimension function* on a fusion category $\mathcal{C}$ is a unital ring homomorphism $K_0(\mathcal{C}) \to \mathbb{C}$. Thus the collection of dimension functions on a fusion category only depends on $K_0(\mathcal{C})$.

A fusion category with a *commutative* Grothendieck ring (e.g. a braided fusion category) has exactly as many dimension functions as it has simple isomorphism classes of objects (see e.g. [**7**, Theorem 2.3]). A non-commutative fusion category may admit only the trivial dimension function: Vec_G for a perfect group G is such an example, since a dimension function for Vec_G is a linear character of G.

We will describe three dimension functions each of a rather distinct nature, and then argue that they are identical in the physically relevant case.

FP-dimension The fusion rules $N_{i,j}^k := \dim \mathrm{Hom}(X_i \otimes X_j, X_k)$ supply us with an explicit realization of the left-regular representation of $K_0(\mathcal{C})$ via $X_i \to N_i$ where $(N_i)_{k,j} := N_{i,j}^k$. Indeed, $X_i \otimes X_j \cong \bigoplus_k N_{i,j}^k X_k$ so that it suffices to check the corresponding matrix equation (exercise) and then extend to $K_0(\mathcal{C})$ linearly $X \to N_X$. The first dimension function is the *Frobenius-Perron (FP) dimension*, defined as $\mathrm{FPdim}(X) := \max \mathrm{Spec}(N_X)$ for simple, i.e. the maximal eigenvalue of the fusion matrix N_i for simple X. The existence of such an eigenvalue follows from the Perron-Frobenius theorem applied to the positive matrix $\sum_{i,j} N_i N_j N_i^T$ that commutes with each N_k. (exercise, see [**18**, Section 8]). The fact that FPdim is a dimension function and is uniquely determined by $\mathrm{FPdim}(\mathbf{1}) = 1$ and $\mathrm{FPdim}(X_i) > 0$ is found in [**18**].

Asymptotic Dimension In the categorical model for topological phases of matter, the state space of n anyons of type X on a disk with boundary labeled by the vacuum anyon $\mathbf{1}$ is $\mathrm{Hom}(\mathbf{1}, X^{\otimes n})$. How does $\dim_{\mathbb{C}} \mathrm{Hom}(\mathbf{1}, X^{\otimes n})$ grow with n? If X were simply a d_X-dimensional vector space over $\mathbb{C}$ and $\mathbf{1} = \mathbb{C}$ then $\dim_{\mathbb{C}} \mathrm{Hom}(\mathbf{1}, X^{\otimes n}) = d_X^n$ for all n. More generally a natural measure of the dimension of an object X in a fusion category is a constant $d_X > 0$ such that $\dim_{\mathbb{C}} \mathrm{Hom}(\mathbf{1}, X^{\otimes n})$ grows like d_X^n, i.e. the asymptotic dimension of $\mathrm{Hom}(\mathbf{1}, X^{\otimes n})$. Of course it can happen that $\dim_{\mathbb{C}} \mathrm{Hom}(\mathbf{1}, X^{\otimes n})$ is 0 for many n, but there is a minimal $k > 0$ such that $\dim_{\mathbb{C}} \mathrm{Hom}(\mathbf{1}, X^{\otimes kn}) > 0$ for all $n > 0$ (see [**15**, Lemma F.6]), so that we may instead define $d_X > 0$ to be a constant such that $\dim_{\mathbb{C}} \mathrm{Hom}(\mathbf{1}, X^{\otimes k(n+1)}) / \dim_{\mathbb{C}} \mathrm{Hom}(\mathbf{1}, X^{\otimes kn}) \approx (d_X)^k$. The Perron-Frobenius theorem applied to the fusion matrix N_X implies that $d_X = \mathrm{FPdim}(X)$ (exercise, compute $\dim_{\mathbb{C}} \mathrm{Hom}(\mathbf{1}, X^{\otimes kn})$ in terms of the entries of N_X). Thus the asymptotic dimension coincides with the FP-dimension and in particular is a dimension function.

We illustrate this with two familar examples:

EXAMPLE 2.2. *Consider* $\mathcal{C} = \mathrm{Fib}$ *the Fibonacci modular category. There are 2 simple objects* $\mathbf{1}, X$ *in* $\mathcal{C}$. *Since* $X \otimes X = \mathbf{1} \oplus X$ *then the fusion matrix is* $N_X = \left(\begin{smallmatrix} 0 & 1 \\ 1 & 1 \end{smallmatrix}\right)$ *and it has eigenvalues* $\frac{1 \pm \sqrt{5}}{2}$. *Therefore,* $\dim(X) = \frac{1+\sqrt{5}}{2}$.

The fusion rules of the category $\mathcal{C}$ *are given by the Fibonacci numbers in the following way* $X^{\otimes i} = F(i-1)\mathbf{1} \oplus F(i)X$. *Then we can compute:*

$$\lim_{i \to \infty} \frac{\dim \mathrm{Hom}_{\mathcal{C}}(X^{\otimes 2i}, \mathbf{1})}{\dim \mathrm{Hom}_{\mathcal{C}}(X^{\otimes 2(i-1)}, \mathbf{1})} = \lim_{i \to \infty} \frac{F(2i-2)}{F(2i-4)} = \left(\frac{1+\sqrt{5}}{2}\right)^2.$$

EXAMPLE 2.3. *An Ising modular category* $\mathcal{I}^\nu$ *is any of the* 8 *non-integral* 4*-dimensional unitary modular categories* [**12**]. *There are three simple objects* $\mathbf{1}$, ψ *and* σ. *The fusion matrix for* σ *is* $\begin{pmatrix} 0 & 0 & 1 \\ 0 & 0 & 1 \\ 1 & 1 & 0 \end{pmatrix}$, *so that* $\dim \sigma = \sqrt{2}$. *Computing,* *we have* $\sigma^{2n} = 2^{n-1}(\mathbf{1} \oplus \psi)$ *so that*

$$\frac{\dim \mathrm{Hom}_{\mathcal{I}^\nu}(\sigma^{\otimes 2i}, \mathbf{1})}{\dim \mathrm{Hom}_{\mathcal{I}^\nu}(\sigma^{\otimes 2(i-1)}, \mathbf{1})} = \frac{2^{i-1}}{2^{i-2}} = (\sqrt{2})^2.$$

Categorical Dimension Now let $\mathcal{C}$ be a ribbon fusion category, and $X \in \mathcal{C}$. Now since $\mathrm{End}(\mathbf{1}) \cong \mathbb{C}$, with basis $\mathrm{Id}_{\mathbf{1}}$, we may define $\dim(X)$ to be the coefficient of $\mathrm{Id}_{\mathbf{1}}$ in $\mathrm{Tr}_{\mathcal{C}}(\mathrm{Id}_X)$. In [**48**] it is shown that $\dim(-)$ is a dimension function (exercise: show that $\mathrm{Tr}_{\mathcal{C}}$ is multiplicative with respect to $\otimes$, using naturality of c and θ.). In fact, one may define a categorical dimension for any spherical fusion category using the pivotal trace, but we will focus on the braided case.

In particular, if $\mathcal{C}$ is unitary, then $(Id_X, Id_X) = \mathrm{Tr}_{\mathcal{C}}(Id_X) = \dim(X) > 0$ for all X. Hence for a unitary category $\dim(X) = \mathrm{FPdim}(X)$ for all X.

Unless otherwise stated, in the rest of this article we will assume that $\dim(X) = \mathrm{FPdim}(X)$ for all objects. In many situations (for example if $\mathrm{FPdim}(X)^2 \in \mathbb{Z}$ for all simple X) it is possible to replace the twists (or equivalently the spherical structure) on a ribbon fusion category by another (unique) choice to ensure that $\dim(X) = \mathrm{FPdim}(X)$ (see [**18**]). On the other hand, there are examples [**43**] of modular categories whose underlying fusion rules do not admit a unitary ribbon categorification. The positivity of dim actually implies a slightly stronger condition:

THEOREM 2.4. *For every object* X, $\dim X \geq 1$.

PROOF. The dimension $\dim(X)$ is strictly positive and dominates all other eigenvalues of N_X in modulus. Thus, if $\dim X < 1$, we must have N_X nilpotent since N_X^n would tend to 0. But $\dim \mathrm{Hom}(X^{\otimes n}, \mathbf{1}) > 0$ for some n by [**15**, Lemma F.6], so that N_X^n must be non-zero. □

We introduce the following standard notation:

- If $X \in \mathcal{C}$ has $\dim(X) = 1$ we say that X is *invertible*. If every simple $X \in \mathcal{C}$ is invertible, we say $\mathcal{C}$ is *pointed*.
- If $X \in \mathcal{C}$ has $\dim(X) \in \mathbb{Z}$ we say that X is *integral*. If every simple $X \in \mathcal{C}$ is integral, we say $\mathcal{C}$ is *integral*.
- If $X \in \mathcal{C}$ is simple and has $\dim(X)^2 \in \mathbb{Z}$ we say that X is *weakly integral*. If every simple $X \in \mathcal{C}$ is weakly integral, we say $\mathcal{C}$ is *weakly integral*. If, moreover, there exists a simple object $X \in \mathcal{C}$ with $\dim(X) \notin \mathbb{Z}$ we say $\mathcal{C}$ is *strictly weakly integral*.

REMARK 2.5. While invertibility is usually defined as $X \otimes X^* \cong \mathbf{1}$, the definition above is equivalent. Indeed, let $\mathcal{C}$ be a fusion category and $X \in \mathcal{C}$. Consider

$$G[X] = \{Y \in \mathcal{C} \mid Y \otimes X \cong X\}.$$

If $Y \in G[X]$, then $\mathrm{FPdim}\, Y = 1$.

Moreover, $X \otimes X^* = \sum_{Y \in G[X]} Y + \sum_{\mathrm{FPdim}\, Z > 1} N_{X,X^*}^Z Z$.

2.2. Some basic modular categories. A significant role is played by pointed modular categories, i.e. modular categories with only invertible simple objects. The classification of pointed modular categories is well-known (going back essentially to [**16**], also see [**15**]): they correspond to pairs (A, q) where A is a finite abelian group and q is a non-degenerate quadratic form $q : A \to \mathbb{Q}/\mathbb{Z}$ i.e. $q(-a) = q(a)$ and the symmetric bilinear form on A defined by $\sigma(a, b) = q(a + b) - q(a) - q(b)$ is non-degenerate. We denote such a category by $\mathcal{C}(A, q)$. In [**15**] these categories are called **metric groups**.

The Semion and Ising categories will also appear in the sequel. The Semion categories, denoted by Sem, are modular categories with the same fusion rules as $\mathrm{Rep}(\mathbb{Z}_2)$, see [**40**] for details. The Ising categories introduced in Example 2.3 are rank 3 modular categories with simple objects: $\mathbf{1}$, ψ (a fermion), and σ (the Ising anyon) of dimension $\sqrt{2}$. The key fusion rules are $\psi^2 = \mathbf{1}$ and $\sigma^2 = \mathbf{1} + \psi$; while the modular datum can be found in [**40**]. There are exactly 8 inequivalent Ising categories. They are the smallest examples of **(generalized) Tambara-Yamagami categories**, *i.e.* non-pointed fusion categories with the property that the tensor product of any two non-invertible simple objects is a direct sum of invertible objects. In fact, by [**37**] any *modular* generalized Tambara-Yamagami category is a (Deligne) product of an Ising category and a pointed modular category.

2.2.1. *Balancing.* Standard arguments show that the entries of the S-matrix are determined by the dimensions, the fusion rules and the twists on the simple objects, giving the following extremely useful balancing relation (see [**2**]):

$$S_{i,j} = \frac{1}{\theta_i\theta_j}\sum_k N_{i^*,j}^k \dim(X_k)\theta_k. \tag{2.1}$$

2.3. Centralizers. For $\mathcal{D} \subset \mathcal{C}$ premodular categories the **centralizer** of $\mathcal{D}$ in $\mathcal{C}$ is denoted by $\mathcal{Z}_{\mathcal{C}}(\mathcal{D})$ and is generated by the objects $\{X \in \mathcal{C} \mid S_{X,Y} = d_X d_Y \ \forall Y \in \mathcal{D}\}$ (see [**4**]. The category $\mathcal{Z}_{\mathcal{C}}(\mathcal{C})$ is called the **Müger center** and is often denoted by $\mathcal{C}'$. Note that a useful characterization of a simple object being outside of the Müger center is that its column in the S-matrix is orthogonal to the first column of S. If $\mathcal{C}' = \mathrm{Vec}$, the category of finite-dimensional vector spaces, then $\mathcal{C}$ is a modular category whereas if $\mathcal{C}' = \mathcal{C}$, then $\mathcal{C}$ is called **symmetric**.

2.3.1. *Symmetric fusion categories.* Symmetric fusion categories are classified in terms of group data:

THEOREM 2.6 ([**14**]). *If $\mathcal{C}$ is a symmetric fusion category, then there exists a finite group G such that $\mathcal{C}$ is equivalent to the super-Tannakian category* $\mathrm{Rep}(G, z)$ *of super-representations (i.e. $\mathbb{Z}_2$ graded) of G where $z \in Z(G)$ is a distinguished central element with $z^2 = 1$ acting as the parity operator.*

Observe that $\mathcal{C} \cong \mathrm{Rep}(G, z)$ with $z = e \in G$ if and only if the $\mathbb{Z}_2$-grading is trivial so that $\mathrm{Rep}(G, z) = \mathrm{Rep}(G)$. In this case we say $\mathcal{C}$ is **Tannakian**, and otherwise we say it is non-Tannakian. The smallest non-Tannakian symmetric category is $\mathrm{Rep}(\mathbb{Z}_2, 1)$ which we will denote by sVec when equipped with the (unique) structure of a ribbon category with $\dim(\chi) = 1$ for a non-trivial object χ.

An invertible object in $\mathcal{C}$ that generates the Tannakian category $\mathrm{Rep}(\mathbb{Z}_2)$ is a boson while an invertible object in $\mathcal{C}$ generating sVec is a fermion. Bosons are useful through a process known as de-equivariantization which we will discuss shortly. A modular category $\mathcal{C}$ with a distinguished fermion $f \in \mathcal{C}$ is called a **spin modular** category (see, eg. [**7**]).

2.4. Gradings. A **grading** of a fusion category $\mathcal{C}$ by a finite group G is a decomposition of the category as direct sum $\mathcal{C} = \oplus_{g \in G}\mathcal{C}_g$, where the components are full abelian subcategories of $\mathcal{C}$ indexed by the elements of G, such that the tensor product maps $\mathcal{C}_g \times \mathcal{C}_h$ into $\mathcal{C}_{gh}$. The **trivial component** $\mathcal{C}_e$ (corresponding to the unit of the group G) is a fusion subcategory of $\mathcal{C}$. The grading is called **faithful** if $\mathcal{C}_g \neq 0$, for all $g \in G$, and in this case all the components are equidimensional with $|G| \dim \mathcal{C}_g = \dim \mathcal{C}$.

Every fusion category $\mathcal{C}$ is faithfully graded by the universal grading group $\mathcal{U}(\mathcal{C})$ and every faithful grading of $\mathcal{C}$ is a quotient of $\mathcal{U}(\mathcal{C})$. Furthermore, the trivial component under the universal grading is the **adjoint subcategory** $\mathcal{C}_{\text{ad}}$, the full fusion subcategory generated by the objects $X \otimes X^*$, for X simple. The universal grading was first studied in [**28**], and it was shown that if $\mathcal{C}$ is modular, then $\mathcal{U}(\mathcal{C})$ is canonically isomorphic to the character group of the group $G(\mathcal{C})$ of isomorphism classes of invertible objects in $\mathcal{C}$. Another useful fact is that $\mathcal{C}'_{ad} = \mathcal{C}_{pt}$, the subcategory generated by the invertible objects [**28**].

When $\mathcal{C}$ is a weakly integral fusion category there is another useful grading called the **GN-grading** first studied in [**28**]:

THEOREM 2.7. [**28**, Theorem 3.10] *Let $\mathcal{C}$ be a weakly integral fusion category. Then there is an elementary abelian 2-group E, a set of distinct square-free positive integers n_x, $x \in E$, with $n_0 = 1$, and a faithful grading $\mathcal{C} = \oplus_{x \in E} \mathcal{C}(n_x)$ such that* $\dim(X) \in \mathbb{Z}\sqrt{n_x}$ *for each $X \in \mathcal{C}(n_x)$.*

The trivial component of the GN-grading is the integral subcategory $\mathcal{C}_{\text{int}}$.

2.5. Equivariantization. Our approach to classification relies upon *equivariantization* and its inverse functor *de-equivariantization* (see for example [**15**]) which we now briefly describe. An *action* of a finite group G on a fusion category $\mathcal{C}$ is a strong tensor functor $\rho : \underline{G} \to \text{End}_{\otimes}(\mathcal{C})$. The G**-equivariantization** of the category $\mathcal{C}$ is the category $\mathcal{C}^G$ of G-equivariant objects and morphisms of $\mathcal{C}$. When $\mathcal{C}$ is a fusion category over an algebraically closed field $\mathbb{K}$ of characteristic 0, the G-equivariantization, $\mathcal{C}^G$, is a fusion category with $\dim \mathcal{C}^G = |G| \dim \mathcal{C}$. The fusion rules of $\mathcal{C}^G$ can be determined in terms of the fusion rules of the original category $\mathcal{C}$ and group-theoretical data associated to the group action [**10**]. De-equivariantization is the inverse to equivariantization. Given a fusion category $\mathcal{C}$ and a Tannakian subcategory $\text{Rep}\, G \subset \mathcal{C}$, consider the algebra $A = \text{Fun}\,(G)$ of functions on G. Then A is a commutative algebra in $\mathcal{C}$. The category of A-modules on $\mathcal{C}$, $\mathcal{C}_G$, is a fusion category called a G**-de-equivariantization** of $\mathcal{C}$. We have $|G| \dim \mathcal{C}_G = \dim \mathcal{C}$, and there are canonical equivalences $(\mathcal{C}_G)^G \cong \mathcal{C}$ and $\left(\mathcal{C}^G\right)_G \cong \mathcal{C}$. An important property of the de-equivariantization is that if the Tannakian category in question is $\mathcal{C}'$, then $\mathcal{C}_G$ is modular and is called the **modularization** of $\mathcal{C}$ [**4, 33**].

2.6. Gauging. Two processes that we employ in our analysis are gauging and de-gauging. First let us describe de-gauging. Let $\mathcal{C}$ be modular and $\text{Rep}(G) \cong \mathcal{D} \subset \mathcal{C}$ a Tannakian subcategory (here a Tannakian category is a symmetric braided fusion category equivalent to $\text{Rep}(G)$ for some finite group G). The G-de-equivariantization $\mathcal{C}_G$ of $\mathcal{C}$ is a faithfully G-graded category (in fact, a braided G-crossed category) with trivial component $[\mathcal{C}_G]_e$ a modular category of dimension $\dim(\mathcal{C})/|G|^2$ (see [**12**]). $[\mathcal{C}_G]_e$ is the G**-de-gauging** of $\mathcal{C}$. The reverse process, G-gauging, is more complicated. Here one starts with a modular category $\mathcal{B}$ and an action of a finite group G by braided tensor autoequivalences: $\rho : G \to \text{Aut}_{\otimes}^{br}(\mathcal{B})$. A G**-gauging** of $\mathcal{B}$, when it exists, is a new modular category obtained by first constructing a G-graded fusion category $\mathcal{D}$ with trivial component $\mathcal{D}_e = \mathcal{B}$ and then equivariantizing $\mathcal{D}^G$. There are obstructions to the existence of a gauging, and when the obstructions vanish there can be many G-gaugings.

Here is a key example: consider the pointed (cyclic) modular category $\mathcal{C}(\mathbb{Z}_N, q)$. The elements of the group $\mathrm{Aut}_{\otimes}^{br}\,\mathcal{C}(\mathbb{Z}_N, q)$ are simply those $\phi \in \mathrm{Aut}(\mathbb{Z}_N)$ that preserve q [**13**]. The following is presumably well-known, but its proof is elementary:

PROPOSITION 2.8. *The* ***particle-hole symmetry*** $\phi(a) = -a$ *is the only non-trivial element of* $\mathrm{Aut}_{\otimes}^{br}\,\mathcal{C}(\mathbb{Z}_N, q)$.

PROOF. By the Chinese remainder theorem $\mathcal{C}(\mathbb{Z}_N, q)$ factors as a product of cyclic modular categories of prime power dimension, so we may assume that $N = p^k$. For $p \neq 2$ the two inequivalent quadratic forms $q : \mathbb{Z}_{p^k} \to \mathbb{Q}/\mathbb{Z}$ are $q_u(a) = \frac{ua^2}{p^k}$ for $u = \pm 1$. An automorphism of $\mathbb{Z}_{p^k}$ is given by $\phi_x(a) = xa$ for some $x \in \mathbb{Z}_{p^k}^*$. So $q_u \circ \phi_x = q_u$ implies that $x^2 \equiv 1 \pmod{p^k}$, which has only $x = \pm 1$ as solutions. For $p = 2$ there are 4 inequivalent quadratic forms (see, e.g. [**26**]): $q_u(a) = \frac{ua^2}{2^{k+1}}$ for $u \in \mathbb{Z}_8^*$. The same computation as above shows that $q_u \circ \phi_x = q_u$ only if $x^2 \equiv 1 \pmod{2^{k+1}}$ with $x \in \mathbb{Z}_{2^k}^*$. Although $x^2 \equiv \pmod{2^{k+1}}$ has 4 solutions, there are only two inequivalent solutions modulo 2^k, namely $x = \pm 1$. □

Furthermore, it can be shown that the obstructions vanish in this case, so the action $\rho : \mathbb{Z}_2 \to \mathrm{Aut}_{\otimes}^{br}\,\mathcal{C}(\mathbb{Z}_N, q)$ defined by $\rho(1) = \phi$ can be gauged.

3. Properties Determined by Dimension

In this section we assume that $\mathcal{C}$ is a unitary braided fusion category, and summarize some results showing that several important properties of any anyon are detected by the dimension of the corresponding simple object.

Important examples of unitary braided fusion categories are obtained from quantum groups at roots of unity (see [**42**] for a survey). Here we briefly outline the construction, mainly for notational purposes.

(1) Let $\mathfrak{g}$ be a simple Lie algebra and $q = e^{\pi i/\ell}$ be a root of unity with $\ell > \check{h}$ (dual Coxeter number of $\mathfrak{g}$) and $\ell \in 2\mathbb{Z}$ for types B, C, F and $\ell \in 3\mathbb{Z}$ for type G.
(2) The representation category of quantum group $U_q\mathfrak{g}$ is a non-semisimple ribbon category. A quotient by the tensor ideal of negligible morphisms (essentially the radical of the trace) of this category gives a braided fusion category $\mathcal{C}(\mathfrak{g}, \ell)$.
(3) The integer $k = (\ell - \check{h})/m$ (where $m = 2$ for types B, C, F and $m = 3$ for type G and $m = 1$ otherwise) is called the *level*, and for classical types A, B, C and D we adopt the alternative notation for $\mathcal{C}(\mathfrak{g}, \ell)$ is $G(N)_k$ for $G = SU, SO$ or Sp of dimension N.

The braid group on n strands $\mathcal{B}_n$ generated by $\sigma_1, \ldots, \sigma_{n-1}$ satisfying

(1) $\sigma_i\sigma_{i+1}\sigma_i = \sigma_{i+1}\sigma_i\sigma_{i+1}$ for $i \leq n-2$
(2) $\sigma_i\sigma_j = \sigma_j\sigma_i$ for $|i - j| > 1$

plays an important role in topological phases of matter: particle exchange induces a representation of $\mathcal{B}_n$ on the state space $\mathrm{Hom}(\mathbf{1}, X^{\otimes n})$ for any object X. More generally we obtain a homomorphism of $\mathcal{B}_n$ into $\mathrm{Aut}(X^{\otimes n})$, which is often lifted to a homomorphism of the group algebra $\rho_X : \mathbb{C}\mathcal{B}_n \to \mathrm{End}(X^{\otimes n})$. Moreover, $\mathcal{H}_n := \bigoplus_i \mathrm{Hom}(X_i, X^{\otimes n})$ is a faithful $\mathrm{End}(X^{\otimes n})$-module and hence we will sometimes abuse notation and regard ρ_X as a $\mathcal{B}_n$-representation on $\mathcal{H}_n$.

3.1. Abelian Anyons. We say that $X \in \mathcal{C}$ (or the corresponding anyon) is *non-abelian* if the $\mathcal{B}_n$-representation $(\rho_X, \mathcal{H}_n)$ has non-abelian image for all $n \geq 3$. This is the first property of $X \in \mathcal{C}$ determined by dimension:

THEOREM 3.1 ([**46**]). *If* $X \in \mathcal{C}$ *is simple and* $\dim X > 1$ *then* X *is non-abelian.*

It is clear that any invertible $Z \in \mathcal{C}$ is abelian: $\dim \mathrm{End}(Z^{\otimes n}) = 1$ in this case so $\rho_Z(\mathcal{B}_n)$ acts by scalars. Thus we have a complete characterization of (simple) abelian anyons as those corresponding to invertible objects. A **boson** b is a particular kind of abelian anyon, characterized by $b^{\otimes 2} \cong \mathbf{1}$ and $c_{b,b} = \mathrm{Id}_{b\otimes b}$ whereas a **fermion** f has $f^{\otimes 2} \cong \mathbf{1}$ but $c_{f,f} = -\mathrm{Id}_{f\otimes f}$. Alternatively, in a unitary ribbon category, an object X with $X^{\otimes 2} \cong \mathbf{1}$ is a boson if $\theta_X = 1$ and a fermion if $\theta_X = -1$.

3.2. Localizable Anyons. The $\mathcal{B}_n$ representations ρ_X are somewhat complicated–they exhibit a *hidden locality* [**23**], but are not explicitly local–rather the action of $\rho_X(\sigma_i)$ act non-trivially on the entire space $\mathcal{H}_n$. Explicitly local representations of $\mathcal{B}_n$ can be obtained from solutions $R \in \mathrm{Aut}(V^{\otimes 2})$ to the Yang-Baxter equation on a vector space V:

$$(R \otimes Id_V)(Id_V \otimes R)(R \otimes Id_V) = (Id_V \otimes R)(R \otimes Id_V)(Id_V \otimes R). \tag{3.1}$$

Such a pair (R, V) is called a *braided vector space.* From a braided vector space we obtain a representation ρ^R of $\mathcal{B}_n$ on $V^{\otimes n}$ via

$$\sigma_i \to Id_V^{\otimes(i-1)} \otimes R \otimes Id_V^{\otimes(n-i-1)}.$$

We say that $X \in \mathcal{C}$ is localizable if there is a braided vector space (R, V) and injective algebra maps $\tau_n : \mathbb{C}\rho_X(\mathcal{B}_n) \to \mathrm{End}(V^{\otimes n})$ such that, for all n, $\rho^R = \tau_n \circ \rho_X$. That is, the following diagram commutes:

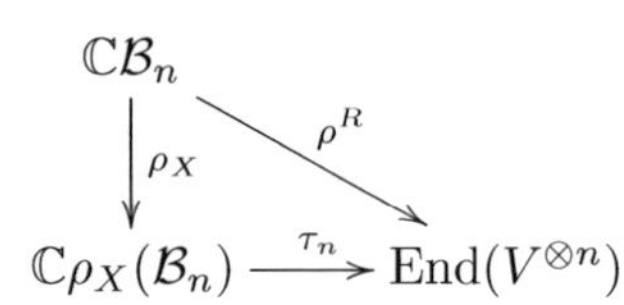

EXAMPLE 3.2. *Consider* $\mathcal{I}^\nu$ *an Ising category (see Example* 2.3*) and the object* σ *of dimension* $\sqrt{2}$. *For* $R = \frac{1}{\sqrt{2}}\begin{pmatrix} 1 & 0 & 0 & 1 \\ 0 & 1 & 1 & 0 \\ 0 & -1 & 1 & 0 \\ -1 & 0 & 0 & 1 \end{pmatrix}$ *the braided vector space* $(R, \mathbb{C}^2)$ *provides a localization of* σ [**22**].

Two slightly less restrictive notions of localizability are studied in [**25**], namely (k, m)-generalized localizations and quasi-localizations. Under some assumptions, localizability is known to be determined by dimension:

THEOREM 3.3 ([**25, 45**]). *Let* $X \in \mathcal{C}$ *be a simple object. If* $\rho_X(\mathbb{C}\mathcal{B}_n) = \mathrm{End}(X^{\otimes n})$ *for* $n \geq 2$ *and* X *is (generalized or quasi-)localizable then* $\dim(X)^2 \in \mathbb{Z}$.

It is believed that this relationship holds more generally:

CONJECTURE 3.4 ([**25,45**]). *Any simple* $X \in \mathcal{C}$ *is (generalized or quasi-)localizable if, and only if* $\dim(X)^2 \in \mathbb{Z}$.

The Gaussian Yang-Baxter operators described in [**27**] provide localizations of the generating (fundamental spinor) objects in $SO(N)_2$.

3.3. Universal Anyons. A given topological model for quantum computation is called (braiding-only) *universal* if any unitary operator can be efficiently approximated up to a phase by braiding anyons [**24**]. We say that $X \in \mathcal{C}$ is *braiding universal* if $\rho_X(\mathcal{B}_n)$ is dense in $SU(W)$ for each irreducible subrepresentation W of $\mathcal{H}_n$. The first step to verify universality is to check that $\rho_X(\mathcal{B}_n)$ is infinite, and very often (see [**24, 31**]) this is sufficient. It is therefore of the utmost importance to determine when $\rho_X(\mathcal{B}_n)$ is finite.

DEFINITION 3.5. An object $X \in \mathcal{C}$ has *property F* if the image $\rho_X(\mathcal{B}_n)$ is finite. A braided fusion category $\mathcal{C}$ has *property F* if the associated braid group representations on the centralizer algebras $\mathrm{End}_{\mathcal{C}}(X^{\otimes n})$ have finite image for all n and all objects X.

CONJECTURE 3.6. [**36**] *A braided fusion category* $\mathcal{C}$ *has property F if, and only if,* $\mathrm{FPdim}(\mathcal{C}) \in \mathbb{Z}$ *(i.e.* $\mathcal{C}$ *is weakly integral).*

Significant progress has been made towards proving this conjecture:

(1) Every $X \in \mathrm{Rep}(D^\omega(G))$ has property F [**21**].
(2) Every weakly integral quantum group category has property F, including $SU(2)_2$, $SU(2)_4$, $SU(3)_3$ and $SO(N)_2$ (see [**24, 30, 31, 41**]).
(3) The standard tensor generators of every non-weakly integral quantum group category do not have property F. (see [**22, 31, 44**]).

From the discussion above, it is clear that weakly integral braided fusion categories (conjecturally) have interesting properties. The main examples we have of such categories are metaplectic categories, which are closely related to $SO(N)_2$. We will explore these categories in some detail in the next sections.

4. Metaplectic modular categories

DEFINITION 4.1. A **metaplectic modular category (of dimension** $4N$**)** is a modular category $\mathcal{C}$ with positive dimensions that is Grothendieck equivalent to $SO(N)_2$, for some integer $N \geq 2$.

These fusion rules of metaplectic categories differ in important ways depending on the value of $N \pmod 4$. Metaplectic categories for N odd were defined in [**29**] and studied in [**1**], while the case $N \equiv 2 \pmod 4$ can be found in [**9**] where the term "even metaplectic" is used to describe metaplectic modular categories of dimension $4N$ with N even. We will simplify the terminology and call them all metaplectic (of dimension $4N$), and specify the value of $N \pmod 4$ when necessary.

The ubiquity of metaplectic categories among weakly integral modular categories as well as their application to topological phases of matter motivate their classification. The complete classification of weakly integral modular categories of dimension $2^n m$, for $n \in \{0, 1, 2, 3\}$ with m a square-free odd integer has been given, employing the classification of metaplectic modular categories of dimension $4N$ with N odd or $N \equiv 2 \pmod 4$ [**6, 9**]. In [**13**] a general description of modular categories of dimension $p^n m$ with m square-free is given in terms of de-equivariantizations. Here we consider metaplectic modular categories of dimension $4N$, recalling the known results for $4 \nmid N$ with new results in the case $4 \mid N$.

4.1. Metaplectic modular categories of dimension $4N$ with $4 \nmid N$. The metaplectic modular categories of dimension $4N$ with $4 \nmid N$ have been studied in [**1, 9**]. Most of the fusion rules for such a category were given in [**36**] with more complete details in [**9, 29**]. For N odd there are 2 invertible objects, $\frac{N-1}{2}$ simple objects of dimension 2 and 2 simple objects of dimension $\sqrt{N}$. All simple objects are self-dual in this case. In the case $N \equiv 2 \pmod 4$ we have 4 invertible objects (including one pair of non-self-dual objects), 4 (non-self-dual) simple objects of dimension $\sqrt{N/2}$, and $\frac{N}{2} - 1$ simple objects of dimension 2 (all of which are self-dual).

We have a complete classification of these categories:

THEOREM 4.2. [**1, 9**] *If $\mathcal{C}$ is a metaplectic category of dimension $4N$, with $4 \nmid N$, then $\mathcal{C}$ is a gauging of the particle-hole symmetry of a $\mathbb{Z}_N$-cyclic modular category $\mathcal{C}(\mathbb{Z}_N, q)$. Moreover, for $N = p_1^{k_1} \cdots p_r^{k_r}$, with p_i distinct primes (where if $p_1 = 2, k_1 = 1$), there are exactly 2^{r+1} many inequivalent such metaplectic modular categories.*

4.2. Metaplectic categories of dimension $4N$, with $4 \mid N$. The modular category $SO(N)_2$ with $N \equiv 0 \pmod 4$ corresponds to Lie type D_k with $2k = N$, has rank $k+7$ and dimension $4N$ [**36**]. The simple objects have dimension $1, 2$ and $\sqrt{k}$.

The simple objects of $SO(N)_2$ are naturally labeled by $\mathfrak{so}_N$ weights μ with $\mu^1 + \mu^2 \leq 2$, i.e. the sum of the first two coordinates can be at most 2. We will provide them with less cumbersome labels, after identifying their weights in terms of the fundamental weights $\lambda_j = (1, \dots, 1, 0, \dots 0)$ with j 1s for $j \leq k-2$, $\lambda_{k-1} = \frac{1}{2}(1, \dots, 1, -1)$, and $\lambda_k = \frac{1}{2}(1, \dots, 1)$. Setting $r = \frac{k}{2} - 1$, we have simple objects as follows:

- objects with weights $\mathbf{0}, 2\lambda_1, 2\lambda_{k-1}, 2\lambda_k$ will be denoted $\mathbf{1}, fg, f, g$,
- simple objects with weights $\lambda_2, \dots, \lambda_{k-2}$ will be denoted $X_0, \dots, X_{r-1}$,
- simple objects with weights $\lambda_1, \lambda_3, \dots, \lambda_{k-3}, \lambda_{k-1} + \lambda_k$ will be denoted $Y_0, \dots, Y_r$,
- simple objects with weights $\lambda_{k-1}, \lambda_1 + \lambda_k$ will be denoted V_1, V_2, and
- simple objects with weights $\lambda_k, \lambda_1 + \lambda_{k-1}$ will be denoted W_1, W_2.

All simple objects are self-dual, and the X_i and Y_i have dimension 2, while V_i, W_i have dimension $\sqrt{k}$. For $k > 2$ the key fusion rules are as follows, where we abuse notation and write $=$ for $\cong$:

- $f^{\otimes 2} = g^{\otimes 2} = \mathbf{1}$, $f \otimes X_i = g \otimes X_i = X_{r-i-1}$ and $f \otimes Y_i = g \otimes Y_i = Y_{r-i}$
- $g \otimes V_1 = V_2, f \otimes V_1 = V_1$ and $f \otimes W_1 = W_2, g \otimes W_1 = W_1$
- $V_1^{\otimes 2} = \mathbf{1} \oplus f \oplus \bigoplus_{i=0}^{r-1} X_i$
- $W_1^{\otimes 2} = \mathbf{1} \oplus g \oplus \bigoplus_{i=0}^{r-1} X_i$
- $W_1 \otimes V_1 = \bigoplus_{i=0}^{r} Y_i$
- $X_i \otimes X_j = \begin{cases} X_{i+j+1} \oplus X_{j-i-1} & i < j \leq \frac{r-1}{2} \\ \mathbf{1} \oplus fg \oplus X_{2i+1} & i = j < \frac{r-1}{2} \\ \mathbf{1} \oplus f \oplus g \oplus fg & i = j = \frac{r-1}{2} < r-1 \end{cases}$
- $Y_i \otimes Y_j = \begin{cases} X_{i+j} \oplus X_{j-i-1} & i < j \leq \frac{r}{2} \\ \mathbf{1} \oplus fg \oplus X_{2i} & i = j \leq \frac{r-1}{2} \\ \mathbf{1} \oplus f \oplus g \oplus fg & i = j = \frac{r}{2}. \end{cases}$

Notice that all other fusion rules may be derived from the above by tensoring with f or g as needed. For example $V_1 \otimes V_2 = g \otimes V_1^{\otimes 2} = f \oplus fg \oplus \bigoplus_{i=0}^{r-1} X_i$.

For $k = 2$, i.e. $SO(4)_2$ we have 9 simple objects: $\mathbf{1}, f, g, fg, Y_0, V_1, V_2, W_1, W_2$ and the fusion rules are the same as $SU(2)_2 \boxtimes SU(2)_2$. The applicable fusion rules above still hold. Indeed, by [**15**, Corollary B.12] any such category is equivalent to a Deligne product of Ising-type modular categories (of which there are 8).

Recall that a metaplectic category of dimension $4N$ with $4 \mid N$ is any modular category $\mathcal{C}$ with the same fusion rules as above. Fix such a category $\mathcal{C}$ and label the simple objects as above. The objects $\mathbf{1}, f, g$ and fg are invertible, and their isomorphism classes form the group $\mathbb{Z}_2 \times \mathbb{Z}_2$ under tensor product. In particular the universal grading group $\mathcal{U}(\mathcal{C}) \cong \mathbb{Z}_2 \times \mathbb{Z}_2$ with graded components labeled as follows:

(1) The adjoint category $\mathcal{C}_0$: with simple objects $\mathbf{1}, fg, f, g$ and $X_0, \ldots, X_{r-1}$
(2) $\mathcal{C}_{(1,1)}$ with simple objects $Y_0, \ldots, Y_r$
(3) $\mathcal{C}_{(1,0)}$ with simple objects V_1, V_2
(4) $\mathcal{C}_{(0,1)}$ with simple objects W_1, W_2.

LEMMA 4.3. *Let $\mathcal{C}$ be a metaplectic modular category of dimension $4N$ with $4 \mid N$, with simple objects labeled as above. Then:*

(1) *fg centralizes every object in $\mathcal{C}_0 \oplus \mathcal{C}_{(1,1)}$.*
(2) *fg is a boson, i.e. $\theta_{fg} = 1$.*

PROOF. Observe that since the pointed subcategory $\mathcal{C}_{pt}$ with simple objects $\mathbf{1}, f, g, fg$ is a subcategory of $\mathcal{C}_{ad} = \mathcal{C}_0$ and $\mathcal{C}_{pt} = \mathcal{C}'_{ad}$ we see that $\mathcal{C}_{pt}$ is symmetric, i.e. self-centralizing. Moreover, since $fg \in \mathcal{C}_{pt} = \mathcal{C}'_{ad}$ it is clear that fg centralizes $\mathcal{C}_0$. For $Y_i \in \mathcal{C}_{(1,1)}$ we have $fg \otimes Y_i = Y_i$ and the balancing equation gives:

$$\theta_{Y_i}\theta_{fg}S_{Y_i,fg} = \theta_{Y_i}\dim(Y_i)$$

so that $S_{Y_i,fg} = \dim(fg)\dim(Y_i)$ if and only if $\theta_{fg} = 1$. If $k > 2$ (so $r > 0$) there is a 2-dimensional object $X_0 \in \mathcal{C}_{ad}$ with $X_0 \otimes fg = X_0$ and $S_{X_0,fg} = 2 = \dim(X_0)\dim(fg)$, so that the balancing equation implies $\theta_{fg} = 1$, and the result follows.

For the case $k = 2$, $\mathcal{C}$ is a product of Ising-type categories, and there is a labeling ambiguity among f, g and fg. Precisely one of these is a boson, since the non-trivial invertible object in any Ising category is a fermion. So we may assume fg is the boson, and the result follows as above. □

In particular, the subcategory $\langle fg \rangle$ generated by fg is equivalent, as a symmetric ribbon category, to $\mathrm{Rep}(\mathbb{Z}_2)$.

We also have the following lemma describing the other invertible objects.

LEMMA 4.4. *Let $\mathcal{C}$ be a metaplectic modular category of dimension $4N$ with $4 \mid N$. If $8 \mid N$ then all the non-trivial invertible objects of $\mathcal{C}$ are bosons, and if $8 \nmid N$ two are fermions (and one is a boson).*

PROOF. From the description above, we know that there are $r = \frac{N}{4} - 1$ 2-dimensional simple objects in the adjoint component $\mathcal{C}_0$, and $r + 1$ 2-dimensional simple objects in $\mathcal{C}_{(1,1)}$. Moreover, the pointed subcategory is symmetric and has fusion rules like $\mathbb{Z}_2 \times \mathbb{Z}_2$.

Since $\mathcal{C}'_0 = \mathcal{C}_{pt}$, if one of the invertibles is a fermion, the action of tensoring with that fermion must be fixed-point free on $\mathcal{C}_0$. Then, the number of 2-dimensional simple objects in $\mathcal{C}_0$, r, must be even in this case. In particular, $N = \dim \mathcal{C} = 4 + 4r$ is not divisible by 8. So if $8 \mid N$, r is odd and so there are no fermions.

On the other hand, if $8 \nmid N$ we must have r even, and so $Y_{\frac{r}{2}}^{\otimes 2} = \mathbf{1} \oplus f \oplus g \oplus fg$. The balancing equation then gives, for example, $-2\theta_{Y_{\frac{r}{2}}}\theta_f = 2\theta_{Y_{\frac{r}{2}}}$, so that $\theta_f = \theta_g = -1$. □

4.3. Proof of Theorem 4.5. We can now prove a partial analogue to Theorem 4.2 for metaplectic modular categories of dimension $4N$ with $4 \mid N$. One exception is the case $N = 4$.

THEOREM 4.5. *If $\mathcal{C}$ is a metaplectic modular category of dimension $4N > 16$ with $4 \mid N$ then the de-equivariantization $\mathcal{D} := \mathcal{C}_{\mathbb{Z}_2}$ by $\langle fg \rangle = \operatorname{Rep}(\mathbb{Z}_2)$ is a generalized Tambara-Yamagami category of dimension $2N$, and, the trivial component $\mathcal{D}_0 := [\mathcal{C}_{\mathbb{Z}_2}]_0 \cong \mathcal{C}(\mathbb{Z}_N, q)$ is a pointed cyclic modular category. Moreover, $\mathcal{C}$ is obtained from $\mathcal{C}(\mathbb{Z}_N, q)$ via a $\mathbb{Z}_2$-gauging of the particle-hole symmetry.*

PROOF. We continue with the notation and labeling as above, with $\mathbf{1}, f, g, fg, X_i, Y_j$ with $0 \le i \le r-1$ and $0 \le j \le r$ where $r = \frac{k}{2} - 1$ and $N = 2k$.

As we noted previously, by Lemma 4.3, $\langle gf \rangle \cong \operatorname{Rep}(\mathbb{Z}_2)$ is a Tannakian subcategory of $\mathcal{C}$. In particular we have the de-equivariantization functor $F : \mathcal{C} \to \mathcal{C}_{\mathbb{Z}_2}$, the image of which is a braided $\mathbb{Z}_2$-crossed fusion category of dimension $2N$ [**15**]. In particular, the trivial component $\mathcal{D}_0$ of $\mathcal{D} = \mathcal{C}_{\mathbb{Z}_2}$ is modular of dimension N [**15**, Proposition 4.56(ii)].

Since gf fixes the 2-dimensional objects X_i, Y_i of $\mathcal{C}$, their images under F give rise to $k-1$ inequivalent pairs $X_i^{(1)}, X_i^{(2)}, Y_i^{(1)}, Y_i^{(2)}$ of distinct invertible objects in $\mathcal{D}$. On the other hand, $gf \otimes \mathbf{1} = gf$ and $g \otimes gf = f$, so that $F(gf) = F(\mathbf{1}) = \mathbf{1}_{\mathcal{D}}$ and $F(g) = F(f) = Z$ are distinct invertible objects in $\mathcal{D}$. In all, we have $2(k-1)+2 = N$ invertible objects, all of which are in the trivial component $\mathcal{D}_0$. Thus $\mathcal{D}_0 \cong \mathcal{C}(A, q)$ for some abelian group of order N.

Since the 4 objects in $\mathcal{C}_{(0,1)} \cup \mathcal{C}_{(1,0)}$ form two orbits under $\_ \otimes fg$, we see that $\mathcal{D}_1$ has two non-integral simple objects which each have dimension $\sqrt{N/2}$. In particular, $\mathcal{D}$ is generalized Tambara-Yamagami category [**32**].

It remains to verify that the classes of simple objects in $\mathcal{D}_0$ form a *cyclic* group, which we do following the inductive idea of [**1**].

First note that $Y_0 \otimes Y_0 = \mathbf{1} \oplus fg \oplus X_0$. Since $F(\mathbf{1}) = F(fg) = \mathbf{1}_{\mathcal{D}}$ is the the trivial object under the de-equivariantization, we have:

$$F(Y_0^{\otimes 2}) = \left(Y_0^{(1)} \oplus Y_0^{(2)}\right)^{\otimes 2} = 2(Y_0^{(1)} \otimes Y_0^{(2)}) \oplus \left(Y_0^{(1)}\right)^{\otimes 2} \oplus \left(Y_0^{(2)}\right)^{\otimes 2}$$
$$= \mathbf{1}_{\mathcal{D}} \oplus X_0^{(1)} \oplus X_0^{(2)}$$

We will show that the class of $Y_0^{(1)}$ generates $[\mathcal{C}_{\mathbb{Z}_2}]_0$. Examining multiplicities we see that we may assume (using the labeling ambiguity $X_0^{(j)}$, $j = 1, 2$) $Y_0^{(1)} \otimes Y_0^{(2)} = \mathbf{1}_{\mathcal{D}}$ while $Y_0^{(j)} \otimes Y_0^{(j)} = X_0^{(j)}$.

Proceeding in a similar way with $Y_0 \otimes X_0 = Y_0 \oplus Y_1$ we must match the 4 simple objects $Y_0^{(i)} \otimes X_0^{(j)}$ for $1 \le i, j \le 2$ with the four simple objects $Y_a^{(b)}$ for $a = 0, 1$ and $b = 1, 2$. Now since $Y_0^{(2)} = (Y_0^{(1)})^*$ we must have

$$Y_0^{(1)} \otimes X_0^{(1)} \oplus Y_0^{(2)} \otimes X_0^{(2)} = Y_1^{(1)} \oplus Y_1^{(2)}.$$

This is again a labeling ambiguity so we may define, without loss of generality, $Y_0^{(j)} \otimes X_0^{(j)} = Y_1^{(j)}$ for $j = 1, 2$.

Now notice that for $n \leq 2r$ the tensor power $Y_0^{\otimes n}$ contains exactly one simple object that has not appeared in lower tensor powers: namely $X_i \subset Y_0^{\otimes(2i+2)}$ and $Y_i \subset Y_0^{\otimes(2i+1)}$. Thus we may proceed inductively and define (using Frobenius reciprocity and the labeling ambiguity) for $j = 1, 2$ and $0 \leq i \leq r-1$:

$$Y_{i+1}^{(j)} := Y_0^{(j)} \otimes X_i^{(j)}, \quad X_i^{(j)} := Y_0^{(j)} \otimes Y_i^{(j)}.$$

Thus we see that all $Y_i^{(1)}$ and $X_i^{(1)}$ are tensor powers of $Y_0^{(1)}$ and all $X_i^{(2)}$ and $Y_i^{(2)}$ are tensor powers of $Y_0^{(2)}$. Next, note that $Y_i^{(2)}$ represents the isomorphism class that is the multiplicative inverse to that of $Y_i^{(1)}$ in the Grothendieck ring of $[\mathcal{C}_{\mathbb{Z}_2}]_0$, since Y_j is self-dual. Thus all $Y_a^{(j)}$ for $0 \leq a \leq r$ and $X_b^{(j)}$ for $0 \leq b \leq r-1$ ($j = 1, 2$) are in the subcategory generated by $Y_0^{(1)}$.

It remains to show that $F(f) = F(g) = Z$ is a tensor power of $Y_0^{(1)}$. For this we compute:

$$Y_0 \otimes Y_r = Y_0 \otimes Y_0 \otimes f = (\mathbf{1} \oplus fg \oplus X_0) \otimes f = X_{r-1} \oplus f \oplus g$$

Applying the functor F and the assignments above we see that $Y_0^{(j)} \otimes Y_r^{(j)} = X_{r-1}^{(\bar{j})}$ for $j = 1, 2$ where $\bar{2} = 1$ and $\bar{1} = 2$. This leaves, for $j = 1, 2$, $Y_0^{(j)} \otimes Y_r^{(\bar{j})} = Z$. In particular, since $Y_r^{(2)}$ is a tensor power of $Y_0^{(1)}$, we have that Z is also. Thus $\mathcal{D}_0 \cong \mathcal{C}(\mathbb{Z}_N, q)$ is a pointed cyclic modular category.

Now we apply the same argument as in [**1**] and [**9**] to see that the only $\mathbb{Z}_2$ gauging on $\mathcal{C}(\mathbb{Z}_N, q)$ that has four invertible objects is the particle-hole symmetry, i.e. inversion on $\mathbb{Z}_N \cong \mathbb{Z}_{2^a} \times \mathbb{Z}_{p_1^{a_1}} \times \cdots \times \mathbb{Z}_{p_b^{a_b}}$ (where $a \geq 2$. Indeed, there can be only two fixed points under the $\mathbb{Z}_2$ action (namely the identity and the unique element of order 2 coming from the $\mathbb{Z}_{2^a}$ factor) as otherwise we obtain $\boxtimes$ factors of the form $\mathcal{C}(\mathbb{Z}_{p_i^{a_i}}, q_i)$ in the gauged category, which is incompatible with the structure of $\mathcal{C}$. □

4.4. Count of inequivalent categories. The following proposition provides a count of inequivalent categories in the $4 \mid N$ case analogous to the count of Theorem 4.2.

PROPOSITION 4.6. *Let $N > 4$ be an integer with $4 \mid N$ and with prime factorization $N = 2^a p_1^{a_1} \cdots p_b^{a_b}$. Then there exist precisely $3(2^{b+2})$ inequivalent metaplectic modular categories of dimension $4N$.*

PROOF. Theorem 4.5 states that $\mathcal{C}$ is obtained from $\mathcal{C}(\mathbb{Z}_N, q)$ for some q via particle-hole gauging. From the prime factorization of N, we find that there are exactly $4 \cdot 2^b$ inequivalent categories of the form $\mathcal{C}(\mathbb{Z}_N, q)$ (see [**26**]). We wish to determine the number of distinct (particle-hole) gaugings of a given fixed $\mathcal{C}(\mathbb{Z}_N, q)$.

Let $\rho : \mathbb{Z}_2 \to \mathrm{Aut}(\mathbb{Z}_N)$ denote the map determined by $\rho(1)(n) = -n$, i.e. the particle-hole symmetry. One known extension $\mathcal{D}$ of $\mathcal{C}(\mathbb{Z}_N, q)$ by ρ has two defects $\sigma_\pm$ with fusion rules determined by

- $\sigma_+ \otimes \sigma_+ = \bigoplus_{a \text{ even}} [a]_N$
- $\sigma_+ \otimes \sigma_- = \bigoplus_{a \text{ odd}} [a]_N$
- $\sigma_\pm \otimes [a] = \begin{cases} \sigma_\pm & [a]_N \text{ even} \\ \sigma_\mp & [a]_N \text{ odd} \end{cases}$,

where $[a]_N$ denotes the simple object corresponding to $a \in \mathbb{Z}_N$ [**3**].

Given a fixed extension (in our case, $\mathcal{D}$), any extension of $\mathcal{C}(\mathbb{Z}_N, q)$ by ρ corresponds to a gauging datum, i.e. a pair $(\alpha, \beta) \in H^2_\rho(\mathbb{Z}_2, \mathbb{Z}_N) \times H^3(\mathbb{Z}_2, U(1)) \cong \mathbb{Z}_2 \times \mathbb{Z}_2$ such that a certain obstruction $O_4(\rho, \alpha)$ vanishes [**11**].

At the level of fusion rules, the action of the nontrivial element $\alpha \in H^2_\rho(\mathbb{Z}_2, \mathbb{Z}_N)$ twists the tensor product of defects by a representative 2-cocycle. One such representative cocycle is the normalized cocycle $\omega \in Z^2(\mathbb{Z}_2, \mathbb{Z}_N)$ determined by $\omega(1,1) = N/2$. The new tensor product $\otimes'$ is given by

$$x_g \otimes' x_h = [\omega(g,h)]_N \otimes x_g \otimes x_h,$$

for x_g in the g-component of the extension and x_h in the h-component [**19**]. Since ω is normalized, the only non-trivial twisting occurs when both g and h are nontrivial. Since $\omega(1,1) = N/2$ is even (because 4 divides N),

$$\sigma_\pm \otimes' \sigma_\pm = [N/2]_N \otimes \sigma_\pm \otimes \sigma_\pm = \bigoplus_{a \text{ even}} [a]_N = \sigma_\pm \otimes \sigma_\pm$$

and

$$\sigma_\pm \otimes' \sigma_\mp = [N/2]_N \otimes \sigma_\pm \otimes \sigma_\mp = \bigoplus_{a \text{ odd}} [a]_N = \sigma_\pm \otimes \sigma_\mp.$$

Hence, the action of α on the fusion rules is trivial. However, since β is an element of a torsor over $H^2_\rho(\mathbb{Z}_2, \mathbb{Z}_N)$, the choice of α is still relevant.

There are two potential obstructions to extending $\mathcal{C}(\mathbb{Z}_N, q)$ by ρ. The first obstruction in $H^3_\rho(\mathbb{Z}_2, \mathbb{Z}_N)$ vanishes since we know that a gauging exists. The second obstruction vanishes since $H^4(\mathbb{Z}_2, U(1)) \cong 0$. There is only one possible set of fusion rules on the $\mathbb{Z}_2$-extension since the action of $H^2_\rho(\mathbb{Z}_2, \mathbb{Z}_N)$ on the fusion rules is trivial. There is a choice of associativity constraints on the $\mathbb{Z}_2$-extension, so that *a priori* we have 4 distinct theories. However, by examining the pairs of FS-indicators of the defects $\sigma_\pm$ as in [**3**, Section X.F] we find that they are $(1,1), (-1,-1), (1,-1)$ and $(-1,1)$, so that the latter two theories can be identified by relabeling. Thus we have 3 distinct gaugings for each fixed $\mathcal{C}(\mathbb{Z}_N, q)$, yielding a count of $3(2^{b+2})$ metaplectic modular categories of dimension $4N > 16$ with $4 \mid N$. □

4.5. Degenerate case: metaplectic modular categories of dimension 16. As we have observed above, metaplectic categories of dimension 16 are of the form $\mathrm{Ising}^{(\nu_1)} \boxtimes \mathrm{Ising}^{(\nu_2)}$ where $\nu_i \in \mathbb{Z}_{16}^\times$. In that case we have $Y_0^{\otimes 2} = \mathbf{1} \oplus f \oplus g \oplus fg$, so that we cannot determine $Y_0^{(1)} \otimes Y_0^{(2)}$ from $(Y_0^{(1)})^{\otimes 2}$ using only the fusion rules–one must be $\mathbf{1}_\mathcal{D}$ and the other Z. The two possiblities lead to $\mathcal{D}_0 \cong \mathcal{C}(\mathbb{Z}_4, q_1)$ or $\mathcal{D}_0 \cong \mathcal{C}(\mathbb{Z}_2 \times \mathbb{Z}_2, q_2)$. Both of these can occur: see [**3**].

The following proposition provides a detailed count for the $N = 4$ case.

PROPOSITION 4.7. *There are precisely 20 inequivalent modular categories of the form* $\mathrm{Ising}^{\nu_1} \boxtimes \mathrm{Ising}^{\nu_2}$*:*

- *12 come from gauging the particle-hole symmetry of a modular category of the form* $\mathcal{C}(\mathbb{Z}_4, q)$
- *8 come from gauging a modular category of the form* $\mathcal{C}(\mathbb{Z}_2 \times \mathbb{Z}_2, q)$ *containing a fermion, 2 from each of the following categories:*
  - *TC (Toric Code),*
  - *3F (3 fermions),*
  - Sem^2 *(semion-squared),*
  - *and* $\overline{\mathrm{Sem}}^2$ *(semion-conjugate-squared).*

Proof. Notice that when $N = 4$ the argument of Proposition 4.6 is still valid if applied only to $\mathcal{C}(\mathbb{Z}_4, q)$: we still obtain 12 theories of the form $\mathrm{Ising}^{\nu_1} \boxtimes \mathrm{Ising}^{\nu_2}$ from these. By counting gaugings as in [**3**], we also get 8 more $\mathrm{Ising}^{\nu_1} \boxtimes \mathrm{Ising}^{\nu_2}$ theories from gauging: two from each of the 4 distinct $\mathcal{C}(\mathbb{Z}_2 \times \mathbb{Z}_2, q)$ modular categories containing a fermion. Thus we obtain a count of 20 metaplectic modular categories of dimension 16. □

Alternatively, we can count T-matrices up to dimension-preserving permutation to get a lower bound on the count of inequivalent categories. As shown below, this lower bound coincides with the gauging-based count, resulting in the following proposition.

Proposition 4.8. *A category of the form* $\mathrm{Ising}^{\nu_1} \boxtimes \mathrm{Ising}^{\nu_2}$ *is determined by its modular data.*

Proof. Any equivalence of modular categories $\widehat{F} : \mathrm{Ising}^{\nu_1} \boxtimes \mathrm{Ising}^{\nu_2} \to \mathrm{Ising}^{\nu_1'} \boxtimes \mathrm{Ising}^{\nu_2'}$ induces a dimension- and twist-preserving map $F : \mathrm{Irr}(\mathrm{Ising}^{\nu_1} \boxtimes \mathrm{Ising}^{\nu_2}) \to \mathrm{Irr}(\mathrm{Ising}^{\nu_1'} \boxtimes \mathrm{Ising}^{\nu_2'})$ on the sets of isomorphism classes of simple objects. Thus, by identifying corresponding simple objects in the domain and codomain of F, the T-matrices of equivalent modular categories are related by dimension-preserving permutations of their basis, the set of isomorphism classes of simple objects. Thus, a lower bound for a count of inequivalent categories is given by counting T-matrices up to dimension-preserving permutation.

To make this count, we use the following table of dimensions and twists of simple objects of $\mathrm{Ising}^{\nu_1} \boxtimes \mathrm{Ising}^{\nu_2}$:

$(\dim_{X\boxtimes Y}, \theta_{X\boxtimes Y})$	$\mathbf{1}$	ψ	σ
$\mathbf{1}$	$(1,1)$	$(1,-1)$	$(\sqrt{2}, e^{\frac{\pi i \nu_2}{8}})$
ψ	$(1,-1)$	$(1,1)$	$(\sqrt{2}, -e^{\frac{\pi i \nu_2}{8}})$
σ	$(\sqrt{2}, e^{\frac{\pi i \nu_1}{8}})$	$(\sqrt{2}, -e^{\frac{\pi i \nu_1}{8}})$	$(2, e^{\frac{\pi i(\nu_1+\nu_2)}{8}})$

We claim that the map $F : \mathrm{Irr}(\mathrm{Ising}^{\nu_1} \boxtimes \mathrm{Ising}^{\nu_2}) \to \mathrm{Irr}(\mathrm{Ising}^{\nu_1'} \boxtimes \mathrm{Ising}^{\nu_2'})$ is determined by $F(\sigma \boxtimes \mathbf{1})$. Since F must preserve simple objects of dimension $\sqrt{2}$, we have $F(\sigma \boxtimes \mathbf{1}) \in \{\sigma \boxtimes \mathbf{1}, \mathbf{1} \boxtimes \sigma, \psi \boxtimes \sigma, \sigma \boxtimes \psi\}$. Thus, $F((\psi \oplus \mathbf{1}) \boxtimes \mathbf{1}) = F(\sigma^{\otimes 2} \boxtimes \mathbf{1}) = (F(\sigma \boxtimes \mathbf{1}))^{\otimes 2} \in \{(\psi \oplus \mathbf{1}) \boxtimes \mathbf{1}, \mathbf{1} \boxtimes (\psi \oplus \mathbf{1})\}$. Hence, $F(\sigma \boxtimes \mathbf{1})$ determines $F(\psi \boxtimes \mathbf{1}) \in \{\psi \boxtimes \mathbf{1}, \mathbf{1} \boxtimes \psi\}$. The value of $F(\psi \boxtimes \mathbf{1})$ also determines $F(\mathbf{1} \boxtimes \psi)$ since these are the only fermions. Hence, $F(\sigma \boxtimes \mathbf{1})$ also determines $F(\sigma \boxtimes \psi)$. Since F must preserve the unique simple object of dimension 2, we have $\nu_1' + \nu_2' = \nu_1 + \nu_2 \pmod{16}$. Since the twist of $F(\sigma \boxtimes \mathbf{1})$ is $\nu_1 \in \{\nu_i', \nu_i' + 8\}$ for some $i \in \{1, 2\}$, this equation determines $\nu_2 \in \{\nu_{3-i}', \nu_{3-i}' + 8\}$, hence the value of $F(\mathbf{1} \boxtimes \sigma)$. The other simple objects are generated by these values, proving the claim.

We thus have the following count of T-matrices up to dimension-preserving permutation. There are $8 \cdot 8 = 64$ ordered pairs (ν_1, ν_2). This splits into 56 pairs of distinct (ν_1, ν_2) and 8 pairs of the form (ν, ν). Since ν is odd, the orbit of (ν, ν) under dimension- and twist-preserving permutation is $\{(\nu, \nu), (\nu + 8, \nu + 8)\}$ of order 2. The 56 pairs of distinct (ν_1, ν_2) further break up into 8 pairs such that $\nu_1 + 8 = \nu_2$ and 48 pairs such that $\nu_1 + 8 \neq \nu_2$. In the case that $\nu_1 + 8 = \nu_2$, the orbit of (ν_1, ν_2) is $\{(\nu_1, \nu_2), (\nu_2, \nu_1)\}$ of order 2. In the case that $\nu_1 + 8 \neq \nu_2$, the orbit is $\{(\nu_1, \nu_2), (\nu_2, \nu_1), (\nu_1 + 8, \nu_2 + 8), (\nu_2 + 8, \nu_1 + 8)\}$ of order 4. Thus, we have $8/2 + 8/2 + 48/4 = 20$ distinct T-matrices up to dimension-preserving permutation.

In light of Proposition 4.7, this implies that the choice of T-matrix determines the category. □

More generally, we would like to know if the modular data distinguishes inequivalent metaplectic modular categories of arbitrary dimension. A first step in this direction would be to determine the modular data:

QUESTION 4.9. *What are the twists of each metaplectic modular category of dimension $4N$ in terms of the twists of the pointed modular category $\mathcal{C}(\mathbb{Z}_N, q)$?*

Formula (412) of [**11**] conjecturally relates twists in the gauged theory to twists in the extension. This still leaves the problem of finding the twists of the defects in the extension. As far as we can tell, the only way to do this is to solve the G-crossed heptagon equations.

5. Modular categories of dimension $16m$, with m odd square-free integer

In this section we will consider $\mathcal{C}$ a strictly weakly-integral modular category of dimension $16m$, with $m > 1$ an odd square-free integer. We will fix this assumptions for all Section 5 unless otherwise specified. We will make a comment on the case $m = 1$, see Remark 5.2 and on the integral case in the discussion below.

We have seen examples, namely metaplectic modular categories of dimension $4N$ with $N = 4m$ with m square-free and odd. The goal is to give a classification of this class of categories similar to the ones in [**6**, **9**] of modular categories of dimension $4m$ and $8m$, respectively. In what follows, we give a classification of modular categories of dimension $16m$ under certain restrictions.

The only *integral* metaplectic modular categories of dimension $16m$ with m square-free are those with $m = 1$. We will disregard this case, pausing only to ask:

QUESTION 5.1. *Are integral modular categories of dimension $16m$ with $m > 1$ and not necessary square-free, pointed?*

REMARK 5.2. The only integral modular categories of dimension $16m$ are for $m = 1$, i.e. dimension 16. But there are strictly weakly-integral modular categories of dimension 16 as well, for example Ising $\boxtimes$ Ising.

REMARK 5.3. While *integral* modular categories of dimension 16 are pointed ([**9**, Lemma 4.3]), it is not the case that all integral modular categories of dimension 2^k are pointed. For example $SO(8)_2$ is a non-pointed integral modular category of dimension 32, which has 4 invertible objects and 7 simple objects of dimension 2.

We can consider $\mathcal{C}$ *prime* (i.e. not of the form $\mathcal{C}_1 \boxtimes \mathcal{C}_2$) since otherwise it reduces to known cases, see [**6**, **9**].

Prime modular categories of dimension $4m$ and $8m$ essentially arise from metaplectic categories when they are not pointed. Thus it is natural to ask:

QUESTION 5.4. *Are prime modular categories of dimension $16m$ Grothendieck equivalent to* $\mathrm{SO}(4m)_2$*?*

Strictly weakly integral modular categories of dimension 16 were classified in [**9**, Lemma 4.9]. These categories are a Deligne product of an Ising category and a pointed modular category, i.e. a Generalized Tamabra-Yamagami category.

5.1. The order of $U(\mathcal{C})$. From [**9**, Lemma 4.1], we know that if 16 divides the order of the universal grading group $U(\mathcal{C})$ of $\mathcal{C}$, then $\mathcal{C}$ is pointed. So we can assume that $16 \nmid |U(\mathcal{C})|$.

As in the proof of Theorem 3.1 in [**6**], using the de-equivariantization process, we can assume that $|U(\mathcal{C})| = 2^k$, for k an integer number.

We will not provide a complete answer to Question 5.4, but we will give a partial response and pose some general questions arising from our analysis.

REMARK 5.5. Let $\mathcal{C}$ have dimension $2^n m$, with $n \geq 3$. Then, the possible dimensions of integral simple objects of $\mathcal{C}$ are 2^k, with $k = 0, \dots, \lfloor \frac{n}{2} \rfloor$ (see e.g. [**13**, Lemma 5.2], [**20**, Theorem 2.11 (i)]).

LEMMA 5.6. *Let $\mathcal{C}$ have dimension $2^n m$, with $n \geq 3$. Then the universal grading group $\mathcal{U}(\mathcal{C})$ has $|\mathcal{U}(\mathcal{C})| \geq 4$.*

PROOF. Since $\mathcal{U}(\mathcal{C})$ is non-trivial and the GN grading group is also non-trivial, it is enough to show that $\mathcal{U}(\mathcal{C}) \ncong \mathbb{Z}_2$. Suppose that $\mathcal{C}$ is as in the statement with $\mathcal{U}(\mathcal{C}) \cong \mathbb{Z}_2$. Notice that the GN-grading group is also $\mathbb{Z}_2$. By faithfulness of the universal grading we have $\dim \mathcal{C}_{ad} = 2^{n-1}m$ (where $n \geq 3$). From the definition of dimension and Remark 5.5, we have $2^{n-1}m = 2 + \sum_{k=1}^{\lfloor \frac{n}{2} \rfloor} 2^{2k} a_k$, where a_k is the number of 2^k-dimensional simple objects in $\mathcal{C}_{ad}$. Since $n > 3$ and $4 \nmid 2$ we have a contradiction. □

LEMMA 5.7. *Let $n \geq 3$. There are no prime self-dual strictly weakly integral modular categories of dimension $2^n m$ whose universal grading group has order 2^{n-1}.*

PROOF. By contradiction, assume that such a category exists. Since $\mathcal{C}$ is self-dual, we must have that $\mathcal{U}(\mathcal{C}) \cong \mathbb{Z}_2^{n-1}$. So let $\mathcal{C}_g$ be a non-integral component. Then $\mathcal{C}_g \otimes \mathcal{C}_g \subset \mathcal{C}_{ad}$. Thus $\mathcal{D} := \mathcal{C}_{ad} \oplus \mathcal{C}_g$ is a fusion category of dimension $4m$. Taking centralizers we have $\mathcal{Z}_{\mathcal{D}}(\mathcal{D}) \subset \mathcal{Z}_{\mathcal{D}}(\mathcal{C}_{ad}) = (\mathcal{C}_{ad})_{pt}$.

Since $\mathcal{C}_{ad}$ has dimension $2m$, we have $2m = \sum_{k=0}^{\lfloor \frac{n}{2} \rfloor} 2^{2k} a_k$, where a_k is the number of 2^k-dimensional simple objects in $\mathcal{C}_{ad}$, by Remark 5.5. Thus $a_0 = 2$ because m is odd. So either $\mathcal{Z}_{\mathcal{D}}(\mathcal{C}_{ad} \oplus \mathcal{C}_g) \cong \mathrm{Vec}$, sVec, or $\mathrm{Rep}(\mathbb{Z}_2)$.

The case $\mathcal{Z}_{\mathcal{D}}(\mathcal{D}) \cong \mathrm{Vec}$ is not possible as $\mathcal{C}$ is prime. Since there are only 2 invertible objects in $\mathcal{C}_{ad}$ and all non-invertibles objecs in $\mathcal{C}_{ad}$ are even dimensional, then they should be fixed by the non-trivial invertible object in $\mathcal{C}_{ad}$, by 2.5. Since the invertible objects are transparent in $\mathcal{C}_{ad}$, the non-trivial invertible object can not be a fermion, i.e. $\mathcal{Z}_{\mathcal{D}}(\mathcal{D}) \ncong \mathrm{sVec}$. If $\mathcal{Z}_{\mathcal{D}}(\mathcal{D}) \cong \mathrm{Rep}(\mathbb{Z}_2)$, we can de-equivariantize to find a $2m$-dimensional modular category. Thus $\mathcal{D}_{\mathbb{Z}_2}$ is pointed (by [**20**, Theorem 2.11 (i)]) and hence $\mathcal{D}$ is integral, an impossibility. □

PROPOSITION 5.8. *If $\mathcal{C}$ is prime and of dimension $16m$, then its universal grading group has order 4. Thus,* $\mathrm{FPdim}\, \mathcal{C}_{pt} = 4$.

PROOF. This statement is a direct consequence of Lemma 5.6 and Lemma 5.7. □

5.2. Self-dual and GN-grading by $\mathbb{Z}_2$ case. We will focus on the self-dual case. Thus, it suffices to consider strictly weakly integral self-dual categories of dimension $16m$ whose universal grading group has order 4.

LEMMA 5.9. *Let $\mathcal{C}$ have dimension $2^n m$, with $n \geq 4$. If $\dim \mathcal{C}_{pt} = 4$, then $\mathcal{Z}_{\mathcal{C}_{ad}}(\mathcal{C}_{ad}) = \mathcal{C}_{pt}$, and $\mathcal{C}_{pt}$ contains a boson.*

PROOF. First note that $\dim \mathcal{C}_{ad} = 2^{n-2}m$. So the dimensions of the simple objects of $\mathcal{C}_{ad}$ are of the form 2^k, $k = 0, \ldots, \lfloor \frac{n}{2} \rfloor$, by Remark 5.5. Thus $2^{n-2}m = \sum_{k=0}^{\lfloor \frac{n}{2} \rfloor} 2^{2k} a_k$. Since $n \geq 4$, then $4 \mid a_0$. Moreover $a_0 = 4$ because $\mathbf{1} \in \mathcal{C}_{ad}$. Thus $\mathcal{C}_{pt} \subset \mathcal{C}_{ad}$. On the other hand, $\mathcal{Z}_{\mathcal{C}_{ad}}(\mathcal{C}_{ad}) = \mathcal{Z}_{\mathcal{C}}(\mathcal{C}_{ad}) \cap \mathcal{C}_{ad} = \mathcal{C}_{pt} \cap \mathcal{C}_{ad} = \mathcal{C}_{pt}$.

Now consider X a non-invertible simple object in $\mathcal{C}_{ad}$ (which exists since $m > 1$). Recall that $2 \mid \text{FPdim}\, Y$, for all non-invertible simple object $Y \in \mathcal{C}_{ad}$. Then, by Remark 2.5, we have that 2 divides the order of the group $G[X]$ of invertible objects that fixes X. Since $\mathbf{1} \in G[X]$, then $G[X] \geq 2$ Then, there exists $g \in G[X] \subseteq \mathcal{C}_{pt} = \mathcal{Z}_{\mathcal{C}_{ad}}(\mathcal{C}_{ad})$ of order 2. Since g fixes X and $g \in \mathcal{C}'_{ad} = \mathcal{C}_{pt}$ then g is not a fermion (see [**34**, Lemma 5.4]), and hence g must be a boson. □

LEMMA 5.10. *If $\mathcal{C}$ has dimension $16m$ with $\dim \mathcal{C}_{pt} = 4$ and GN-grading $\mathbb{Z}_2$, then the simple objects in the integral non-adjoint component are all 2-dimensional. Moreover, all the non-invertible simple objects in $\mathcal{C}_{ad}$ are also 2-dimensional.*

PROOF. Let a_0, b_0, and c_0 denote the number of 1-dimensional, 2-dimensional, and 4-dimensional simples in $\mathcal{C}_{ad}$ respectively. Then $4m = \dim \mathcal{C}_{ad} = a_0 + 4b_0 + 16c_0$. In particular, $4 \mid a_0$. Since $\mathbf{1}$ is in $\mathcal{C}_{ad}$ then $a_0 > 0$. Thus $a_0 = 4$. Consequently, $\mathcal{C}_g$, the integral non-adjoint component must contain only 2-dimensional and 4-dimensional simple objects.

Moreover, $4m = 4b_1 + 16c_1$ where b_1 and c_1 are the number of 2- and 4-dimensionals in the integral non-adjoint component. Since m is odd we know that $b_1 \neq 0$, but $m = b_1 + 4c_1$, so b_1 is odd.

Since the GN-grading group and the universal grading group are not the same then $\mathcal{C}_{ad} \subsetneq \mathcal{C}_{int}$. So we have $\mathcal{Z}_{\mathcal{C}}(\mathcal{C}_{int}) \subsetneq \mathcal{Z}_{\mathcal{C}}(\mathcal{C}_{ad}) = \mathcal{C}_{pt}$. Notice that $\mathcal{C}_{int}$ is not modular, otherwise, by [**35**, Theorem 4.2], [**15**, Theorem 3.13], the category $\mathcal{C} \cong \mathcal{C}_{int} \boxtimes \langle g \rangle$ would be integral which contradicts the assumption of $\mathcal{C}$ being strictly weakly integral. Thus $\mathcal{Z}_{\mathcal{C}}(\mathcal{C}_{int}) = \langle g \rangle$ for some invertible g.

We saw above that there is an odd number of 2-dimensional simple objects in the non-adjoint integral component. Then, at least one these 2-dimensional objects are fixed by g. It follows from [**34**, Lemma 5.4] that g is a boson. Thus $\mathcal{C}_{int}$ is modularizable.

So $(\mathcal{C}_{int})_{\mathbb{Z}_2}$ is a $4m$-dimensional integral modular category and thus is pointed, by [**5**, Theorem 3.1]. In particular, $\mathcal{C}_{int}$ has character degrees 1 and 2. □

PROPOSITION 5.11. *If $\mathcal{C}$ is a self-dual prime modular category of dimension $16m$ with GN-grading $\mathbb{Z}_2$, then $\mathcal{C}$ is spin modular.*

PROOF. Suppose that the non-trivial invertibles, g, h, and gh, are all bosons.

In the same way as in the proof of Lemma 5.10 it can be shown that $\mathcal{Z}_{\mathcal{C}}(\mathcal{C}_{int}) = \langle g \rangle$, for some invertible g. Since the GN-grading group is $\mathbb{Z}_2$, the possible non-integral dimensions of simple objects are $\sqrt{t}$ and $2\sqrt{t}$, for some $t \in \mathbb{N}$.

Denote by $\mathcal{D}_a$, the subcategory $\mathcal{C}_{ad} \oplus \mathcal{C}_a$. Then $\mathcal{Z}_{\mathcal{C}}(\mathcal{D}) \subset \mathcal{Z}_{\mathcal{C}}(\mathcal{C}_{ad}) = \mathcal{C}_{pt}$. By primality and taking double centralizers we can conclude that $\mathcal{Z}_{\mathcal{C}}(\mathcal{D}) \neq \text{Vec}, \mathcal{C}_{pt}$. In particular, there is a boson b such that $\mathcal{Z}_{\mathcal{C}}(\mathcal{D}) = \langle b \rangle$. Once again, by double centralizing, $b \neq g$. Similarly, denoting $\mathcal{D}_c$, the subcategory $\mathcal{C}_{ad} \oplus \mathcal{C}_c$, where $\mathcal{C}_c$ is the other non-integral component, we get $\mathcal{Z}_{\mathcal{C}}(\mathcal{D}) = \langle bg \rangle$. With out lost of generality, we can assume that $b = h$.

Since g does not centralize X we have $S_{g,X} \neq d_X$ and so by balancing and orthogonality of the S-matrix, g must move X. Similarly, g must move Y. On the other hand, g fixes the 2-dimensional simple objects in $\mathcal{C}_{ad}$.

Now, assume there exist a simple $X \in \mathcal{D}_a$ of dimension $\sqrt{t}$ and a simple $Y \in \mathcal{D}_c$ of dimension $2\sqrt{t}$. Then $X \otimes Y$ is in the non-adjoint integral component of $\mathcal{C}$. So $X \otimes Y$ is a sum of 2-dimensional simple objects. But if Z is a 2-dimensional simple object such that $Z \otimes X \cong Y$ then $Y \cong Z \otimes Y \cong g \otimes Z \otimes X \cong g \otimes Y$. This is a contradiction since g does not fix Y.

Then, there is a fermion on $\mathcal{C}$. Moreover, both h and gh are fermions in $\mathcal{C}$. □

COROLLARY 5.12. *If $\mathcal{C}$ is a self-dual prime modular category of dimension $16m$ with GN-grading $\mathbb{Z}_2$. Let h and gh be the fermions in $\mathcal{C}$. Denote by $\mathcal{C}_a$ and $\mathcal{C}_b$ the non-integral components, then $\mathcal{Z}_{\mathcal{C}}(\mathcal{C}_{ad} \oplus \mathcal{C}_a) = \langle h \rangle$ and $\mathcal{Z}_{\mathcal{C}}(\mathcal{C}_{ad} \oplus \mathcal{C}_b) = \langle gh \rangle$. Moreover, there is no component containing only objects of dimension $2\sqrt{t}$ for some square-free integer t.*

PROOF. The proof of the first statement is contained in the proof of Proposition 5.11.

By [**28**, Theorem 3.10], the only possible dimensions of non-integral objects are $\sqrt{t}$ and $2\sqrt{t}$ for a square-free integer t.

Suppose to the contrary that there is some component containing only simples of dimension $2\sqrt{t}$. Without loss of generality we take this to be the $\mathcal{C}_a$ component. Then tensoring with h must permute these simples in a fixed point free manner. Then $4m = \mathrm{FPdim}\,\mathcal{C}_a = 4tk$, where k is the number of simples in $\mathcal{C}$. This is a contradiction since k is even and m odd. □

5.3. Classification for dimension $16m$.

LEMMA 5.13. *Let $\mathcal{C}$ be a self-dual prime modular category of dimension $16m$ with GN-grading $\mathbb{Z}_2$. Then the non-integral objects have dimension $\sqrt{t}$ and $2\sqrt{t}$ for some* square-free even *integer t. Moreover each non-integral component contains an even number of objects of dimension $\sqrt{t}$ and an even number of objects of dimension $2\sqrt{t}$.*

PROOF. By Corollary 5.12 we know that there is at least one object of dimension $\sqrt{t}$ in each non-integral component, let X and Y be such objects in $\mathcal{C}_a$ and $\mathcal{C}_b$ (under the notation of Corollary 5.12), respectively. Then $X \otimes Y$ must be in the integral non-adjoint component under the universal grading. In particular, by Lemma 5.10, $t = \dim X \otimes Y = 2s$ for some $s \in \mathbb{Z}$. The t is even. Furthermore, t is square-free since t divides m.

By Proposition 5.11, $\mathcal{C}$ is a spin modular category. Denote $\mathcal{D} = \mathcal{C}_{ad} \oplus \mathcal{C}_a$, then $\mathcal{Z}_{\mathcal{D}}(\mathcal{D}) \cong \mathrm{sVec}$ and even multiplicity follows from the fact that the action of the fermion is fixed-point-free. The same holds for $\mathcal{C}_{ad} \oplus \mathcal{C}_b$. □

COROLLARY 5.14. *Let $\mathcal{C}$ be a self-dual prime modular category of dimension $16m$ with GN-grading $\mathbb{Z}_2$. Then all of the non-integral simple objects have dimension $\sqrt{2m}$.*

PROOF. By Proposition 5.11, $\mathcal{C}$ must be spin modular.

Suppose there are objects X and Y in the same component of dimension $2\sqrt{t}$ and $\sqrt{t}$ respectively. Then $X \otimes Y$ is in $\mathcal{C}_{ad}$. Since X and Y have different dimensions, no invertibles can appear as subobjects of $X \otimes Y$. Thus $X \otimes Y = \bigoplus N_{X,Y}^{Z_i} Z_i$ where

Z_i are the 2-dimensional simple objects in $\mathcal{C}_{ad}$. But $N_{X,Y}^{Z_i} = N_{X,Z_i}^{Y}$. If we consider $Z_i \otimes X$ there are 3 possibilities: it is the direct sum of 2 simple objects of dimension $\sqrt{t}$, it is equal to a simple object of dimension $2\sqrt{t}$, or it equal to 2 copies of a simple object of dimension $\sqrt{t}$. If we tensor $Z \otimes X$ by the boson g, since Z is fixed by g, we get that $Z \otimes X = X_1 \oplus gX_1$, where X_1 is a $\sqrt{t}$-dimensional simple object in $\mathcal{C}_a$. This implies that $N_{X,Y}^{Z_i} = N_{X,Z_i}^{Y} = 0$ for all X, Y, and Z as above. But this is a contradiction because $X \otimes Y$ has dimension $2t$. It follows that the non-integral objects all have dimension $\sqrt{t}$ for some even square-free integer t, see Corollary 5.12.

Now suppose there exist non-integral simple objects $W \neq V$ from the same component such that $W \otimes V$ does not contain any invertible objects. Then the fermion that doesn't centralize W and V must fix them. On the other hand, this fermion must permute all elements of the adjoint subcategory since it is transparent in $\mathcal{C}_{ad}$. Hence $W \otimes V$ must decompose into an even number of 2-dimensional simples in the adjoint. Computing dimensions we see $4 \mid \dim W \otimes V = t$, which is impossible since t is square-free. Thus for any non-integral W, V in the same component, $W \otimes V$ contains an invertible, say a. Thus $a \otimes W = V$. In particular, this component can only contain two non-integral objects. The result now follows by equidimensionality of the universal grading. □

We see that under the assumptions above, the category $\mathcal{C}$ has the following structure, where we take $\mathbb{Z}_2 \times \mathbb{Z}_2$ grading:

- $\mathcal{C}_0$ has 1 boson, 2 fermions and $m-1$ objects of dimension 2.
- $\mathcal{C}_{(1,1)}$ has m objects of dimension 2
- $\mathcal{C}_{(0,1)}$ and $\mathcal{C}_{(1,0)}$ each have 2 objects of dimension $\sqrt{2m}$.

The equation together with Corollary 5.12 shows that the twists of the simple objects in $\mathcal{C}_{(1,0)}$ and $\mathcal{C}_{(0,1)}$ come in pairs of the form $(\theta, -\theta)$. Similarly, the $m-1$ simple objects of dimension 2 in $\mathcal{C}_0$ come in pairs with twists of the form $(\theta, -\theta)$. In particular the central charge of $\mathcal{C}$ only depends on the twists in $\mathcal{C}_{(1,1)}$. This is not quite sufficient to conclude that $\mathcal{C}$ is metaplectic, but for certain the trivial component of the $\mathbb{Z}_2$-de-equivariantization has $[\mathcal{C}_{\mathbb{Z}_2}]_0 \cong \mathcal{C}(A, q)$ where A is an abelian group of order $4m$, and the corresponding action of $\mathbb{Z}_2$ by braided tensor autoequivalences on $\mathcal{C}(A, q)$ has exactly 2 fixed points.

Although we do not get as sharp a result in the case of $16m$ as we do in the cases $4m$ and $8m$, we can still give a significant amount of the structure under some assumptions:

THEOREM 5.15. *Let $\mathcal{C}$ be a prime modular category of dimension $2^n m$ with m odd and square free.*

- *If $n = 1$, then $\mathcal{C}$ is pointed.*
- **[6, 9]** *If $n = 2$ or 3, then $\mathcal{C}$ is a metaplectic modular category.*
- *If $n = 4$ and $\mathcal{C}$ is self-dual and its GN grading is $\mathbb{Z}_2$, then*
  - *$\mathcal{C}$ is metaplectic, or*
  - *$\mathcal{C}$ is obtained as a $\mathbb{Z}_2$ gauging of $\mathcal{C}(\mathbb{Z}_2 \times \mathbb{Z}_{2m}, q)$ via an action of $\mathbb{Z}_2$ with exactly 2 fixed points.*

We do not know if the $\mathbb{Z}_2$ gauging of $\mathcal{C}(\mathbb{Z}_2 \times \mathbb{Z}_{2m}, q)$ is self-dual or prime, so it is possible that a sharper result can be obtained. Moreover, we do not know what happens if the GN-grading is $\mathbb{Z}_2 \times \mathbb{Z}_2$.

5.4. Related questions.

QUESTION 5.16. *How large can the GN-grading group be for a prime strictly weakly integral modular category?*

There is an upper bound, given in [**13**, Theorem 3.1], for weakly-integral modular categories of dimension $2^n m$, where $n \geq 0$ and m odd. In this case, the order of the GN-grading group is at most $2^{\frac{n}{2}}$.

It is of course straight forward to develop non-prime categories with large GN-grading via Deligne product of $\mathbb{Z}_2$-graded categories. A natural place to look for large GN-grading with fewer prime factors is through equivariantization. That is, perhaps one can find an equivariantization of one of these Deligne products that has a diverse prime factor. This of course leads to the following related question:

QUESTION 5.17. *How do Deligne products behave under (de-)equivariantization?*

References

[1] Eddy Ardonne, Meng Cheng, Eric C. Rowell, and Zhenghan Wang, *Classification of metaplectic modular categories*, J. Algebra **466** (2016), 141–146, DOI 10.1016/j.jalgebra.2016.08.001. MR3541678

[2] Bojko Bakalov and Alexander Kirillov Jr., *Lectures on tensor categories and modular functors*, University Lecture Series, vol. 21, American Mathematical Society, Providence, RI, 2001. MR1797619

[3] M. Barkeshli, P. Bonderson, M. Cheng, and Z. Wang. Symmetry, defects, and gauging of topological phases.

[4] Alain Bruguières, *Catégories prémodulaires, modularisations et invariants des variétés de dimension 3* (French, with English summary), Math. Ann. **316** (2000), no. 2, 215–236, DOI 10.1007/s002080050011. MR1741269

[5] Paul Bruillard, César Galindo, Seung-Moon Hong, Yevgenia Kashina, Deepak Naidu, Sonia Natale, Julia Yael Plavnik, and Eric C. Rowell, *Classification of integral modular categories of Frobenius-Perron dimension pq^4 and p^2q^2*, Canad. Math. Bull. **57** (2014), no. 4, 721–734, DOI 10.4153/CMB-2013-042-6. MR3270794

[6] Paul Bruillard, César Galindo, Siu-Hung Ng, Julia Y. Plavnik, Eric C. Rowell, and Zhenghan Wang, *On the classification of weakly integral modular categories*, J. Pure Appl. Algebra **220** (2016), no. 6, 2364–2388, DOI 10.1016/j.jpaa.2015.11.010. MR3448800

[7] Paul Bruillard, César Galindo, Siu-Hung Ng, Julia Y. Plavnik, Eric C. Rowell, and Zhenghan Wang, *On the classification of weakly integral modular categories*, J. Pure Appl. Algebra **220** (2016), no. 6, 2364–2388, DOI 10.1016/j.jpaa.2015.11.010. MR3448800

[8] Paul Bruillard, Siu-Hung Ng, Eric C. Rowell, and Zhenghan Wang, *Rank-finiteness for modular categories*, J. Amer. Math. Soc. **29** (2016), no. 3, 857–881, DOI 10.1090/jams/842. MR3486174

[9] Paul Bruillard, Julia Yael Plavnik, and Eric C. Rowell, *Modular categories of dimension p^3m with m square-free*, Proc. Amer. Math. Soc. **147** (2019), no. 1, 21–34, DOI 10.1090/proc/13776. MR3876728

[10] Sebastian Burciu and Sonia Natale, *Fusion rules of equivariantizations of fusion categories*, J. Math. Phys. **54** (2013), no. 1, 013511, 21, DOI 10.1063/1.4774293. MR3059899

[11] Shawn X. Cui, César Galindo, Julia Yael Plavnik, and Zhenghan Wang, *On gauging symmetry of modular categories*, Comm. Math. Phys. **348** (2016), no. 3, 1043–1064, DOI 10.1007/s00220-016-2633-8. MR3555361

[12] Alexei Davydov, Michael Müger, Dmitri Nikshych, and Victor Ostrik, *The Witt group of non-degenerate braided fusion categories*, J. Reine Angew. Math. **677** (2013), 135–177. MR3039775

[13] Alexei Davydov and Dmitri Nikshych, *The Picard crossed module of a braided tensor category*, Algebra Number Theory **7** (2013), no. 6, 1365–1403, DOI 10.2140/ant.2013.7.1365. MR3107567

[14] P. Deligne, *Catégories tensorielles* (French, with English summary), Mosc. Math. J. **2** (2002), no. 2, 227–248, DOI 10.17323/1609-4514-2002-2-2-227-248. Dedicated to Yuri I. Manin on the occasion of his 65th birthday. MR1944506

[15] Vladimir Drinfeld, Shlomo Gelaki, Dmitri Nikshych, and Victor Ostrik, *On braided fusion categories. I*, Selecta Math. (N.S.) **16** (2010), no. 1, 1–119, DOI 10.1007/s00029-010-0017-z. MR2609644

[16] Samuel Eilenberg and Saunders Mac Lane, *On the groups $H(\Pi, n)$. II. Methods of computation*, Ann. of Math. (2) **60** (1954), 49–139, DOI 10.2307/1969702. MR0065162

[17] Pavel Etingof, Shlomo Gelaki, Dmitri Nikshych, and Victor Ostrik, *Tensor categories*, Mathematical Surveys and Monographs, vol. 205, American Mathematical Society, Providence, RI, 2015. MR3242743

[18] Pavel Etingof, Dmitri Nikshych, and Viktor Ostrik, *On fusion categories*, Ann. of Math. (2) **162** (2005), no. 2, 581–642, DOI 10.4007/annals.2005.162.581. MR2183279

[19] Pavel Etingof, Dmitri Nikshych, and Victor Ostrik, *Fusion categories and homotopy theory*, Quantum Topol. **1** (2010), no. 3, 209–273, DOI 10.4171/QT/6. With an appendix by Ehud Meir. MR2677836

[20] Pavel Etingof, Dmitri Nikshych, and Victor Ostrik, *Weakly group-theoretical and solvable fusion categories*, Adv. Math. **226** (2011), no. 1, 176–205, DOI 10.1016/j.aim.2010.06.009. MR2735754

[21] Pavel Etingof, Eric Rowell, and Sarah Witherspoon, *Braid group representations from twisted quantum doubles of finite groups*, Pacific J. Math. **234** (2008), no. 1, 33–41, DOI 10.2140/pjm.2008.234.33. MR2375313

[22] Jennifer M. Franko, Eric C. Rowell, and Zhenghan Wang, *Extraspecial 2-groups and images of braid group representations*, J. Knot Theory Ramifications **15** (2006), no. 4, 413–427, DOI 10.1142/S0218216506004580. MR2221527

[23] Michael H. Freedman, Alexei Kitaev, and Zhenghan Wang, *Simulation of topological field theories by quantum computers*, Comm. Math. Phys. **227** (2002), no. 3, 587–603, DOI 10.1007/s002200200635. MR1910832

[24] Michael H. Freedman, Michael Larsen, and Zhenghan Wang, *A modular functor which is universal for quantum computation*, Comm. Math. Phys. **227** (2002), no. 3, 605–622, DOI 10.1007/s002200200645. MR1910833

[25] César Galindo, Seung-Moon Hong, and Eric C. Rowell, *Generalized and quasi-localizations of braid group representations*, Int. Math. Res. Not. IMRN **3** (2013), 693–731, DOI 10.1093/imrn/rnr269. MR3021796

[26] César Galindo and Nicolás Jaramillo, *Solutions of the hexagon equation for abelian anyons*, Rev. Colombiana Mat. **50** (2016), no. 2, 273–294, DOI 10.15446/recolma.v50n2.62213. MR3605649

[27] César Galindo and Eric C. Rowell, *Braid representations from unitary braided vector spaces*, J. Math. Phys. **55** (2014), no. 6, 061702, 13, DOI 10.1063/1.4880196. MR3390645

[28] Shlomo Gelaki and Dmitri Nikshych, *Nilpotent fusion categories*, Adv. Math. **217** (2008), no. 3, 1053–1071, DOI 10.1016/j.aim.2007.08.001. MR2383894

[29] Matthew B. Hastings, Chetan Nayak, and Zhenghan Wang, *On metaplectic modular categories and their applications*, Comm. Math. Phys. **330** (2014), no. 1, 45–68, DOI 10.1007/s00220-014-2044-7. MR3215577

[30] Michael J. Larsen and Eric C. Rowell, *An algebra-level version of a link-polynomial identity of Lickorish*, Math. Proc. Cambridge Philos. Soc. **144** (2008), no. 3, 623–638, DOI 10.1017/S0305004107000424. MR2418707

[31] Michael J. Larsen, Eric C. Rowell, and Zhenghan Wang, *The N-eigenvalue problem and two applications*, Int. Math. Res. Not. **64** (2005), 3987–4018, DOI 10.1155/IMRN.2005.3987. MR2206918

[32] J. Liptrap. Generalized tambara-yamagami categories. *J. Algebra*, 2010. to appear.

[33] Michael Müger, *Galois theory for braided tensor categories and the modular closure*, Adv. Math. **150** (2000), no. 2, 151–201, DOI 10.1006/aima.1999.1860. MR1749250

[34] Michael Müger, *Galois theory for braided tensor categories and the modular closure*, Adv. Math. **150** (2000), no. 2, 151–201, DOI 10.1006/aima.1999.1860. MR1749250

[35] Michael Müger, *On the structure of modular categories*, Proc. London Math. Soc. (3) **87** (2003), no. 2, 291–308, DOI 10.1112/S0024611503014187. MR1990929

[36] Deepak Naidu and Eric C. Rowell, *A finiteness property for braided fusion categories*, Algebr. Represent. Theory **14** (2011), no. 5, 837–855, DOI 10.1007/s10468-010-9219-5. MR2832261

[37] Sonia Natale, *Faithful simple objects, orders and gradings of fusion categories*, Algebr. Geom. Topol. **13** (2013), no. 3, 1489–1511, DOI 10.2140/agt.2013.13.1489. MR3071133

[38] Chetan Nayak, Steven H. Simon, Ady Stern, Michael Freedman, and Sankar Das Sarma, *Non-abelian anyons and topological quantum computation*, Rev. Modern Phys. **80** (2008), no. 3, 1083–1159, DOI 10.1103/RevModPhys.80.1083. MR2443722

[39] Victor Ostrik, *Module categories, weak Hopf algebras and modular invariants*, Transform. Groups **8** (2003), no. 2, 177–206, DOI 10.1007/s00031-003-0515-6. MR1976459

[40] Eric Rowell, Richard Stong, and Zhenghan Wang, *On classification of modular tensor categories*, Comm. Math. Phys. **292** (2009), no. 2, 343–389, DOI 10.1007/s00220-009-0908-z. MR2544735

[41] Eric C. Rowell and Hans Wenzl, $SO(N)_2$ *braid group representations are Gaussian*, Quantum Topol. **8** (2017), no. 1, 1–33, DOI 10.4171/QT/85. MR3630280

[42] Eric C. Rowell, *From quantum groups to unitary modular tensor categories*, Representations of algebraic groups, quantum groups, and Lie algebras, Contemp. Math., vol. 413, Amer. Math. Soc., Providence, RI, 2006, pp. 215–230, DOI 10.1090/conm/413/07848. MR2263097

[43] Eric C. Rowell, *Unitarizability of premodular categories*, J. Pure Appl. Algebra **212** (2008), no. 8, 1878–1887, DOI 10.1016/j.jpaa.2007.11.004. MR2414692

[44] Eric C. Rowell, *Braid representations from quantum groups of exceptional Lie type*, Rev. Un. Mat. Argentina **51** (2010), no. 1, 165–175. MR2681262

[45] Eric C. Rowell and Zhenghan Wang, *Localization of unitary braid group representations*, Comm. Math. Phys. **311** (2012), no. 3, 595–615, DOI 10.1007/s00220-011-1386-7. MR2909757

[46] Eric C. Rowell and Zhenghan Wang, *Degeneracy and non-Abelian statistics*, Phys. Rev. A **93** (2016), no. 3, 030102(R), 4, DOI 10.1103/physreva.93.030102. MR3855013

[47] Vladimir Turaev and Hans Wenzl, *Semisimple and modular categories from link invariants*, Math. Ann. **309** (1997), no. 3, 411–461, DOI 10.1007/s002080050120. MR1474200

[48] Vladimir G. Turaev, *Modular categories and 3-manifold invariants*, Internat. J. Modern Phys. B **6** (1992), no. 11-12, 1807–1824, DOI 10.1142/S0217979292000876. Topological and quantum group methods in field theory and condensed matter physics. MR1186845

Expedia Group, Bellevue, WA 98004
Email address: pbruillard@expedia.com

Department of Electrical Engineering, Wright State University, Dayton, OH 45435
Email address: paul.gustafson@wright.edu

Department of Mathematics, Indiana University, Bloomington, IN 47405
Email address: jplavnik@iu.edu

Department of Mathematics, Texas A&M University, College Station, TX 77843-3368
Email address: rowell@math.tamu.edu

Contemporary Mathematics
Volume **747**, 2020
https://doi.org/10.1090/conm/747/15041

The rank of G-crossed braided extensions of modular tensor categories

Marcel Bischoff

Abstract. We give a short proof for a well-known formula for the rank of a G-crossed braided extension of a modular tensor category.

1. Introduction

G-crossed braided extensions of modular tensor categories are an important ingredient for the process of gauging in topological phases of matter in the framework of modular tensor categories [**BBCW14**]. Let $\mathcal{C}$ be a modular tensor category, which for the purpose of this note is a semisimple, $\mathbb{C}$-linear abelian ribbon category with simple tensor unit, such that the set of isomorphism classes of simple objects $\operatorname{Irr}(\mathcal{C})$ is finite and the braiding is non-degenerate, see e.g. [**BK01**]. A global symmetry [**BBCW14**] of $\mathcal{C}$ is a pair (G,ρ) consisting of a finite group G and homomorphism $\rho\colon G\to \operatorname{Aut}^{\mathrm{br}}_{\otimes}(\mathcal{C})$, where $\operatorname{Aut}^{\mathrm{br}}_{\otimes}(\mathcal{C})$ is the group of isomorphism classes of braided autoequivalences of $\mathcal{C}$. Gauging a global symmetry (G,ρ) of $\mathcal{C}$ [**CGPW16**] is a two step process which eventually produces a new modular tensor category and is given as follows:

(1) Construct a G-crossed braided extension $\mathcal{F}=\bigoplus_{g\in G}\mathcal{F}_g$ of $\mathcal{C}$.
(2) Consider the equivariantization $\mathcal{F}^G$ of $\mathcal{F}$ to obtain a new modular tensor category, which therefore contains $\operatorname{Rep}(G)$ as a symmetric subcategory.

A G-crossed braided fusion category is a fusion category $\mathcal{F}$ equipped with the following structure:

- $\mathcal{F}$ is faithfully G-graded, i.e. $\mathcal{F}=\bigoplus\mathcal{F}_g$ and $X_g\otimes Y_h\in\mathcal{F}_{gh}$ for every $g,h\in G$, $X_g\in\mathcal{F}_g$, and $Y_h\in\mathcal{F}_h$.
- There is an G-action $X\mapsto {}^gX=\rho(g)(X)$ given by a monoidal functor $\underline{\rho}\colon \underline{G}\to\underline{\operatorname{Aut}_{\otimes}}(\mathcal{F})$, where $\underline{\operatorname{Aut}_{\otimes}}(\mathcal{F})$ is the categorical group of monoidal autoequivalences.
- There is a natural family of isomorphisms
$$c_{X,Y}\colon X\otimes Y\to {}^gY\otimes X\,,\qquad\qquad g\in G, X\in\mathcal{F}_g, Y\in\mathcal{F}$$
with $\rho(g)(\mathcal{F}_h)\subset\mathcal{F}_{ghg^{-1}}$ fulfilling certain coeherence diagrams, see e.g. [**Tur10**].

2010 *Mathematics Subject Classification.* Primary 18D10; Secondary 81T40, 81T45.
Supported in part by NSF Grant DMS-1700192/1821162.

©2020 American Mathematical Society

We note that $\mathcal{F}_e$ is a braided fusion category and we get a global symmetry $\rho\colon G \to \mathrm{Aut}^{\mathrm{br}}_{\otimes}(\mathcal{F}_e)$ given by restriction and truncation of $\underline{\rho}$. We are only interested in the case where $\mathcal{F}_e$ is a modular tensor category. We refer to [**DGNO10**, **Tur10**] for more details.

We call a G-crossed braided fusion category $\mathcal{F} = \bigoplus_{g\in G} \mathcal{F}_g$ a G-crossed braided extension of a modular tensor category $\mathcal{C}$ if $\mathcal{F}_e$ is braided equivalent to $\mathcal{C}$. The above data of $\mathcal{F}$ gives by restriction a global symmetry (G,ρ) on $\mathcal{C} \cong \mathcal{F}_e$. We note that given a modular tensor category $\mathcal{C}$ and a global action (G,ρ) as an input one has to check that certain obstructions have to vanish in order for a G-crossed braided to exist, see [**ENO10**].

G-crossed braided extensions arise naturally in rational conformal field theory. Let $\mathcal{A}$ be a (completely) rational conformal net, then the category of representations $\mathrm{Rep}(\mathcal{A})$ is a unitary modular tensor category [**KLM01**]. Let $G \leq \mathrm{Aut}(\mathcal{A})$ be a finite group of automorphisms of the net $\mathcal{A}$. There is a category $G\text{–}\mathrm{Rep}(\mathcal{A})$ of G-twisted representations of $\mathcal{A}$, which is a G-crossed braided extension of the unitary modular tensor category $\mathrm{Rep}(\mathcal{A})$. The fixed point or "orbifold net" $\mathcal{A}^G$ is again completely rational and $\mathrm{Rep}(\mathcal{A}^G)$ is a gauging of $\mathrm{Rep}(\mathcal{A})$ by G, i.e. $\mathrm{Rep}(\mathcal{A}^G)$ is braided equivalent to the G-equivariantization $(G\text{–}\mathrm{Rep}(\mathcal{A}))^G$, see [**Müg05**] for the original reference and [**Bis18**] for a review.

Starting with a modular tensor category $\mathcal{C}$ we want to compute certain invariants of possible G-crossed braided extensions $\mathcal{F} = \bigoplus_{g\in G} \mathcal{F}_g$ with $\mathcal{F}_e \cong \mathcal{C}$. The simplest invariant of $\mathcal{F}$ is its rank $\mathrm{rk}(\mathcal{F}) := |\operatorname{Irr}(\mathcal{F})|$.

We note that the global symmetry (G,ρ) associated with a G-crossed braided extension $\mathcal{F}$ of $\mathcal{C}$ equips $\operatorname{Irr}(\mathcal{C})$ with a G-action. This G-space already contains all the information about the rank of possible G-crossed braided extensions associated with (G,ρ). Namely, the following well-know formula, cf. [**BBCW14**, Eq. (348)] holds.

PROPOSITION 1.1. *Let $\mathcal{F} = \bigoplus_{g\in G} \mathcal{F}_g$ be a (faithful) G-crossed braided extension of a modular tensor category $\mathcal{C} \cong \mathcal{F}_e$. Then the rank of $\mathcal{F}_g$ is equal to the size of the stabilizer* $\operatorname{Irr}(\mathcal{C})^g$, *i.e.*

$$\mathrm{rk}(\mathcal{F}_g) = |\operatorname{Irr}(\mathcal{C})^g|\,.$$

In particular, the rank of $\mathcal{F}$ is determined via the G-action on $\operatorname{Irr}(\mathcal{C})$ *by*

$$\mathrm{rk}(\mathcal{F}) = \sum_{g\in G} |\operatorname{Irr}(\mathcal{C})^g| = |G||\operatorname{Irr}(\mathcal{C})/G|\,.$$

The goal of this note is to provide a short proof of this statement using modular invariants, which were used in the operator algebra literature, e.g. [**BEK99**, **BEK00**].

2. The rank of module categories and G-crossed braided extensions

Let $\mathcal{C}$ be a modular tensor category and A a simple non-degenerate algebra object in $\mathcal{C}$. Here non-degenerate means that the trace pairing $\Phi_A\colon A \to \overline{A}$ given by

$$\Phi_A = ((\mathrm{ev}_A)\circ(\mathrm{id}_{\bar{A}}\otimes m)\otimes \mathrm{id}_{\bar{A}})\circ(\mathrm{id}_{\bar{A}}\otimes(\mathrm{id}_A\otimes \mathrm{coev}_A)\circ m)\circ(\mathrm{coev}_{\bar{A}}\otimes \mathrm{id}_A)$$

is invertible, which implies that A has the structure of a special symmetric Frobenius algebra object, see [**KR08**, Lemma 2.3].

We denote the category with the opposite braiding by $\overline{\mathcal{C}}$. The full center $Z(A)$ is a Lagrangian algebra object in $Z(\mathcal{C}) \cong \mathcal{C} \boxtimes \overline{\mathcal{C}}$ [**FFRS06**, **KR08**, **DMNO13**]. We note that the forgetful functor $F\colon \mathcal{C} \boxtimes \overline{\mathcal{C}} \to \mathcal{C}$ maps $X \boxtimes Y \mapsto X \otimes Y$ and has an adjoint $I\colon \mathcal{C} \to \mathcal{C} \boxtimes \overline{\mathcal{C}}$. The image $I(\mathbb{1})$ has the canonical structure of a Lagrangian algebra in $\mathcal{C} \boxtimes \overline{\mathcal{C}}$. If $X \in \mathcal{C}$ we denote by $\bar{X} \in \mathcal{C}$ a dual object. The full center $Z(A)$ can be realized as the left center of the braided product of $(A \boxtimes \mathbb{1}) \otimes^+ I(\mathbb{1})$ [**FFRS06**, **KR08**]. Given a simple non-degenerate algebra A there are two functors $\alpha^\pm$ from $\mathcal{C}$ to the category ${}_A\mathcal{C}_A$ of A-bimodules in $\mathcal{C}$. They are given by equipping the image of the free right module functor $-\otimes A$ with a left action using the braiding or opposite braiding, respectively, see [**FFRS06**].

The following proposition is well-known to experts.

PROPOSITION 2.1. *Let $\mathcal{C}$ be a modular tensor category, A a simple non-degenerate algebra object in $\mathcal{C}$, and Z the associated modular invariant matrix, i.e. $Z = (Z_{X,Y})_{X,Y\in\mathrm{Irr}(\mathcal{C})}$ is the square matrix given by*

$$Z_{X,Y} = \dim_{\mathbb{C}} \mathrm{Hom}_{{}_A\mathcal{C}_A}(\alpha^+(X), \alpha^-(\bar{Y}))\,.$$

Then

(1) *The full center $Z(A)$ of A as an object in $\mathcal{C} \boxtimes \overline{\mathcal{C}}$ is given by*

$$Z(A) \cong \bigoplus_{X,Y\in\mathrm{Irr}(\mathcal{C})} Z_{X,Y} \cdot X \boxtimes \bar{Y}\,.$$

(2) *The rank $\mathrm{rk}(\mathcal{C}_A)$ of the category of right A-modules $\mathcal{C}_A$ is given by the trace of the matrix Z:*

$$\mathrm{rk}(\mathcal{C}_A) = \mathrm{tr}(Z) = \sum_{X\in\mathrm{Irr}(\mathcal{C})} Z_{X,X}\,.$$

PROOF. The first statement is [**FFRS06**, Remark 3.7]. The second statement is [**BEK99**, Corollary 6.1] in the case that A is a Q-system and $\mathcal{C}$ is a unitary modular tensor category realized as endomorphisms of a type III factor. In the more general setting, it follows from [**KR08**, Eq. (4.4) and Prop. 4.3]. Namely, $F(Z(A)) \cong \bigoplus_{X,Y\in\mathrm{Irr}(\mathcal{C})} Z_{X,Y} X \otimes \bar{Y}$ thus $\dim \mathrm{Hom}(\mathbb{1}, F(Z(A)) = \mathrm{tr}(Z)$ equals the number of isomorphism classes of simple modules in $\mathcal{C}_A$, since $F(Z(A))$ is equivalent to $\bigoplus_{M\in\mathrm{Irr}(\mathcal{C}_A)} M \otimes_A \bar{M}$ and since $M \otimes_A \bar{M}$ is a connected algebra by [**KR08**, Proof of Prop. 4.10]. □

For $\phi \in \mathrm{Aut}^{\mathrm{br}}_{\otimes}(\mathcal{C})$ there is a canonical Lagrangian algebra $L_\phi = (\mathrm{id} \boxtimes \phi) I(1)$ in $\mathcal{C} \boxtimes \overline{\mathcal{C}}$, see e.g. [**DNO13**, Sec 3.2-3.3]. The algebra L_ϕ corresponds to the full center of a $\mathcal{C}$-module category, which we denote by $\mathcal{C}_\phi$, cf. [**KR08**, **DMNO13**]. This module category can be obtained, using e.g. [**KR08**], by taking any connected summand A of the algebra $F(L_\phi)$ (which always fulfills $Z(A) = L_\phi$) and considering the $\mathcal{C}$-module category $\mathcal{C}_\phi := \mathcal{C}_A$, see [**KR08**], and also [**ENO10**, Lemma 5.1]. Since $L_\phi \cong \bigoplus_{X\in\mathrm{Irr}(\mathcal{C})} X \boxtimes \phi(\bar{X})$, i.e. $Z_{X,Y} = \delta_{\phi(X),Y}$, Proposition 2.1 immediately gives the following statement.

LEMMA 2.2. *The rank of $\mathcal{C}_\phi$ equals $|\,\mathrm{Irr}(\mathcal{C})^\phi|$, i.e. the number of isomorphism classes of simple objects in $\mathcal{C}$ which are fixed by ϕ.*

PROOF OF PROPOSITION 1.1. Let $\mathcal{F} = \bigoplus_{g\in G} \mathcal{F}_g$ be a G-crossed braided extension of a modular tensor category $\mathcal{C}$. By [**ENO10**, Sec. 5.4] the category $\mathcal{F}_g$ is equivalent as a left $\mathcal{C}$-module category to the category $\mathcal{C}_g$ in Lemma 2.2 and the statement follows. □

2.1. The rank of permutation extensions. Let $\mathcal{C}$ be a modular tensor category. Gannon and Jones announced **[GJ19]** that certain cohomological obstructions vanish and thus that there is always an S_n-crossed braided extension of $\mathcal{C}^{\boxtimes n}$ which we denote by $\mathcal{C} \wr S_n$, where the action of S_n is given by permuations of objects. More general, for any subgroup $G \leq S_n$ there is a G-crossed braided extension $\mathcal{C} \wr G$.

For the cyclic subgroup $\mathbb{Z}_n \cong \langle (12\cdots n)\rangle \subset S_n$ with n prime we obtain

$$\mathrm{rk}(\mathcal{C} \wr \mathbb{Z}_n) = \mathrm{rk}(\mathcal{C})^n + (n-1)\,\mathrm{rk}(\mathcal{C})\,,$$

which in the case $n = 2$ is also true by construction for the examples considered in **[EMJP18]**.

In general, for the full symmetric group, we get that $\mathrm{rk}((\mathcal{C} \wr S_n)_g)$ depends only on the conjugacy class C_g of $g \in S_n$. Namely, $\mathrm{rk}(\mathcal{C} \wr S_n)_g) = \mathrm{rk}(\mathcal{C}))^{|a|}$ where $a = (a_1, \ldots, a_n)$ and a_j is the number of j-cycles (counting 1-cycles) in the cycle decomposition of g and $|a| = \sum_j a_j$. Thus summing over all elements of the $p(n)$ conjugacy classes we get

$$\mathrm{rk}(\mathcal{C} \wr S_n) = \sum_{a=(a_1,\ldots,a_n)} c_a\,\mathrm{rk}(\mathcal{C})^{|a|}\,, \qquad c_a = \frac{n!}{\prod_j (j)^{a_j}(a_j!)}\,,$$

where $c_a = |C_g|$ the size of the conjugacy class C_g of an element $g \in S_n$ with cycle type $a = (a_1, \ldots, a_n)$ and the sum runs over all unordered partitions of $\{1, \ldots, n\}$ with a_j the number of partitions of length j. Explicitly,

$$\begin{aligned}
\mathrm{rk}(\mathcal{C} \wr S_3) &= \mathrm{rk}(\mathcal{C})^3 + 3\,\mathrm{rk}(\mathcal{C})^2 + 2\,\mathrm{rk}(\mathcal{C})\\
\mathrm{rk}(\mathcal{C} \wr S_4) &= \mathrm{rk}(\mathcal{C})^4 + 6\,\mathrm{rk}(\mathcal{C})^3 + 11\,\mathrm{rk}(\mathcal{C})^2 + 6\,\mathrm{rk}(\mathcal{C})\\
\mathrm{rk}(\mathcal{C} \wr S_5) &= \mathrm{rk}(\mathcal{C})^5 + 10\,\mathrm{rk}(\mathcal{C})^4 + 35\,\mathrm{rk}(\mathcal{C})^3 + 50\,\mathrm{rk}(\mathcal{C})^2 + 24\,\mathrm{rk}(\mathcal{C})\\
&\cdots
\end{aligned}$$

Acknowledgments. I would like to thank Corey Jones, David Penneys, and Julia Plavnik for discussions.

References

[BBCW14] M. Barkeshli, P. Bonderson, M. Cheng, and Z. Wang, *Symmetry, defects, and gauging of topological phases*, arXiv preprint arXiv:1410.4540 (2014).

[BEK00] J. Böckenhauer, D. E. Evans, and Y. Kawahigashi, *Chiral structure of modular invariants for subfactors*, Comm. Math. Phys. **210** (2000), no. 3, 733–784, DOI 10.1007/s002200050798. MR1777347

[BEK99] J. Böckenhauer, D. E. Evans, and Y. Kawahigashi, *On α-induction, chiral generators and modular invariants for subfactors*, Comm. Math. Phys. **208** (1999), no. 2, 429–487, DOI 10.1007/s002200050765. MR1729094

[Bis18] M. Bischoff, *Conformal net realizability of Tambara-Yamagami categories and generalized metaplectic modular categories*, arXiv preprint arXiv:1803.04949 (2018).

[BK01] B. Bakalov and A. Kirillov Jr., *Lectures on tensor categories and modular functors*, University Lecture Series, vol. 21, American Mathematical Society, Providence, RI, 2001. MR1797619

[CGPW16] S. X. Cui, C. Galindo, J. Y. Plavnik, and Z. Wang, *On gauging symmetry of modular categories*, Comm. Math. Phys. **348** (2016), no. 3, 1043–1064, DOI 10.1007/s00220-016-2633-8. MR3555361

[DGNO10] V. Drinfeld, S. Gelaki, D. Nikshych, and V. Ostrik, *On braided fusion categories. I*, Selecta Math. (N.S.) **16** (2010), no. 1, 1–119, DOI 10.1007/s00029-010-0017-z. MR2609644

[DMNO13] A. Davydov, M. Müger, D. Nikshych, and V. Ostrik, *The Witt group of non-degenerate braided fusion categories*, J. Reine Angew. Math. **677** (2013), 135–177. MR3039775

[DNO13] A. Davydov, D. Nikshych, and V. Ostrik, *On the structure of the Witt group of braided fusion categories*, Selecta Math. (N.S.) **19** (2013), no. 1, 237–269, DOI 10.1007/s00029-012-0093-3. MR3022755

[EMJP18] C. Edie-Michell, C. Jones, and J. Plavnik, *Fusion rules for $\mathbb{Z}/2\mathbb{Z}$ permutation gauging*, arXiv preprint arXiv:1804.01657 (2018).

[ENO10] P. Etingof, D. Nikshych, and V. Ostrik, *Fusion categories and homotopy theory*, Quantum Topol. **1** (2010), no. 3, 209–273, DOI 10.4171/QT/6. With an appendix by Ehud Meir. MR2677836

[FFRS06] J. Fröhlich, J. Fuchs, I. Runkel, and C. Schweigert, *Correspondences of ribbon categories*, Adv. Math. **199** (2006), no. 1, 192–329. MR2187404 (2007b:18007)

[GJ19] T. Gannon and C. Jones, *Vanishing of Categorical Obstructions for Permutation Orbifolds*, Comm. Math. Phys. **369** (2019), no. 1, 245–259, DOI 10.1007/s00220-019-03288-9. MR3959559

[KLM01] Y. Kawahigashi, R. Longo, and M. Müger, *Multi-interval subfactors and modularity of representations in conformal field theory*, Comm. Math. Phys. **219** (2001), no. 3, 631–669, DOI 10.1007/PL00005565. MR1838752

[KR08] L. Kong and I. Runkel, *Morita classes of algebras in modular tensor categories*, Adv. Math. **219** (2008), no. 5, 1548–1576, DOI 10.1016/j.aim.2008.07.004. MR2458146

[Müg05] M. Müger, *Conformal orbifold theories and braided crossed G-categories*, Comm. Math. Phys. **260** (2005), no. 3, 727–762, DOI 10.1007/s00220-005-1291-z. MR2183964

[Tur10] V. Turaev, *Homotopy quantum field theory*, EMS Tracts in Mathematics, vol. 10, European Mathematical Society (EMS), Zürich, 2010. Appendix 5 by Michael Müger and Appendices 6 and 7 by Alexis Virelizier. MR2674592

Department of Mathematics, Morton Hall 321, 1 Ohio University, Athens, OH 45701, USA

Email address: bischoff@ohio.edu

Email address: marcel@localconformal.net

Contemporary Mathematics
Volume **747**, 2020
https://doi.org/10.1090/conm/747/15042

Symmetry defects and their application to topological quantum computing

Colleen Delaney and Zhenghan Wang

Abstract. We describe the mathematical theory of topological quantum computing with symmetry defects in the language of fusion categories and unitary representations. Symmetry defects together with anyons are modeled by G-crossed braided extensions of unitary modular tensor categories. The algebraic data of these categories afford projective unitary representations of the braid group. Elements in the image of such representations correspond to quantum gates arising from exchanging anyons and symmetry defects in topological phases of matter with symmetry. We provide some small examples that highlight features of practical interest for quantum computing. In particular, symmetry defects show the potential to generate non-abelian statistics from abelian topological phases and to be used as as a tool to enlarge the set of quantum gates accessible to an anyonic device.

1. Introduction

Topological order of a 2-dimensional topological phase of matter can be modeled by a unitary modular tensor category $\mathcal{C}$. If a 2-dimensional topological phase of matter has a symmetry group G, then $\mathcal{C}$ inherits an action of G. A nontrivial action of G on $\mathcal{C}$ is a form of symmetry breaking, called weak symmetry breaking in [**17**]. It follows that the associated topological defects, which manifest the topological properties of symmetry breaking patterns, provide a window to the world of symmetry and topology. A practical reason to study such symmetry defects is as a new mechanism to generate non-abelian statistics from a purely abelian topological phase. For example, we show that the symmetry defects arising from the electromagnetic duality of the $\mathbb{Z}_2$ toric code behave almost identically to Majorana zero modes or Ising anyons.

The symmetry defects we consider here are extrinsic point-like objects that are in general confined. Like anyons, they lead to topological degeneracy and braid group representations, though only projectively. In this paper, we investigate the mathematical modeling of symmetry defects and their application to topological quantum computing.

In Section 2 we review the algebraic description of symmetry defects via G-crossed extensions of unitary modular tensor categories following [**2**], where this mathematical model for topological order with symmetry was first introduced.

2010 *Mathematics Subject Classification.* Primary 18D10, 81P68, 57R56.

©2020 American Mathematical Society

Some examples are presented to illustrate the theory and are further developed in Section 3, where we describe the (projective) unitary representations associated to unitary G-crossed extensions and discuss their interpretation as quantum gates for a topological quantum computer built by braiding symmetry defects and anyons. In Section 4, we give a mathematical analysis of an example of topological order with $\mathbb{Z}_2$ symmetry that shows allowing projective measurement and braiding with symmetry defects can enlarge an anyonic gate set with finite image in the unitary group to one that is universal for quantum computation. This feature of two layers of the Ising theory with a layer-exchange symmetry $\left(\mathbf{Ising}^{(1)} \boxtimes \mathbf{Ising}^{(1)}\right)^{\times}_{\mathbb{Z}_2}$ was proposed in [**6**] based on hypothetical data for the $\mathbb{Z}_2$-crossed extension and a physical argument. We conclude in Section 5 with some comments about future directions, physical and mathematical.

2. Algebraic theory of symmetry defects

Symmetry defects are point-like objects that are not intrinsic quasiparticle excitations of a topological phase of matter, but can nevertheless be used for topological quantum computing (TQC). In analogy with the mathematical definition of an anyon as a simple object in a unitary modular tensor category (UMTC), a symmetry defect can be defined as a simple object in a non-trivial sector of a unitary G-crossed extension $\mathcal{C}_G^{\times}$ of a UMTC $\mathcal{C}$ [**2**]. The mathematical model of symmetry defects and anyons is then a *unitary G-crossed braided fusion category*, UGxBFC for short. Since in physics symmetry defect models arise from anyon models $\mathcal{C}$ with symmetry G, we begin with the construction of UGxBFCs as extensions of UMTCs and delay an abstract definition of a UGxBFC until later in the section. Both notions will be useful to have on hand when we discuss how symmetry defect models can also come from a topological phase $\mathcal{D}$ in a different way: as the first step in a two-step "topological Bose condensation" procedure (de-equivariantization in the language of fusion categories) [**15**].

2.1. G-crossed extensions of UMTCs.

Here we briefly recall the construction of a G-crossed category from a UMTC $\mathcal{C}$ and identify symmetry defects and their mathematical properties. Our discussion follows [**2**] but our notation is slightly different in that we use standard font for group elements.

2.1.1. *Preliminaries.* In what follows we assume G is a finite group. Although G can be non-abelian we will consistently use 0 to denote the identity element. We will work with a certain "categorification" of G, a categorical group $\mathcal{G}$. The objects of $\mathcal{G}$ are group elements and morphism sets $\mathrm{Hom}(g,h) = \delta_{gh}$ of two objects g and h are empty if $g \neq h$ and contain only the identity if $g = h$. That is, there are no nontrivial morphisms between distinct group elements, and the only morphism between identical group elements is the identity. With respect to the tensor product given by group multiplication $g \otimes h = gh$, $\mathcal{G}$ forms a strict monoidal category.

We will use $\mathcal{C}$ to denote an arbitrary UMTC. We denote the set of objects of $\mathcal{C}$ by $\mathrm{Obj}(\mathcal{C})$. Recall that an anyon is a simple object in $\mathcal{C}$. We call the isomorphism class of an anyon its type. When we work with concrete examples we always pass to a skeletonization of $\mathcal{C}$, in which case there is only one object per isomorphism class and we conflate anyons and anyon types. We call an anyon abelian if it has quantum dimension one. This definition is equivalent to the others in use, for example that in terms of images of associated braid group representations by the work of [**22**].

The subset of $\mathrm{Obj}(\mathcal{C})$ consisting of abelian anyon types forms a subgroup under the tensor product, which we denote by $\mathcal{A}$. To see why this is true, observe that the tensor unit (the vacuum) has quantum dimension one and the quantum dimension function $d_{\_} : \mathrm{Obj}(\mathcal{C}) \to \mathbb{R}_{\geq 0}$ is linear and multiplicative, in the sense that

$$a \otimes b = \bigoplus_c N_c^{ab} c \implies d_a \cdot d_b = \sum_c N_c^{ab} d_c.$$

In particular, if $d_a = d_b = 1$, then $a \otimes b = c$ where $d_c = 1$. It also follows that the dual a^* of an abelian anyon a is also abelian, since $a \otimes a^* = 1$ implies $d_a^* = 1$. It follows that the subset of abelian anyon types contains the vacuum and is closed under multiplication and inverses. Since fusion is associative on the level of isomorphism classes, $\mathcal{A}$ forms a group.

We denote by $\mathrm{Aut}_{br}^{\otimes}(\mathcal{C})$ the group of braided-tensor autoequivalences of $\mathcal{C}$ up to natural isomorphism under composition. We will also need the categorical group $\underline{\mathrm{Aut}_{br}^{\otimes}(\mathcal{C})}$, with group elements of $\mathrm{Aut}_{br}^{\otimes}(\mathcal{C})$, tensor product given by composition, and morphisms given by natural isomorphisms of functors.

2.1.2. *From group symmetry to group extension theory of MTCs.* To construct a UGxBFC, we start with a UMTC $\mathcal{C}$ and a group homomorphism $\rho : G \to \mathrm{Aut}_{br}^{\otimes}(\mathcal{C})$, called a *global symmetry*. When this group homomorphism ρ can be promoted to a monoidal functor $\underline{\rho} : \mathcal{G} \to \underline{\mathrm{Aut}_{br}^{\otimes}(\mathcal{C})}$, it is called a *categorical global symmetry* of $\mathcal{C}$. The existence of such a lift $\underline{\rho}$ depends on the vanishing of a certain cohomological obstruction in $H^3(G, \mathcal{A})$, where $\mathcal{A}$ is the subgroup of abelian anyons under fusion. When this obstruction vanishes, the different extensions are "classified" by $H_\rho^2(G, \mathcal{A})$ in the sense that the distinct liftings of the global symmetry ρ to the categorical global symmetry $\underline{\rho}$ form a torsor over $H_\rho^2(G, \mathcal{A})$.

With a categorical global symmetry in hand, the next ingredient for constructing a G-extension of $\mathcal{C}$ is an isomorphism of the categorical groups $\underline{\mathrm{Aut}_{br}^{\otimes}(\mathcal{C})}$ and $\underline{\mathrm{Pic}(\mathcal{C})}$, the categorical Picard group consisting of invertible bimodule categories of $\mathcal{C}$ [**14**]. This associates to each categorical braided tensor autoequivalence $\underline{\rho_g}$ an invertible $\mathcal{C}$-bimodule category $\mathcal{C}_g$, a correspondence which is at the heart of the relationship between symmetry in the abstract and symmetry defects. When an additional cohomological obstruction in $H^4(G; U(1))$ vanishes, consistent fusion of invertible bimodule categories $\mathcal{C}_g \boxtimes \mathcal{C}_h$ can be defined so that elements of $\mathcal{C}_g$ satisfy the pentagon equations. The different solutions to the pentagon equations then form a torsor over $H^3(G, U(1))$.

Each such solution corresponds noncanonically to a unitary fusion category denoted by $\mathcal{C}_G^{\times} = \bigoplus_{g \in G} \mathcal{C}_g$, where $\mathcal{C}_0 = \mathcal{C}$. The inequivalent G-crossed braided extensions of a UMTC $\mathcal{C}$ are classified by the data (ρ, α, β), where $\alpha \in H_\rho^2(G, \mathcal{A})$ and $\beta \in H^3(G, U(1))$ [**2, 14**]. We will see in the next subsection that $\mathcal{C}_G^{\times}$ has the structure of a unitary *G-crossed braided fusion category.*

Now the anyons of $\mathcal{C}$ are identified as simple objects in $\mathcal{C}_0$ and the symmetry defects as simple objects in $\bigoplus_{g \neq 0} \mathcal{C}_g$:

DEFINITION 2.1. A symmetry defect with flux g is a simple object in the invertible $\mathcal{C}_0$-bimodule category $\mathcal{C}_g$.

We will sometimes call the bimodule category $\mathcal{C}_g$ the g-flux sector, or a g-sector.

2.1.3. *Physical interpretation of UGxBFCs.* We note that in the physics literature, especially in the early development of TQC, the word "anyon" was used in a broad sense to describe both intrinsic topological order and extrinsic defects [**13**]. Here we use the term "anyon" for simple objects in a UMTC, and reserve the term "symmetry defects" for simple objects in a nontrivial sector of a UGxBFC. While in this framework anyons could be considered as trivial defects, the physics of extrinsic defects differs in important ways from quasiparticle excitations of a topological phase of matter and justifies a distinction in the terminology.

The mathematical formalism describing symmetry defects, namely unitary G-crossed extensions of UMTCs, holds for any finite group G acting by a categorical global symmetry (when the H^4 obstruction vanishes). However, there are further considerations to make about the nature of the categorical group action $\underline{\rho}$ when trying to identify them as models for quantum systems. Apart from unitarity, generally one is interested in symmetries that are local in the following sense: operators that effect the symmetry transformations can be written as a product of local operators that remain localized in the same region under the symmetry action. These are called "on-site" symmetries. When the group action is unitary and on-site, the resulting class of UGxBFCs characterize what are known in condensed matter theory as (2+1)D symmetry-enriched topological (SET) phases of matter. However, not all physically meaningful symmetries are unitary and "on-site": the authors of [**2**] make sense of anti-unitary and "quasi-on-site" symmetries, which can include time-reversal, translational, rotational, and other spacetime symmetries.

2.2. G-crossed braided fusion categories. We will see that a GxBFC is a fusion category together with three compatible structures: a G-action, a G-grading, and a G-braiding. Below $\mathcal{C}$ will denote a fusion category, not a UMTC as in the previous subsection.

Recall from the previous subsection that a G-action is a monoidal functor $\underline{\rho}$: $\mathcal{G} \to \underline{\mathrm{Aut}_{br}^{\otimes}(\mathcal{C})}$. We write the G-action on objects of $\mathcal{C}$ as $\rho_g(X) = {}^gX$.

In the fusion category setting, a G-grading is a map $\partial : \mathrm{Obj}(\mathcal{C}_{hom}) \to G$, where $\mathcal{C}_{hom}$ is a full fusion subcategory of $\mathcal{C}$ with the property that it generates the objects of $\mathcal{C}$ by direct sum [**20**]. The full subcategories $\mathcal{C}_g = \partial^{-1}(g)$ given by the G-grading are the g-flux sectors seen previously.

Definition 2.2. A G-crossed braided fusion category $\mathcal{C}_G^{\times}$ is a fusion category with

- a G-action ${}^g(\cdot) : \mathcal{G} \to \mathrm{Aut}_{\otimes}(\mathcal{C}_G^{\times})$
- a G-grading ∂
  such that $\partial({}^gX) = g\,(\partial X)\,g^{-1}$ for all $X \in C_{hom}$
- a G-crossed braiding:
  natural isomorphisms $\{c_{X,Y} : X \otimes Y \to {}^gY \otimes X\}$ for all $X \in \mathcal{C}_g$, $Y \in \mathcal{C}_0$, $g \in G$

satisfying compatibility conditions with one another and the associators [**15**], [**20**].

The following theorem from [**2**], adapted from [**14**], says that the G-crossed extension of a UMTC from Section 2.1 satisfies the above definition.

Theorem 2.3. *The unitary G-crossed extension $\mathcal{C}_G^{\times}$ of a UMTC $\mathcal{C}$ has a canonical G-braiding and categorical G-action that make $\mathcal{C}_G^{\times}$ into a unitary G-crossed braided fusion category.*

In particular, the G-action on a unitary G-crossed extension of a UMTC $\mathcal{C}$ comes from extending the categorical action from $\mathcal{C}_0$ to all of the sectors $\mathcal{C}_g$ and the G-grading satisfies the property that the action of g on an object in the sector $\mathcal{C}_h$ sends it to an object in the sector $\mathcal{C}_{ghg^{-1}}$.

To describe topological phases of matter with symmetry we needed G-crossed braided fusion categories which were unitary. We have omitted the abstract definition of unitarity in fusion categories and instead will simply state what it means for a skeletonization of a UMTC or UGxBGC to be unitary. This is sufficient for our purposes since in order to work out concrete elements in the image of the projective braid group representations associated to a GxBFC $\mathcal{C}_G^\times$ one must pass to a skeletonization of $\mathcal{C}_G^\times$. Next we discuss these skeletonizations, which we refer to as working with the algebraic data of a UMTC or UGxBFC.

2.2.1. *Algebraic data of a GxBFC.* It is well established that a UMTC is characterized by the algebraic data defining its skeletonization, and can effectively be defined as a collection of numbers $\{N_{ab}^c, [F_d^{abc}]_{ef}, R_c^{ab}\}$ satisfying certain compatibility conditions. Then an MTC is unitary if there is a basis in which the matrices $[F_d^{abc}]$ and R_c are unitary. Unitarity of the R- and F- symbols then implies that the associated braid group representations are unitary, endowing them with an interpretation as quantum mechanical processes.

In [**2**] this result is extended to GxBFCs, and hence it is justified to define a UGxBFC as a collection of numbers $\{N_{ab}^c, [F_d^{abc}]_{ef}, R_c^{ab}, U_k(a,b;c), \eta_x(g,h)\}$ which form unitary matrices and satisfy various consistency equations. Here the R-symbols keep track of the G-crossed braiding, which we recall is not a braiding in the usual sense like it is in a MTC.

When the topological charges are all anyons (labeled by the trivial element of G), the corresponding symbols will obey the usual consistency equations for a UMTC, for example the hexagon equation that encodes compatibility of the associator and braiding in $\mathcal{C}_0$. The symbols carrying symmetry defect labels now obey more complicated G-crossed consistency equations. There is an enlarged set of pentagon equations that must be satisfied by the F-symbols, and now compatibility of the associator and G-crossed braiding is given by a "heptagon" equation see Section 2.4. The additional symbols U and η arise from the categorical symmetry, and give rise to additional compatibility relations that the symbols must satisfy. Restricted to the C_0 sector, the U-symbols encode how the symmetry acts on morphism spaces and the *eta*-symbols encode the natural isomorphism identifying ρ_{gh} with $\rho_g\rho_h$ [**2**].

2.2.2. *Computational methods for finding algebraic data.* If one has an anyon model with symmetry, the direct way to determine the G-crossed categorical data is to look for unitary solutions to the G-crossed consistency equations. This is how the algebraic data for the examples in Section 3 were found [**2**]. Of course in general it is a difficult problem to solve the consistency equations for a UMTC, let alone for a G-extension.

Another approach to deriving the algebraic data for a symmetry defect model is to make use of "two-step" gauging [**9**]. For a UMTC $\mathcal{C}$ with a symmetry G, there is a two-part procedure to gauge the symmetry G and produce another UMTC $\mathcal{D}$. The first step in this process is to take a G-crossed extension $\mathcal{C}_G^\times$ of $\mathcal{C}$; the second is to G-equivariantize $\mathcal{C}_G^\times$ to get the category $\mathcal{D} = \left(\mathcal{C}_G^\times\right)^G$. Equivariantization is a functor of fusion categories for which there is an inverse functor, called de-equivariantization. De-equivariantization is the process of passing from a fusion category $\mathcal{D}$ containing $\text{Rep}(G)$ as a symmetric fusion subcategory to the category of A-modules over $\mathcal{D}$, where $A = \text{Fun}(G, \mathbb{C})$. It is known that equivariantization and de-equivariantization establish a bijection between certain equivalence classes of GxBFCs and MTCs containing $\text{Rep}(G)$ as a symmetric fusion subcategory [**15**,**20**].

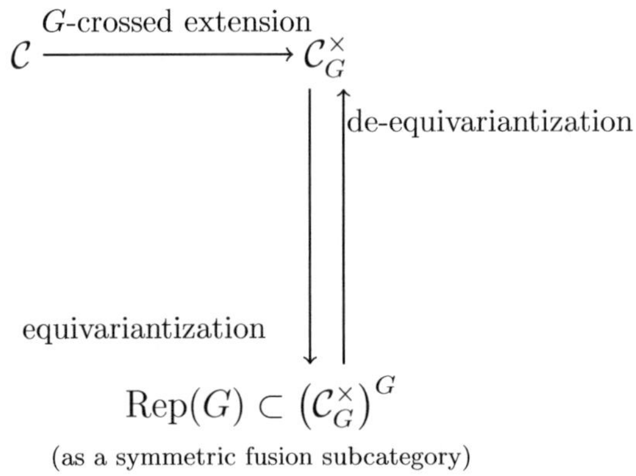

FIGURE 1. The relationship between the gauging of an MTC $\mathcal{C}$ and the GxBFC $(\mathcal{C})_G^{\times}$.

As a consequence, if one knows the gauged UMTC $\mathcal{D}$ or can calculate it from the original UMTC $\mathcal{C}$, then one can de-equivariantize $\mathcal{D}$ and calculate the algebraic data for a G-crossed theory $\mathcal{C}_G^{\times}$ explicitly [**10**]. Depending on the category this can be an easier approach to finding solutions to the G-crossed consistency equations [**12**].

2.3. Small examples for $G = \mathbb{Z}_2$. A $\mathbb{Z}_2$-extension has two graded components, the trivial sector and a nontrivial sector, $\mathcal{C}_{\mathbb{Z}_2}^{\times} = \mathcal{C}_0 \oplus \mathcal{C}_1$. It is known that the number of defects in the g-flux sector $\mathcal{C}_g$ is given by the number of fixed points of $\mathcal{C}_0$ under the action of g [**19**]. As a consequence one can adopt a convention for labeling g-defects in $\mathcal{C}_g$ by topological charges in $\mathcal{C}_0$ which are fixed under the action of g. We will use this convention with the exception of the toric code symmetry defects, for reasons that will become clear. In Section 3 we will return to these examples and demonstrate how they can be applied for quantum information processing.

2.3.1. *Toric code with electromagnetic duality.* As a UMTC the toric code has four anyon types, given by $\{1, e, m, \psi\}$. The fusion rules are given by the group $\mathbb{Z}_2 \times \mathbb{Z}_2$ with the identification $1 = (0,0)$, $e = (1,0)$, $m = (0,1)$, $\psi = (1,1)$ and have a $\mathbb{Z}_2$ symmetry coming from exchanging e and m. One can check that the action lifts to a categorical action. Then the $\mathbb{Z}_2$-grading is

$$\text{TC}_{\mathbb{Z}_2}^{\times} = \{1, e, m, \psi\} \oplus \{\sigma_+, \sigma_-\},$$

with symmetry defects $\sigma_{\pm}$. All anyons have quantum dimension equal to one, i.e. are abelian anyons, while the symmetry defects satisfy $d_{\sigma_\pm} = \sqrt{2}$ and have fusion rules with degeneracy.

The table below shows the fusion for the $\mathbb{Z}_2$-crossed category.

$\otimes$	e	m	ψ	σ_+	σ_-
e	1	ψ	m	σ_-	σ_+
m	ψ	1	e	σ_-	σ_+
ψ	m	e	1	σ_+	σ_-
σ_+	σ_-	σ_-	σ_+	$1 \oplus \psi$	$e \oplus m$
σ_-	σ_+	σ_+	σ_-	$e \oplus m$	$1 \oplus \psi$

where we have ommitted the vacuum fusion rules.

The $\mathbb{Z}_2$-crossed R-, F-, U- and η-symbols were found by solving the G-crossed consistency equations in [**2**]. We will see in the next section that braiding of the symmetry defects leads to non-abelian statistics.

2.3.2. *$\mathbb{Z}_3$ anyons with charge conjugation duality.* The UMTC $\mathbb{Z}_3$ has three anyon types, labeled by $1, \omega$, and ω^*, with fusion rules given by the group multiplication in $\mathbb{Z}_3$. The category has a $\mathbb{Z}_2$ symmetry given by interchanging ω and ω^*, which lifts to a categorical $\mathbb{Z}_2$-action and hence admits a $\mathbb{Z}_2$-crossed extension which we write

$$(\mathbb{Z}_3)^{\times}_{\mathbb{Z}_2} = \{1, \omega, \omega^*\} \oplus \{X_1\}.$$

As a fusion category, $(\mathbb{Z}_3)^{\times}_{\mathbb{Z}_2}$ is a Tambara-Yamagami (TY) category, whose algebraic data is known and relatively easy to describe [**23**]. Briefly, given a finite group A a TY category has simple objects $A \sqcup \{m\}$ and fusion rules

$$a \otimes b = ab, \quad a \otimes m = m, \quad \text{and} \quad m \otimes m = \bigoplus_{a \in A} a.$$

A choice of F-symbols for TY categories with this fusion are determined by a choice of nondegenerate symmetric bicharacter $\chi : A \times A \to \mathbb{C}^{\times}$ and a choice of square root of $|A|$. One can check that in order to correspond to a unitary $\mathbb{Z}_2$-crossed BFC the F-symbols are determined by a choice of a primitive cube root of unity. For calculations in the next section we choose χ to take the value $\xi = e^{2\pi i/3}$ on the non-identity diagonal elements of $\mathbb{Z}_3 \times \mathbb{Z}_3$ and $\bar{\xi}$ on the off-diagonal elements.

For $A = \mathbb{Z}_3$ the fusion is given by

$\otimes$	1	ω	ω^*	X_1
1	1	ω	ω^*	X_1
ω	ω	ω^*	1	X_1
ω^*	ω^*	1	ω	X_1
X_1	X_1	X_1	X_1	$1 \oplus \omega \oplus \omega^*$

Formulas for $\mathbb{Z}_2$-crossed R-symbols were found for general $\mathbb{Z}_n$ anyons in [**2**].

2.4. The G-crossed graphical calculus. Before we can apply UGxBFCs for computing we need to introduce the graphical calculus associated to their skeletonizations. The graphical calculus for defects generalizes the graphical calculus for UMTCs in the sense that whenever the topological charges involved in a diagram are all anyons, we recover the graphical calculus of a UMTC. We refer the reader to [**2, 8**] for a review of the diagrammatic calculus for anyons. As with anyons, the consistency equations that ensure the various categorical structures are compatible with one another are encoded in the topological properties of the diagrams. Here we point out only the parts of the G-crossed graphical calculus that differ from the usual one for MTCs.

Diagrams are read from bottom to top and interpreted as quantum mechanical processes between anyons and symmetry defects. For simplicity, in what follows we assume our theory is multiplicity-free and suppress the vertex labels. Recall the notation ${}^h a_g$ is shorthand for $\rho_h(a_g)$, and $\bar{g} = g^{-1}$.

Trivalent vertices encode admissible splitting/fusion processes between anyons and defects, implicitly encoding the N-matrix entries $N_{c_{gh}}^{a_g b_g}$.

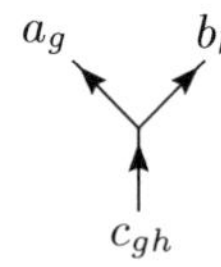

The F-symbols have the same diagrammatic definition as for anyons, except now the topological charges carry group element labels.

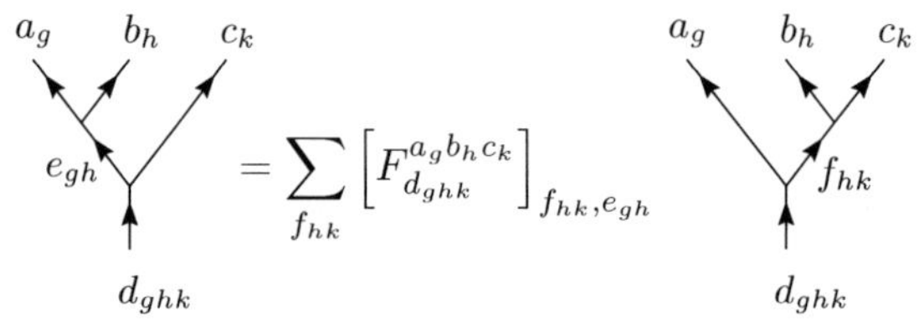

As we saw above, the algebraic data of a GxBFC differs in three main ways: the R-symbols are extended to become the matrix entries of a G-braiding, and new symbols U and η arise.

The G-crossed R-symbols are written in the same way as in a UMTC, although they are no longer the matrix entries of a braiding isomorphism $c_{X,Y} : X \otimes Y \to Y \otimes X$ but a G-crossed braiding, $c_{X,Y} : X \otimes Y \to {}^g Y \otimes X$.

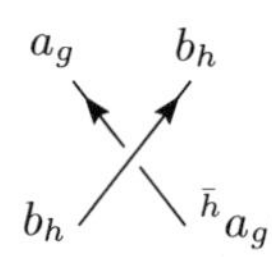

Crossings of defect lines are resolved in the same manner as for anyons, except now the R-symbols are G-crossed.

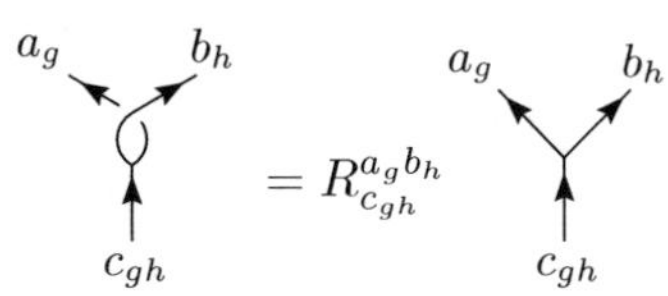

The picture is similar for a left-handed crossing, but with a factor of $\left(R_{c_{gh}}^{a_g b_h}\right)^{-1}$.

Now instead of satisfying a hexagon, the G-crossed R-symbols satisfy G-crossed heptagon equations [**2**].

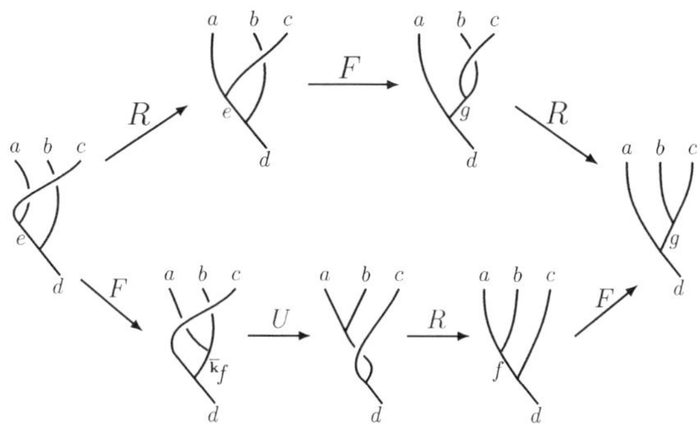

A similar picture encodes compatibility of the associators with R^{-1}.

While for anyons strands can be freely moved over and under vertices, U- and η- symbols are picked up when a strand labeled by a symmetry defect passes over or under a trivalent vertex, respectively. Sliding a defect over a splitting vertex results in a U-symbol as follows.

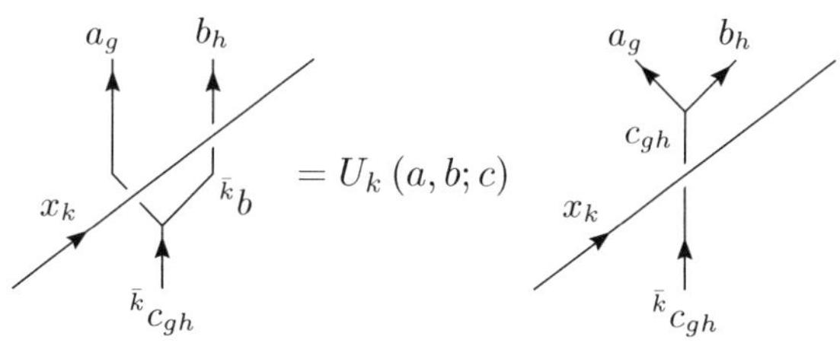

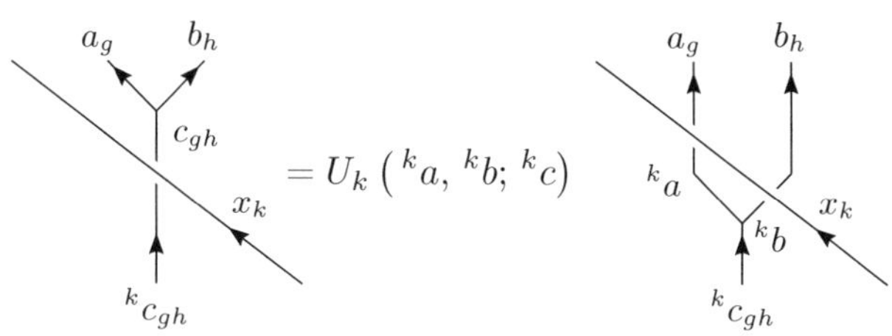

The picture is similar for fusion vertices, see [**2**]. Defect lines can be slid under splitting vertices at the cost of an η-symbol as follows.

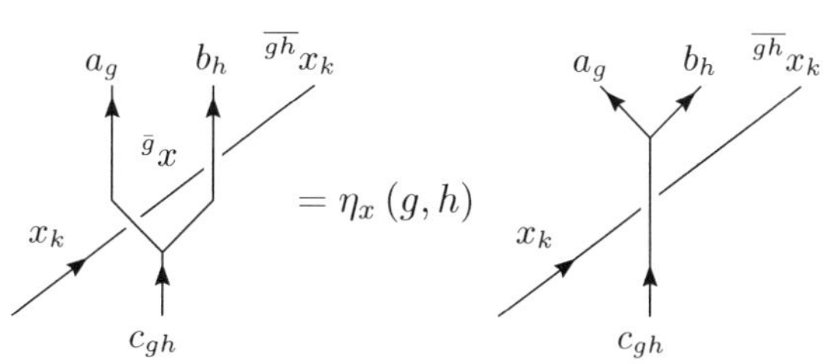

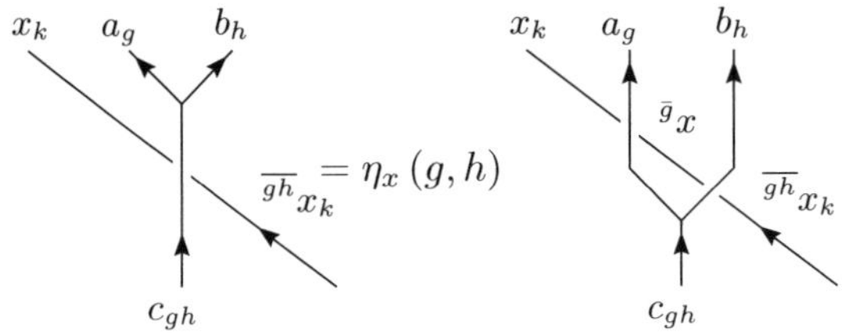

As with the U-symbols there is a similar picture for sliding defect lines under fusion vertices. Notice that the orientation of the g-defect with respect to the the trivalent vertex results in different arguments to the U- and η-symbols.

In practice, understanding these moves of diagrams is sufficient to work in the G-extended graphical calculus. The G-crossed consistency equations enforcing compatibility of all of these structures are quite complicated, and the full set of such equations is listed in [**2**], although there is some redundancy. In particular, many of the diagrams involving compatibility of the U- and η- symbols can be derived from the G-crossed pentagons and heptagons, and hence the consistency equations do not encode additional information [**7**].

3. Topological quantum computing with symmetry defects

The mathematical formalism of topological quantum computation using UGxBFCs proceeds along the same lines as for UMTCs. Qudits are encoded in the same manner, namely as subspaces of Hom spaces of the category, and the associated braid group representations have essentially the same definition, only now they are projective. In this section we describe how to encode quantum information in the state spaces of collections of symmetry defects and how to produce quantum gates from braiding symmetry defects. To illustrate the theory we analyze some quantum gate sets arising from the examples of Section 2.

3.1. Qudit encodings. Let $\mathcal{C}_G^\times = \bigoplus_{g\in G}\mathcal{C}_g$ denote a UGxBFC. Let x_{g_i} denote $n+1$ simple objects in $\mathcal{C}_{g_i}$, with $0 \le i \le n$. Then $V_{x_{g_0},\ldots,x_{g_n}} = \mathrm{Hom}\left(x_{g_0}, \bigotimes_{i=1}^n x_{g_i}\right)$ is a finite-dimensional vector space that represents the state space of the objects (anyons and defects) $\{x_{g_i}\}_{i\ge 1}$ with total charge x_{g_0}. Typically for applications to quantum computing one takes all topological charges to be identical, and considers vector spaces of the form $V_{n,x_g}^{y_h} := \mathrm{Hom}(y_h, x_g^{\otimes n})$.

Then $V_{n,x_g}^{y_h}$ is the Hilbert space of states associated to a collection of n symmetry defects $x_g \in \mathcal{C}_g$ with total charge $y_h \in \mathcal{C}_h$ on a surface with trivial topology. We identify a d-dimensional qudit $\mathbb{C}^d$ with $V_{n,x_g}^{y_h}$.

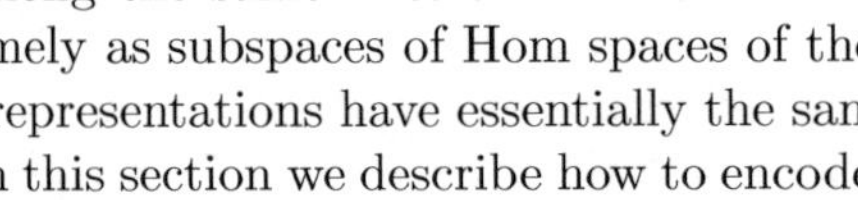

As we will see in the next subsection these vector spaces give rise to a projective representation of the n-strand braid group $\mathcal{B}_n$.

$$\mathcal{B}_n = \left\langle \sigma_1, \sigma_2, \ldots, \sigma_{n-1} \;\middle|\; \begin{array}{l} \sigma_i\sigma_j = \sigma_j\sigma_i \text{ for } |i-j| \ge 2 \\ \sigma_i\sigma_{i+1}\sigma_i = \sigma_{i+1}\sigma_i\sigma_{i+1}, i = 1, 2, \ldots, n-1 \end{array} \right\rangle.$$

In general $V_{x_{g_0},\ldots,x_{g_n}}$ will not be a representation of the braid group, or even the pure braid group. However, it is a representation of the subgroup of $\mathcal{B}_n$ that preserves the data of the topological charges x_{g_i}.

As with anyons, for physical reasons one is typically interested in state spaces with trivial total charge. When all n simple objects x_{g_i} live in $\mathcal{C}_0$, i.e. are anyons, then the formalism reduces to TQC with anyons, see [**8**].

3.2. Projective braid group representations. Next we define the representations associated to n symmetry defects x_g with total charge y_h. When the n topological charges are identical, there is an action of the n-strand braid group on $\mathrm{Hom}(y_h, x_g^{\otimes n})$. An orthonormal basis of qudit states is given by fusion trees whose internal edges are admissibly labeled by simple objects in $\mathcal{C}_G^\times$, normalized with respect to the Markov trace inner product.

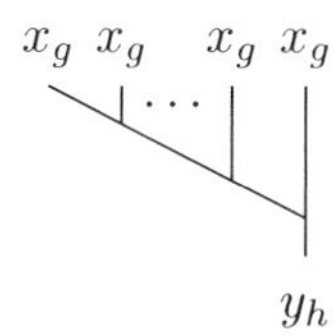

In the remainder of the paper we suppress the arrows indicating the orientations when drawing fusion trees, and assume all diagrams are oriented from the bottom up.

The action of the braid group is defined exactly the same as for anyons. Abstractly, each element of the braid group is interpreted as a morphism $b : x_g^{\otimes n} \to x_g^{\otimes n}$. Then elements of the braid group act on a fusion tree basis element $f \in V_{n,x_g}^{y_h}$ by postcomposition:

$$b \cdot f = b \circ f.$$

On the level of diagrams, elements of the braid group $b \in \mathcal{B}_n$ act on an admissibly labeled fusion trees by stacking the diagram $\boxed{b}$ of b on the fusion tree.

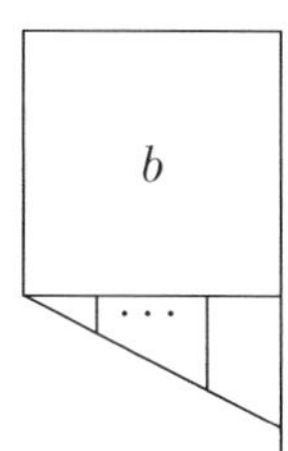

By extending linearly to all of $V_{n,x_g}^{y_h}$ this gives a projective representation ρ_{n,x_g,y_h}, which we refer to as the representation corresponding to n symmetry defects of type x_g with total charge y_h.

The matrix representation in the fusion tree basis is then given by resolving $b \cdot f$ as a linear combination of admissibly labeled fusion trees using the graphical calculus. (Or in practice, by taking inner products as in Section 4.) However, this defines a braid group representation which is not linear in general, only projective. This is due to symmetry defects being confined, see [**2**]. However, since quantum mechanics is insensitive to a global phase already, even when analyzing representations coming from UMTCs one is only interested in the projective image. That means we can meaningfully compare quantum computers designed using either anyons or symmetry defects.

3.2.1. *Universality.* For applications to quantum computing one desires to be able to approximate any unitary matrix in $U(d)$ to arbitrary accuracy with only polynomially many matrices in some small gate set $S \subset U(d)$.

For gate sets coming from braiding n symmetry defects x_g with total charge y_h, this translates into the question of whether the projective image of the braid group representation $V^{y_h}_{n,x_g}$ is dense in the unitary group. That is, one wants $\overline{\rho_{n,x_g,y_h}(\mathcal{B}_n)} \supset SU(d)$ projectively.

Thus the problem of analyzing the computational power of a symmetry defect model boils down to studying the images of its associated projective braid group representations in the unitary group. As a result all of the techniques and results used to study quantum gates in general quantum information apply. In particular, standard results about optimality of encodings and leakage out of computational subspaces carry over without modification [**1**].

For application to quantum information one also needs to know how to represent non-unitary physical processes, for example projective measurement of topological charge.

3.2.2. *Projective measurement.* Not only is projective measurement the way to perform read-out of a topological quantum computer, but it is also used in the bulk of computation as a way to force measurement outcomes of subroutines. Projective measurement can also be leveraged to simulate physical braiding along with the help of ancillary qudits [**4**]. This may be especially relevant for symmetry defects, which behave differently energetically than anyons and whose physical realizations may not be as conducive to physical braiding [**2, 6**]. As it stands many state-of-the-art proposals for anyonic hardware rely on the measurement-only formalism, see for example [**16**].

On the level of diagrams, a projective measurement of symmetry defects a_g and b_h onto c_{gh} is given by

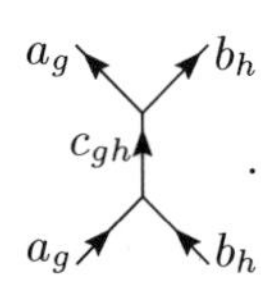

.

The matrix representation of this morphism can be calculated by specifying a basis of fusion trees, then stacking the diagram and resolving it as a linear combination of basis elements via the rules of the graphical calculus, exactly as one would for a unitary process. The example in Section 4 incorporates this technique.

Having covered the construction of quantum gates from symmetry defects we return to the small examples introduced in the previous section.

3.3. Toric code with electromagnetic duality. The F-, G-crossed R-, U-, and η-symbols were found in [**2**] by making a gauge choice and solving the G-crossed consistency equations. It was also observed that the fusion subcategories $\{1, \psi, \sigma_+\}$ and $\{1, \psi, \sigma_-\}$ have Ising fusion rules. We show that this relationship extends even further: a toric code defect qubit is identical to an Ising qubit.

We consider the projective representation of the four-strand braid group $\mathcal{B}_4$ afforded by braiding four symmetry defects σ_+ with trivial total charge. The fusion space $V^1_{4,\sigma_+} = \mathrm{Hom}(1, \sigma_+^{\otimes 4})$ is two-dimensional, corresponding to the admissible labelings of the following fusion tree.

$|a\rangle =$

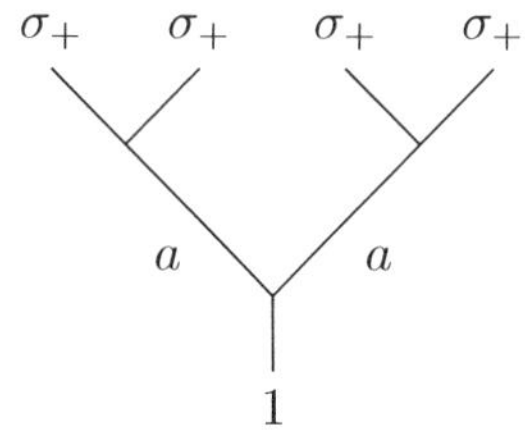

Thus the symmetry defects encode a qubit with basis states $|1\rangle$ and $|\psi\rangle$.

We find

$$\rho(\sigma_1) = \rho(\sigma_3) = e^{-\pi i/8}\begin{pmatrix}1 & 0\\ 0 & i\end{pmatrix}$$

and

$$\rho(\sigma_2) = \frac{e^{-\pi i/8}}{2}\begin{pmatrix}1+i & i-1\\ 1-i & 1+i\end{pmatrix}.$$

We remark that this generates the same image as the Ising qubit [**8**]. With the gauge choice made here the generating gates are actually equal on the nose to the Ising qubit gates. The Ising qubit is a Jones representation of $\mathcal{B}_4$, whose projective image is finite and given by $\mathbb{Z}_2^2 \rtimes S_3$ [**18**].

3.4. $(\mathbb{Z}_3)^X_{\mathbb{Z}_2}$ symmetry defect qutrit. We consider the projective representation of the four-strand braid group $\mathcal{B}_4$ afforded by braiding four symmetry defects X_ω with trivial total charge. The fusion space $V^1_{4,X_\omega} = \mathrm{Hom}(1, X_\omega^4)$ is three-dimensional, corresponding to the admissible labelings of the following fusion tree.

$|ab\rangle =$

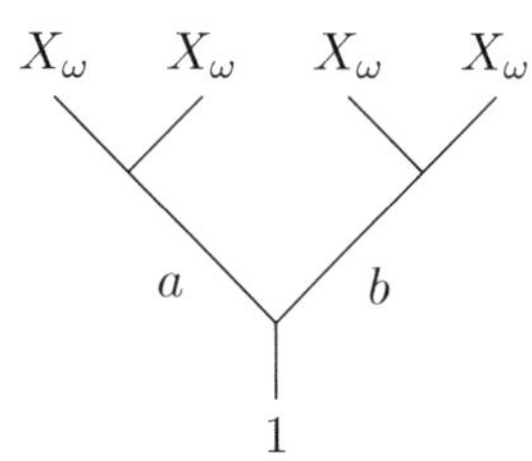

We work in the ordered basis given by $\{|11\rangle, |\omega\omega^*\rangle, |\omega^*\omega\rangle\}$. We find

$$\rho(\sigma_1) = \rho(\sigma_3) \sim \begin{pmatrix}R_1^{X_\omega X_\omega} & 0 & 0\\ 0 & R_\omega^{X_\omega X_\omega} & 0\\ 0 & 0 & R_{\omega^*}^{X_\omega X_\omega}\end{pmatrix}$$

and

$$\rho(\sigma_2) \sim \frac{1}{3\sqrt{3}}\begin{pmatrix}1 & \frac{1}{2}(-1+i\sqrt{3}) & \frac{1}{2}(-1+i\sqrt{3})\\ \frac{1}{2}(-1+i\sqrt{3}) & 1 & \frac{1}{2}(-1+i\sqrt{3})\\ \frac{1}{2}(-1+i\sqrt{3}) & \frac{1}{2}(-1+i\sqrt{3}) & 1\end{pmatrix}.$$

We remark that these matrices form a subset of the matrices generating the Jones representation of $\mathcal{B}_4$ coming from anyon 1 and total charge 2 in the $\mathbb{Z}_2$-gauged theory $\mathbb{Z}_3^{\mathbb{Z}_2} = TL_4$, the Temperley-Lieb-Jones algebroid at $A = ie^{-\pi i/12}$. In particular the image of the projective representation is finite [**18**].

So far all of our examples have been concerned with what happens when one is interested in defect-defect braiding alone. Next we go more in depth with an example that shows how to incorporate anyon-defect braiding and measurement into the mathematical formalism.

4. Bilayer Ising with $\mathbb{Z}_2$ layer-exchange symmetry

In this section we outline how to get a T-gate from a $\mathbb{Z}_2$-extension of the category $\mathbf{Ising}^{(1)} \boxtimes \mathbf{Ising}^{(1)}$. This follows the work of physicists who argued that certain types of topological defects called "genons" that arise in bilayer Ising anyon models can have non-abelian statistics [**6**]. It is important to note that there are eight inequivalent MTCs of rank 3 with nonintegral quantum dimensions which all go by the name of Ising. These are the categories $\mathbf{Ising}^{(\nu)}$ with isomorphism classes of simple objects $\{1, \sigma, \psi\}$, where ν parametrizes the quantum twist of the anyon σ. These lead to 20 inequivalent MTCs of the form $\mathbf{Ising}^{(\nu_1)} \boxtimes \mathbf{Ising}^{(\nu_2)}$ [**5**]. Here we take $\nu_1 = \nu_2 = 1$, and by **Ising** we mean $\mathbf{Ising}^{(1)}$, the category with modular data given in [**8**].

4.1. Towards a universal quantum computer based on Ising anyons. Prior to the description of symmetry defects using UGxBFCs [**2**], physicists argued that certain types of symmetry defects called *genons* could give rise to non-abelian braiding statistics and thus could be used together with anyons for quantum information processing [**6**].

Braiding Ising anyons alone is not universal for quantum computation: in particular, the T-gate is missing for universal single-qubit operations. The authors of [**6**] designed a protocol that addressed this shortcoming in another way, by constructing a T-gate from braiding and measurement of $\mathbb{Z}_2$ symmetry defects associated to two layers of the Ising theory with layer-exchange symmetry. More precisely, they furnished a universal gate set $\{T = G_1, G_2, G_3\}$ which can be realized by braiding anyons and defects. Mathematically, this is described by a $\mathbb{Z}_2$-extension of the Deligne product of the category **Ising** with itself, $(\mathbf{Ising} \boxtimes \mathbf{Ising})_G^\times$.

It is an open question to find a full solution to the G-crossed consistency equations for $(\mathbf{Ising} \boxtimes \mathbf{Ising})_G^\times$. However, under certain physically motivated assumptions about the algebraic data defining this $\mathbb{Z}_2$-extension, we use the mathematical formalism of fusion categories to verify that the physical protocol of [**6**] produces a T-gate.

A mathematical treatment of this proposal proceeds by formulating the hypothetical data as a conjecture about a symmetry defect X_1 in the category $(\mathbf{Ising} \boxtimes \mathbf{Ising})_{\mathbb{Z}_2}^\times$. For the motivation behind the conjecture see [**6**].

CONJECTURE 4.1. *There exists a set of solutions to the $\mathbb{Z}_2$-crossed consistency equations associated to* ***Ising*** $\boxtimes$ ***Ising*** *such that*

$$R_{a\boxtimes a}^{X_1 X_1} = \theta_a \quad \text{for } a \in \textbf{\textit{Ising}}$$

and

$$[F_{X_1}^{X_1 X_1 X_1}] = S_{\textbf{\textit{Ising}}} = \frac{1}{2}\begin{pmatrix} 1 & \sqrt{2} & 1 \\ \sqrt{2} & 0 & -\sqrt{2} \\ 1 & -\sqrt{2} & 1 \end{pmatrix}.$$

Our analysis etablishes the following corollary of the conjecture.

COROLLARY 4.2 (To Conjecture 4.1). *Braiding and projective measurement of bilayer Ising anyons together with $\mathbb{Z}_2$ symmetry defects is universal for quantum computation.*

Should Conjecture 4.1 hold, our analysis provides mathematical justification for including symmetry defects in the quantum information scientist's toolkit for producing universal gate sets from topological phases of matter.

In the remainder of this section we outline the proof of the corollary, which involves encoding the protocol in the morphism space of the UGxBFC given by $(\mathbf{Ising} \boxtimes \mathbf{Ising})^{\times}_{\mathbb{Z}_2}$ and calculating its matrix representation. Due to the projective measurement, the protocol is not a braid, and hence the T-gate is not in the projective braid group image. But the algebraic data of the G-crossed category nevertheless affords a well-defined matrix representation corresponding to the physical process.

4.2. The category $(\mathbf{Ising} \boxtimes \mathbf{Ising})^{\times}_{\mathbb{Z}_2}$. As a $\mathbb{Z}_2$-graded category,

$$\left(\mathbf{Ising}^{\boxtimes 2}\right)^{\times}_{\mathbb{Z}_2} = \bigoplus_{a,b\in\{1,\sigma,\psi\}} \{a \boxtimes b\} \bigoplus \{X_1, X_\sigma, X_\psi\}.$$

That is, there are 12 isomorphism classes of simple objects: 9 anyon types and 3 symmetry defect types. The defects are labeled X_a for the fixed points $a \boxtimes a$ of the layer-exchange symmetry.

The quantum dimensions are given by

$$d_{11} = d_{1\psi} = d_{\psi 1} = d_{\psi\psi} = 1,\ d_{1\sigma} = d_{\sigma 1} = d_{\sigma\psi} = d_{\psi\sigma} = \sqrt{2},\ d_{\sigma\sigma} = 2$$

and

$$d_{X_1} = d_{X_\psi} = 2,\ d_{X_\sigma} = 2\sqrt{2}.$$

Fusion in the trivial sector is given by **Ising** fusion in each factor of the Deligne product. Among the defects the fusion channels are by given by

$\otimes$	X_1	X_σ	X_ψ
X_1	$11 \oplus \sigma\sigma \oplus \psi\psi$	$1\sigma \oplus \sigma 1 \oplus \sigma\psi \oplus \psi\sigma$	$1\psi \oplus \sigma\sigma \oplus \psi 1$
X_σ	$1\sigma \oplus \sigma 1 \oplus \sigma\psi \oplus \psi\sigma$	$11 \oplus 1\psi \oplus 2\,\sigma\sigma \oplus \psi 1 \oplus \psi\psi$	$1\sigma \oplus \sigma 1 \oplus \sigma\psi \oplus \psi\sigma$
X_ψ	$1\psi \oplus \sigma\sigma \oplus \psi 1$	$1\sigma \oplus \sigma 1 \oplus \sigma\psi \oplus \psi\sigma$	$11 \oplus \sigma\sigma \oplus \psi\psi$

while between the bilayer anyons and defects they are given by

$\otimes$	11	1σ	1ψ	$\sigma 1$	$\sigma\sigma$	$\sigma\psi$	$\psi 1$	$\psi\sigma$	$\psi\psi$
X_1	X_1	X_σ	X_ψ	X_σ	$X_\psi \oplus X_1$	X_σ	X_ψ	X_σ	X_1
X_σ	X_σ	$X_\psi \oplus X_1$	X_σ	$X_\psi \oplus X_1$	$2X_\sigma$	$X_\psi \oplus X_1$	X_σ	$X_\psi \oplus X_1$	X_σ
X_ψ	X_ψ	X_σ	X_1	X_σ	$X_\psi \oplus X_1$	X_σ	X_1	X_σ	X_ψ

In what follows we will focus on the defect X_1 and the fusion channel

$$X_1 \otimes X_1 = 11 \oplus \sigma\sigma \oplus \psi\psi.$$

4.3. Qubit and T-gate encoding. The single-qubit T-gate will be realized as the logical operation enacted on a two-dimensional subspace $\mathcal{H}_{log}$ of a physical Hilbert space $\mathcal{H}_{phys}$ associated to the fusion space of a collection of four monolayer Ising anyons $\sigma 1$ and four symmetry defects X_1 with trivial total charge. That is,

$$\mathcal{H}_{phys} = \mathrm{Hom}\left(11, \sigma 1^{\otimes 4} \otimes X_1^{\otimes 4}\right)$$

as a morphism space in the category $(\mathbf{Ising} \boxtimes \mathbf{Ising})^{X}_{\mathbb{Z}_2}$.

We fix a basis $|xy\rangle$ of the full fusion space to be the set of admissibly labeled fusion trees, where $x \in \{11, \psi 1\}$ and $y \in \{11, \sigma\sigma, \psi\psi\}$.

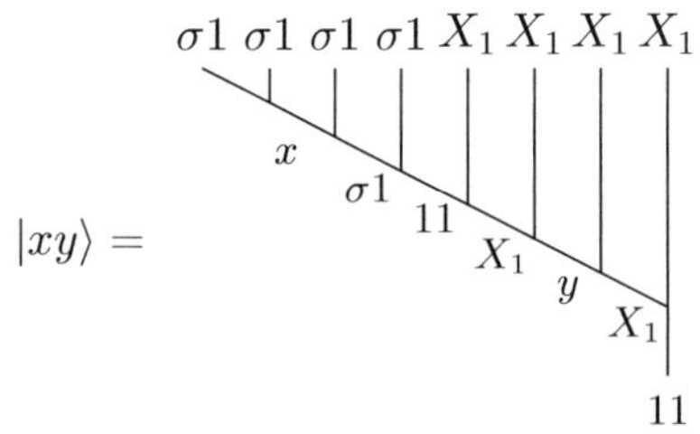

Following [**6**] the logical qubit is encoded in the $\mathbb{C}$-span of $\{|11, 11\rangle, |\psi 1, 11\rangle\}$. In general there will be leakage out of the logical subspace. In what follows we simplify the notation and suppress both the index label of the bottom layer and symmetry defect fusion channel and write $|11, 11\rangle = |1\rangle, |\psi 1, 11\rangle = |\psi\rangle$.

$$\mathcal{H}_{log} = \mathrm{Span}_{\mathbb{C}}\{|1\rangle, |\psi\rangle\}.$$

The physical protocol to initialize into this logical subspace is as follows:

(1) Create two pairs of monolayer Ising anyons $\sigma 1$ from the vacuum.
(2) Create two pairs of symmetry defects X_1 from the vacuum, using projective measurement if necessary to fix the total charge of each pair to be trivial.

The protocol to enact the logical T-gate is given by the following sequence of steps.

(3) Select one of the monolayer Ising anyons from a pair and braid it around the middle two genons in the counterclockwise direction.
(4) Fix the pair of Ising anyons from (3) to have trivial total charge with a projective measurement.
(5) Perform a full exchange of the middle two defects.
(6) Braid the anyon from (3) with the middle two defects in the clockwise direction.

Readout of the computation then proceeds by pair annihilation and projective measurement.

(7) Annihilate the pairs of Ising anyons and the pairs of defects.

Steps (3)-(6) determine a morphism in Hom $\left(\sigma^{\otimes 4} \otimes X_1^{\otimes 4}, \sigma^{\otimes 4} \otimes X_1^{\otimes 4}\right)$. In the graphical calculus the diagram for this morphism takes the form depicted in Figure 4.1.

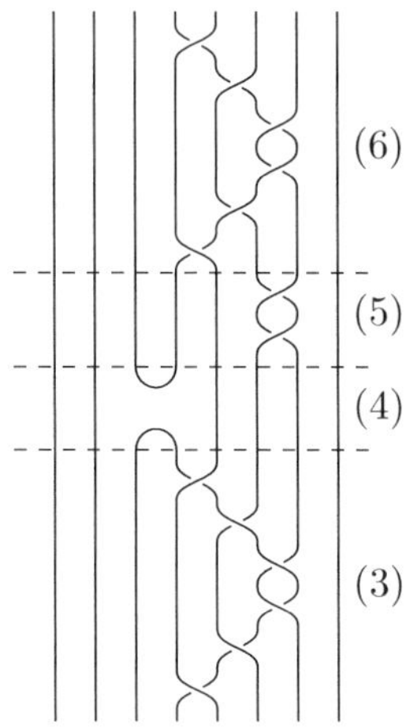

FIGURE 2. Diagrammatic representation of T-gate protocol.

The matrix entries T_{ij} of the protocol T restricted to the logical subspace can be found by calculating $\langle v_i|T|v_j\rangle$, which gives a morphism $11 \to 11$ and hence by Schur's Lemma is a scalar times the identity.

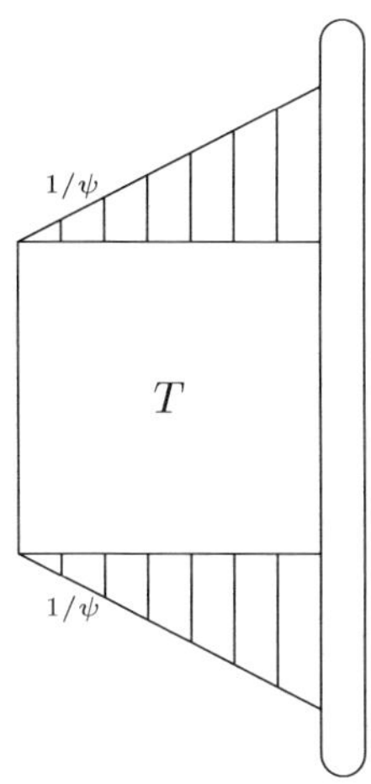

FIGURE 3. Diagrammatic representation of matrix entries.

The matrix entries are determined by using the graphical calculus to find these scalars, yielding

$$\rho(T) = \begin{pmatrix} \langle 1|T|1\rangle & \langle \psi|T|1\rangle \\ \langle 1|T|\psi\rangle & \langle \psi|T|\psi\rangle \end{pmatrix}$$

up to a normalization.

We briefly recall the tools needed for these calculations.

4.3.1. *Preliminaries for calculating matrix entries.* In addition to the S-matrix, R- and F- symbols for the UMTC **Ising ⊠ Ising**, one needs (at minimum) the R- and F-symbols specified in Conjecture 4.1. Of course there are many choices of sequences of diagrammatic moves to resolve diagrams in the graphical calculus and in general one needs all of the G-crossed data: $\{R, F, U, \eta\}$. We note however that for this specific example we were able to calculate the matrix of T in a way that is independent of U- and η- symbols.

The formulas for the matrix entries involve the S-matrix, which is defined diagrammatically by

$$S_{ab} = \frac{1}{\mathcal{D}}$$

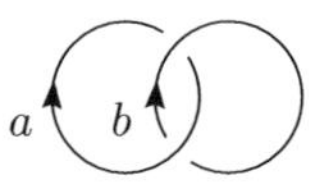

where $\mathcal{D} = \sqrt{\sum_{a \in \mathrm{Irr}(\mathcal{C})} d_a^2}$.

4.3.2. *The diagonal entries of $\rho(T)$.* By observing that the leftmost two strands in Figure 2 are through strands, it follows from Schur's lemma that the off-diagonal entries of $\rho(T)$ vanish. That is,

$$\langle \psi | T | 1 \rangle = \langle 1 | T | \psi \rangle = 0.$$

As for the diagonal entries, they can be found in terms of the algebraic data of the $\mathbb{Z}_2$-crossed category. We find that up to a normalization

$$\langle 1 | T | 1 \rangle = d_{\sigma 1} d_{X_1}^2 \sum_c \left(R_c^{X_1 X_1} \right)^2 \left| \left[F_{X_1}^{X_1 X_1 X_1} \right]_{c,11} \right|^2 \left(\frac{S_{\sigma 1, c}}{S_{11,c}} \right)^2$$

and

$$\langle \psi | T | \psi \rangle = d_{\sigma 1} d_{X_1}^2 \sum_c \left(R_c^{X_1 X_1} \right)^2 \left| \left[F_{X_1}^{X_1 X_1 X_1} \right]_{c,11} \right|^2 \left(1 - \left(\frac{S_{\psi 1, c}}{S_{11,c}} \right) \right),$$

where $c \in \{11, \sigma\sigma, \psi\psi\}$. The details of the diagrammatic calculations that produce these formulas are (1) somewhat subtle due to the additional rules of the G-crossed graphical calculus and (2) constrained by a desire to make as few assumptions as possible about the solutions to the G-crossed consistency equations, not yet being available. As a result the derivation of these formulas is a bit involved, so we have included it as an appendix.

Assuming that Conjecture 4.1 holds, one can check that

$$\langle \psi | T | \psi \rangle = e^{\pi i/4} \langle 1 | T | 1 \rangle.$$

This would confirm that the protocol implements a phase shift of $\pi/4$ between the two qubit states, effecting a T-gate

$$\rho(T) \sim \begin{pmatrix} 1 & 0 \\ 0 & e^{\pi i/4} \end{pmatrix}$$

on the logical qubit. This proves the corollary to the conjecture.

5. Discussion

When one is interested in topologically protected gate sets coming from braiding defects alone, one can analyze the image of the projective braid group representation associated to a UGxBFC as outlined in section 3. It is expected that braiding symmetry defects behaves like braiding anyons in several important ways. First, it is expected that the image of the braid group representations coming from braiding defects will be finite when the quantum dimensions are weakly integral (this is a generalization of the property-F conjecture to UGxBFCs). It is also expected that if the underlying anyons are not universal then the symmetry defects will not be universal either.

The example of the bilayer Ising model with $\mathbb{Z}_2$ symmetry defects in Section 4 indicates how to analyze the logical operations that can be realized through braiding and measurement of anyons and symmetry defects in general. In particular, our calculation supports the previous work of physicists who argued that symmetry defects could be used along with measurement to enlarge deficient gate sets to universal ones [**6**]. The calculation provides further proof-of-concept for the inclusion of symmetry defects in the quantum information theorist's toolkit, along with projective measurement, magic state distillation, and other ways to supplement deficient anyonic gate sets. However, encoding physical protocols in the language of G-crossed extensions of UMTCs is evidently quite technical.

The efforts to build a topological qubit and ultimately a scalable universal topological quantum computer are still underway. The majority of such efforts currently focus on measurement-based schemes involving Majorana zero modes (MZMs), Ising anyon-like objects that appear at the ends of nanowires placed on a superconductor in the presence of a magnetic field [**16**]. But it is within the realm of possibility that engineering a topological phase with symmetry defects could present a shorter path to a universal quantum computer.

On this note, bilayer systems are interesting to study because any topological phase can be stacked into a bilayer phase that then possesses a $\mathbb{Z}_2$ bilayer-exchange symmetry. The best understood examples of topological phases are fractional quantum Hall liquids, so it is realistic to consider what experiments can be performed to probe symmetry defects in bilayer fractional quantum Hall systems. Since fractional quantum Hall systems are electron systems, strictly speaking one needs a theory of G-extensions of super-modular categories in order to formulate the problem mathematically. One important exception is the $\nu = \frac{1}{3}$ fractional quantum Hall liquid, where the associated super-modular category splits into $SU(3)_1$ and the physical electron. In this case the system can be studied via a bilayer $SU(3)_1$ topological order. An interesting question is then how to use symmetry defects to generate a universal gate set for quantum computing from this abelian topological phase, which is available in almost every fractional quantum Hall lab.

Apart from the physical motivations, symmetry defects in the context of TQC are interesting for mathematical reasons alone. Going forward, it is highly desirable to understand the precise mathematical relationship between the (projective) braid group representations afforded by symmetry defects and the braid group representations afforded by the underlying anyons. Relatedly, it is an open question to determine the precise relationship of the braid group representations arising from a UMTC $\mathcal{C}$ and its gauging $(\mathcal{C}_G^\times)^G$. An answer to this question could help resolve the Property-F conjecture [**21**], one of most well-known open problems in modular tensor category theory.

A systematic and detailed study of TQC with symmetry defects is to appear in [**12**].

Acknowledgements

We would like to thank Maissam Barkeshli, Parsa Bonderson, and Meng Cheng for contributing figures, Xingshan Cui for sharing Mathematica code, as well as Eric Samperton and the referees for helpful feedback on the draft.

Appendices

In the appendices we provide a derivation of the matrix entries of the T-gate protocol from Section 4.2, with data and graphical formulas for reference.

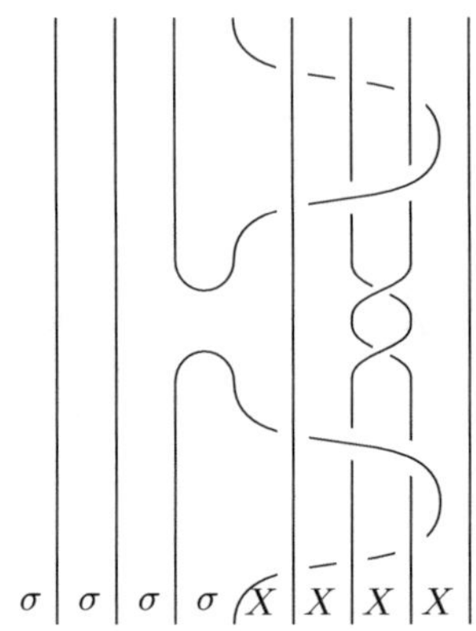

In Appendix A we list all the data needed to compute the matrix entries, including the conjectured G-crossed braiding and F-symbols for the $\mathbb{Z}_2$-extension of **Ising** ⊠ **Ising**. In Appendix B we provide a list of helpful moves in the graphical calculus for UGxBFCs (and the UMTCs they contain) for reference. In Appendix C, we demonstrate a sequence of moves in the graphical calculus that produces the formulas for the matrix entries. We find that the phase difference induced on the basis states is indeed $e^{\pi i/4}$ as anticipated by the physical argument given in [**6**].

Appendix A. Data

A.1. Algebraic data for Ising. There are three isomorphism classes of simple objects in **Ising**, labeled $\{1, \sigma, \psi\}$.

Quantum dimensions $d_1 = 1,\ d_\sigma = \sqrt{2},\quad d_\psi = 1$

Twists $\theta_1 = 1,\ \theta_\sigma = e^{2\pi i/16},\ \theta_\psi = -1$

S-matrix $$\frac{1}{2}\begin{pmatrix} 1 & \sqrt{2} & 1 \\ \sqrt{2} & 0 & -\sqrt{2} \\ 1 & -\sqrt{2} & 1 \end{pmatrix}$$

F-symbols $$F_\sigma^{\sigma\sigma\sigma} = \frac{1}{\sqrt{2}}\begin{pmatrix} 1 & 1 \\ 1 & -1 \end{pmatrix}$$

A.2. Algebraic data for Ising ⊠ Ising. We also use the quantum dimensions and S-matrix of the doubled theory **Ising ⊠ Ising**. We write objects in the bilayer category as $ab := a \boxtimes b$ for simplicity. Under the Deligne product quantum dimensions simply multiply

$$d_{ab} = d_a d_b$$

and the S-matrix respects the tensor product

$$S_{\mathbf{Ising}\boxtimes\mathbf{Ising}} = S_{\mathbf{Ising}} \otimes S_{\mathbf{Ising}}$$

but we nevertheless provide the values for convenience. The S-matrix is written with respect to the basis $\{11, 1\sigma, 1\psi, \sigma 1, \sigma\sigma, \sigma\psi, \psi 1, \psi\sigma, \psi\psi\}$.

Quantum dimensions $d_{11} = d_{1\psi} = d_{\psi 1} = d_{\psi\psi} = 1,\ d_{\sigma 1} = d_{1\sigma} = d_{\sigma\psi} = d_{\psi\sigma} = \sqrt{2},$
$d_{\sigma\sigma} = 2$

S-matrix
$$\frac{1}{4}\begin{pmatrix} 1 & \sqrt{2} & 1 & \sqrt{2} & 2 & \sqrt{2} & 1 & \sqrt{2} & 1 \\ \sqrt{2} & 0 & -\sqrt{2} & 2 & 0 & -2 & \sqrt{2} & 0 & -\sqrt{2} \\ 1 & -\sqrt{2} & 1 & \sqrt{2} & -2 & \sqrt{2} & 1 & -\sqrt{2} & 1 \\ \sqrt{2} & 2 & \sqrt{2} & 0 & 0 & 0 & -\sqrt{2} & -2 & -\sqrt{2} \\ 2 & 0 & -2 & 0 & 0 & 0 & -2 & 0 & 2 \\ \sqrt{2} & -2 & \sqrt{2} & 0 & 0 & 0 & -\sqrt{2} & 2 & -\sqrt{2} \\ 1 & \sqrt{2} & 1 & -\sqrt{2} & -2 & -\sqrt{2} & 1 & \sqrt{2} & 1 \\ \sqrt{2} & 0 & -\sqrt{2} & -2 & 0 & 2 & \sqrt{2} & 0 & -\sqrt{2} \\ 1 & -\sqrt{2} & 1 & -\sqrt{2} & 2 & -\sqrt{2} & 1 & -\sqrt{2} & 1 \end{pmatrix}$$

A.3. Algebraic data for $(\mathbf{Ising} \boxtimes \mathbf{Ising})^{\times}_{\mathbb{Z}_2}$. While the full set of $\mathbb{Z}_2$-crossed R-symbols, F-, U-, and η- symbols is not yet known, we will only use the quantum dimensions and fusion rules (which are known) along with Conjecture 4.1 which we restate below for convenience.

Quantum dimensions of defects $d_{X_1} = d_{X_\psi} = 2,\ d_{X_\sigma} = 2\sqrt{2}$
Fusion rules $X_1 \otimes X_1 = 11 \oplus \sigma\sigma \oplus \psi\psi$
(see Section 4)

Hereafter we write $X := X_1$.

CONJECTURE A.1. *There exists a set of solutions to the $\mathbb{Z}_2$-crossed consistency equations associated to **Ising ⊠ Ising** such that*

$$R^{XX}_{aa} = \theta_a \quad \text{for } a \in \textbf{\textit{Ising}}$$

and

$$[F^{XXX}_X] = S_{\textbf{\textit{Ising}}} = \frac{1}{2}\begin{pmatrix} 1 & \sqrt{2} & 1 \\ \sqrt{2} & 0 & -\sqrt{2} \\ 1 & -\sqrt{2} & 1 \end{pmatrix}.$$

Appendix B. Graphical calculus

We recall several moves involving the G-crossed R-, U-, and η-symbols from [**2**] in the case where the fusion rules of $(\mathcal{C})_G^\times$ are multiplicity-free.

(1) $R^{a_g b_h} =$ 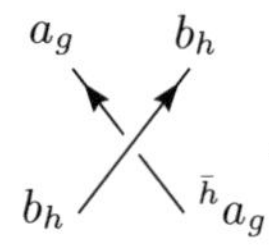

$= \sum_c \sqrt{\frac{d_{c_{gh}}}{d_{a_g} d_{b_h}}} R^{a_g b_h}_{c_{gh}}$ 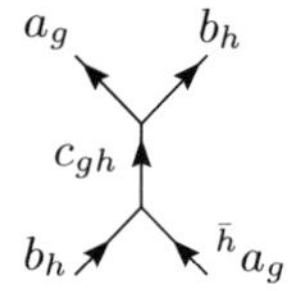

(2) $\left(R^{a_g b_h}\right)^{-1} =$ 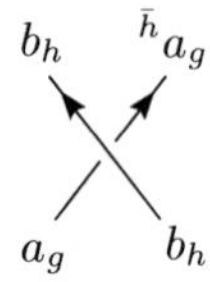

$= \sum_c \sqrt{\frac{d_{c_{gh}}}{d_{a_g} d_{b_h}}} \left(R^{a_g b_h}_{c_{gh}}\right)^{-1}$ 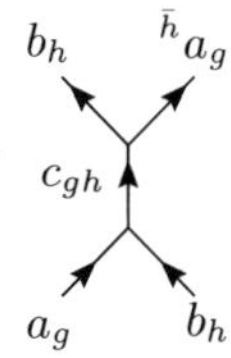

(3) 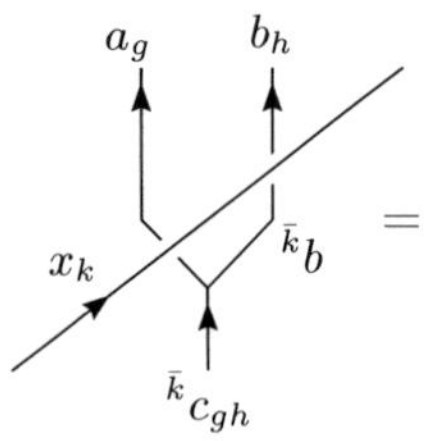

$= U_k\left(a, b; c\right)$ 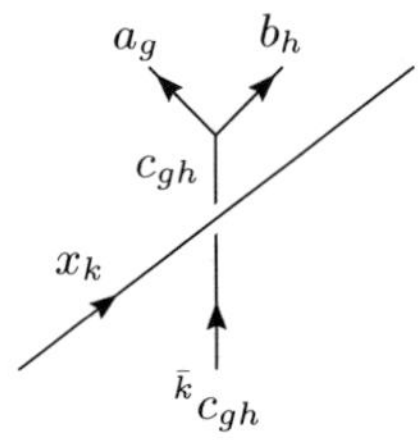

(4) 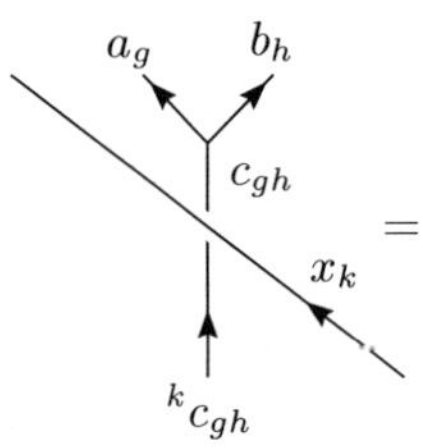

$= U_k\left({}^k a, {}^k b; {}^k c\right)$ 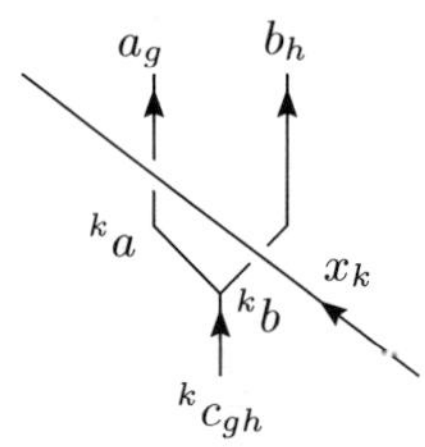

(5) 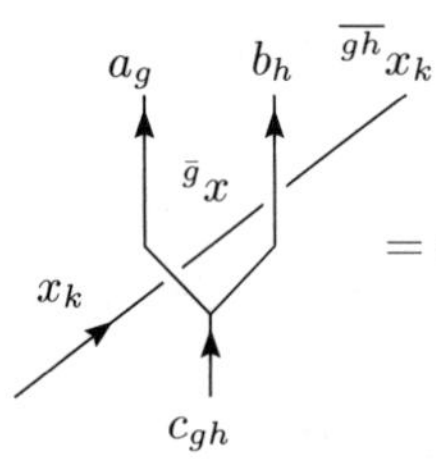

$= \eta_x\left(g, h\right)$ 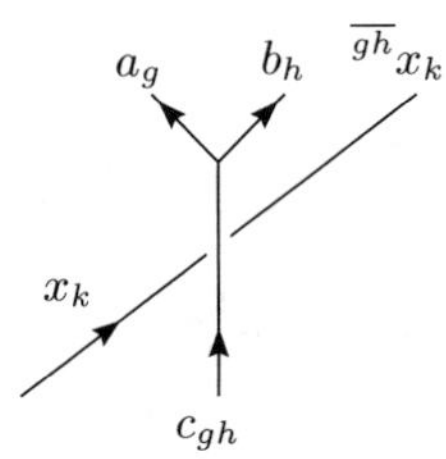

(6) 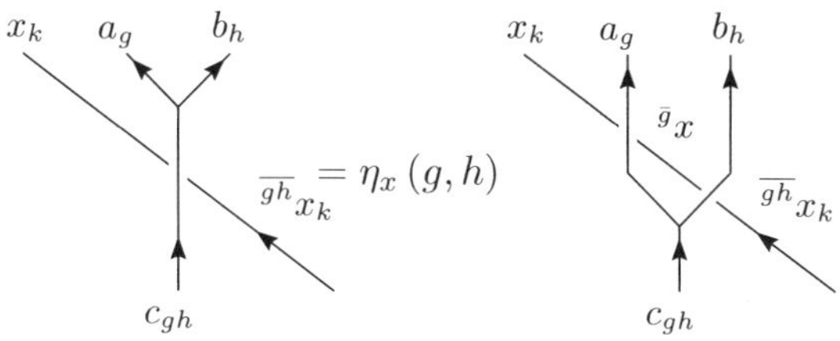

(7) 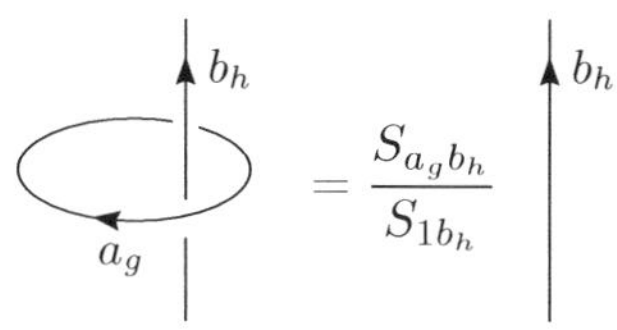

(8) 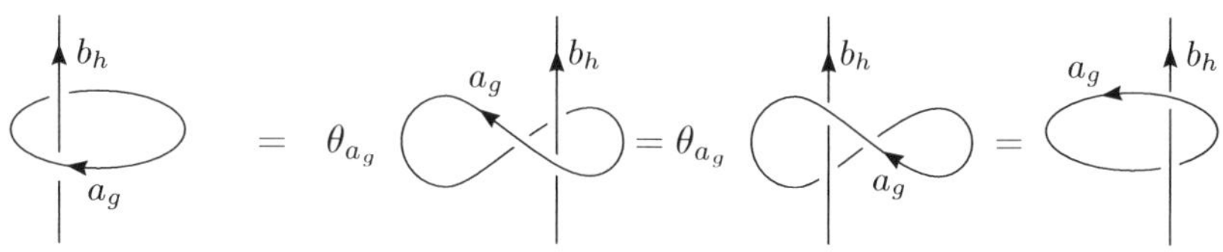

(9) 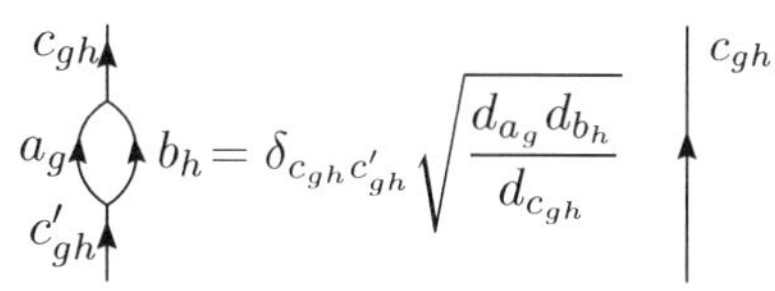

Appendix C. Calculation of matrix entries of the T-gate protocol

C.1. Qudit and protocol encoding. In terms of the graphical calculus, the qubit is given by the $\mathbb{C}$-span of the (normalized) fusion tree basis $|x\rangle$, which has the dual basis $\langle x|$.

$|x\rangle =$ 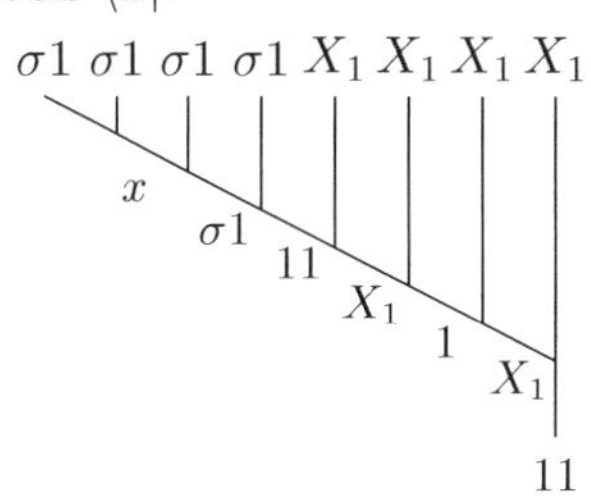

$\langle x| =$ 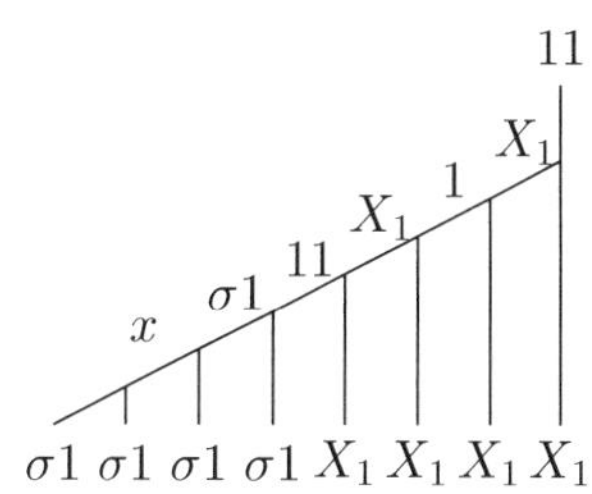 .

In terms of diagrams the matrix entries $\langle y|T|x\rangle$ are given by stacking and taking a trace. Below we show the general diagram for the matrix entries alongside the shorthand notation that we use in what follows. In particular, we suppress the index on the symmetry defect, writing $X := X_1$, and since all anyons live in a single layer, we write $a := a1$ for a a simple object in **Ising**. Moreover, since the total charge of the fusion trees are trivial, we suppress the labeling of the tracial strand.

As was mentioned in Section 4.3.2, the off diagonal matrix elements vanish by a Schur's Lemma argument, so it remains to calculate $\langle 1|T|1\rangle$ and $\langle\psi|T|\psi\rangle$.

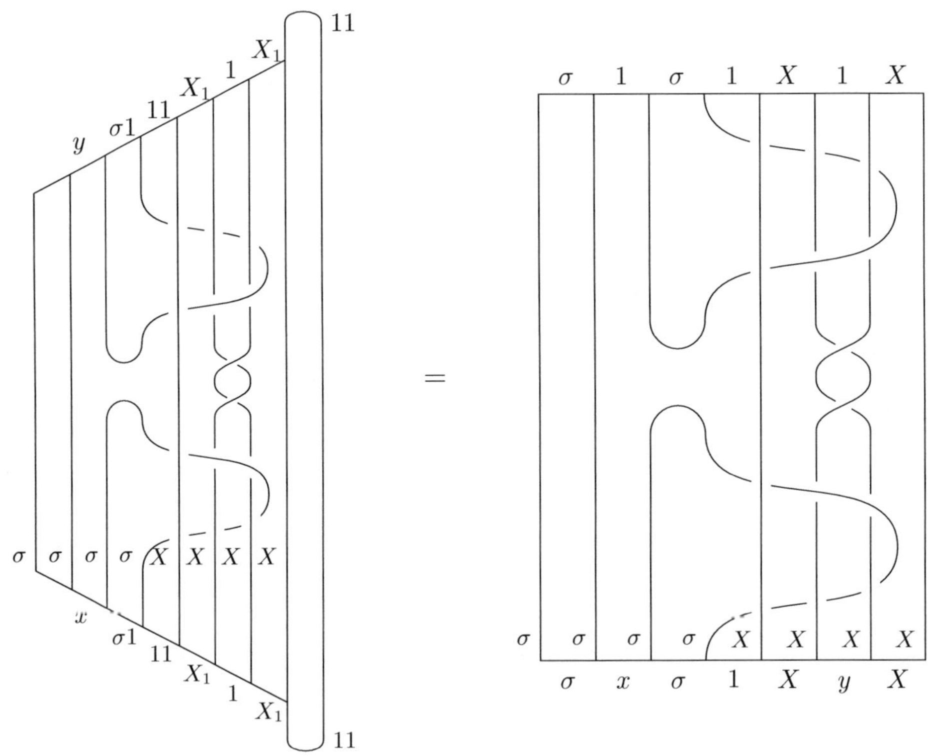

FIGURE 4. On the left is the diagram representing the matrix entries of the protocol. On the right is the shorthand notation we use in the following section.

C.2. Calculation of $\langle 1 \mid T \mid 1\rangle$. We break down the calculation of the first matrix entry (up to normalization) into five steps. Below is a formula that shows the contributions from each step, before simplification.

$$\langle 1|T|1\rangle = \underbrace{\frac{1}{d_X}\sum_c \sqrt{d_c}(R_c^{XX})^2}_{\text{Step 1}} \underbrace{\left(\frac{S_{\sigma 1,c}}{S_{11,c}}\right)^2}_{\text{Step 3}} \underbrace{\left|[F_X^{XXX}]_{c,11}\right|^2 \frac{d_X^2}{d_c}}_{\text{Step 4}} \underbrace{d_{\sigma 1} d_X \sqrt{d_c}}_{\text{Step 5}}$$

The dashed rectangles in each equation indicate the region of the diagram where the graphical calculus is being applied.
Step 1: Unbraid the symmetry defects using equation (1) twice.

(10) 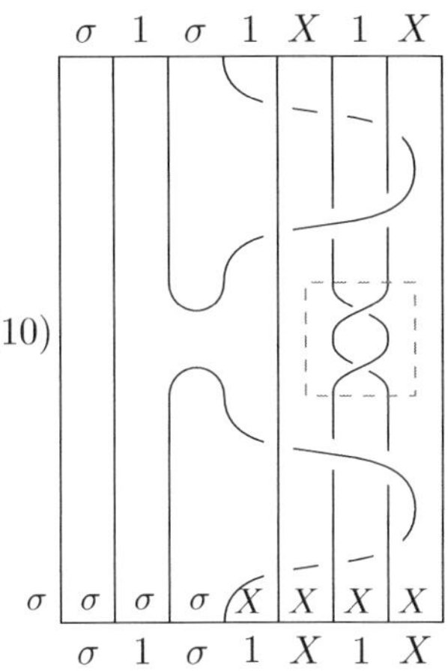

$$= \sum_c \sqrt{\frac{d_c}{d_X d_X}} \left(R^{XX}_c\right)^2$$

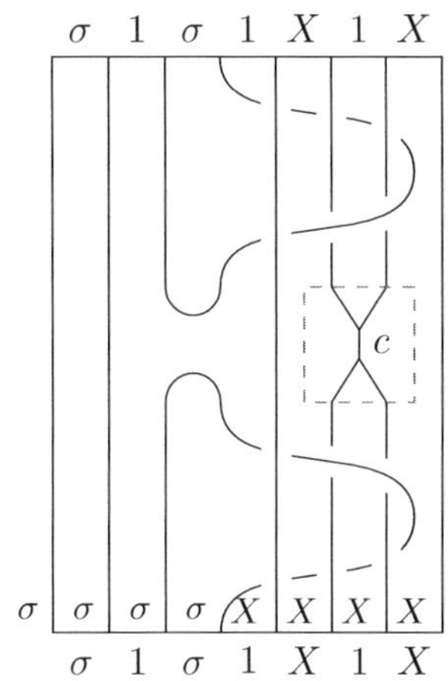

Note that in this case there is a loop labeled by σ that we can replace with d_σ right away. However, we have left it unsimplified so that the same picture applies for the calculation of $\langle\psi|T|\psi\rangle$.

Step 2: Slide the σ loops under the first defect charge line so that they encircle pairs of defect charge lines using equations (3)-(6). The U- and η- symbols this introduces all cancel, so that the anyons and defects can be moved past each other at no cost.

(11) 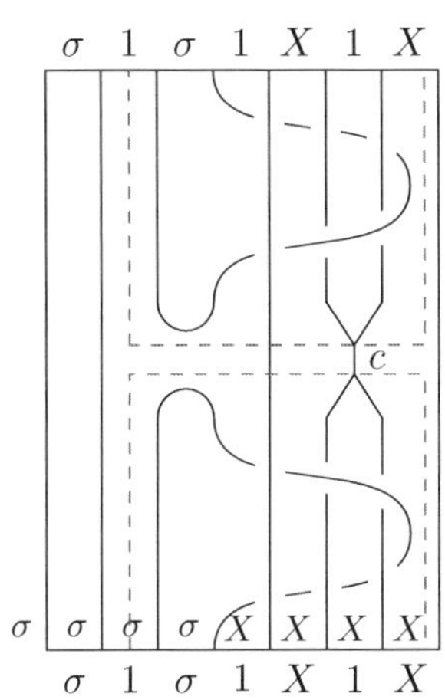

$$=$$

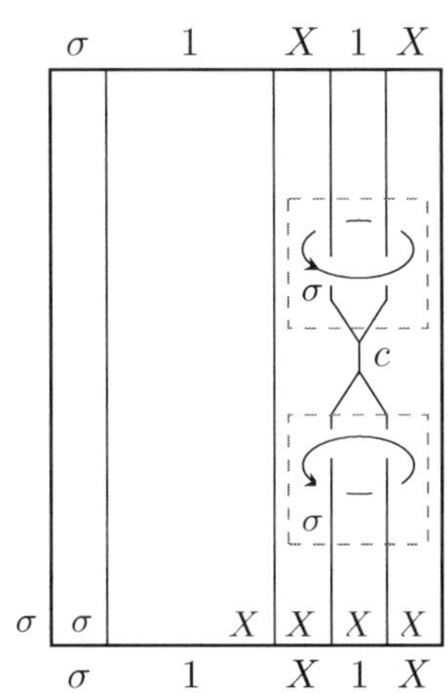

Step 3: Slide the loops over the trivalent vertices so that they encircle the fusion channel labeled by c, flip the tilt of the bottom loop using (6), then use the loop removal relation (7) twice.
In general with loops encircling charge lines in the G-crossed graphical calculus we need to pay careful attention to the relative positions of charge lines, see equations (306) and (307) in [**2**]. However, in this case the subdiagram involves only anyons, so we can use the graphical calculus of a UMTC and ignore them. Nevertheless, a careful account of the factors using the full G-extended calculus gives the same result: any factors of U- and η- cancel.

(12)

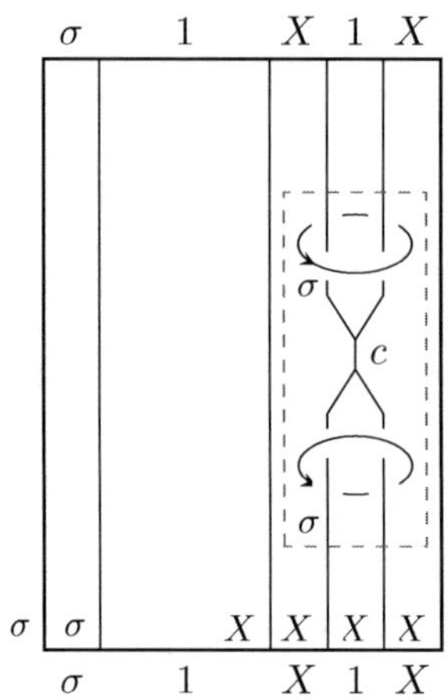

=

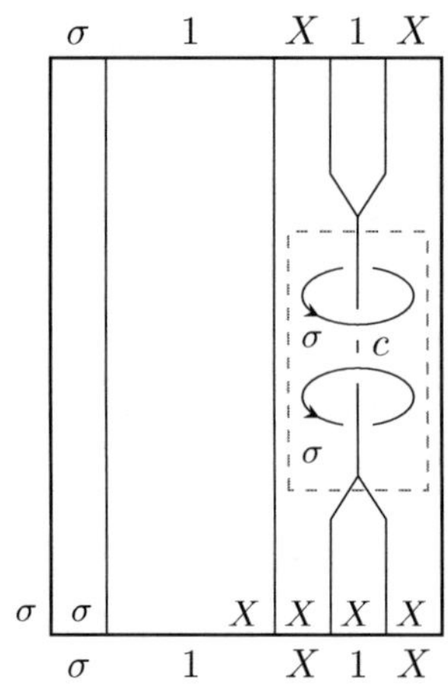

(13)

$$= \left(\frac{S_{\sigma 1,c}}{S_{11,c}}\right)^2$$

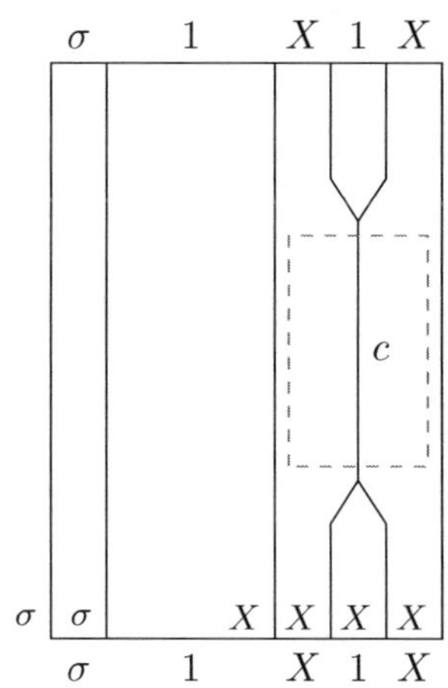

Step 4: Perform a sequence of F-moves to change the basis states, then use the bubble-popping relation (9).

(14)

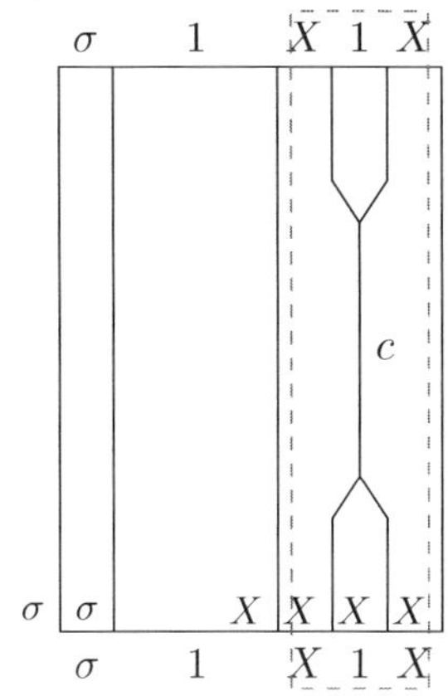

$$= \sum_{d,d'} \left[F^{XXX}_{X}\right]_{d,11} \left[F^{XXX}_{X}\right]^{*}_{d',11}$$

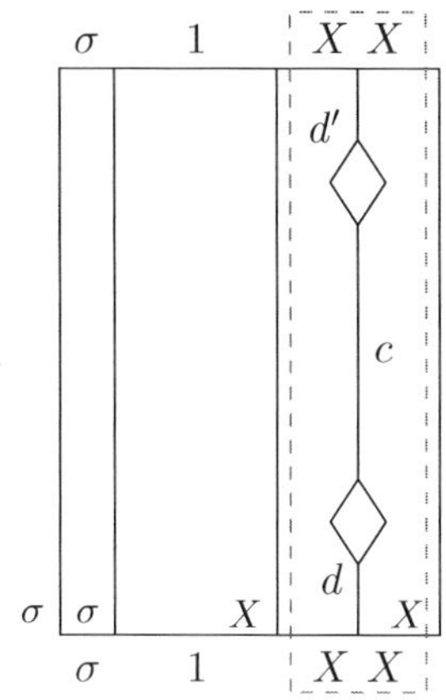

(15) $$= \sum_{d,d'} \left[F^{XXX}_{X}\right]_{d,11} \left[F^{XXX}_{X}\right]^{*}_{d',11} \left(\sqrt{\frac{d_X d_X}{d_c}}\right)^2 \delta_{dc}\delta_{d'c}$$

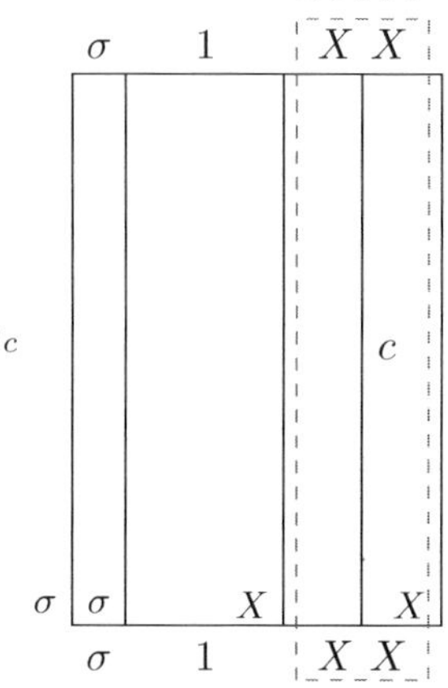

Step 5: The diagram that remains contributes a scalar factor that can be calculated using equations (4) and (5) repeatedly until the empty diagram is obtained.

(16)

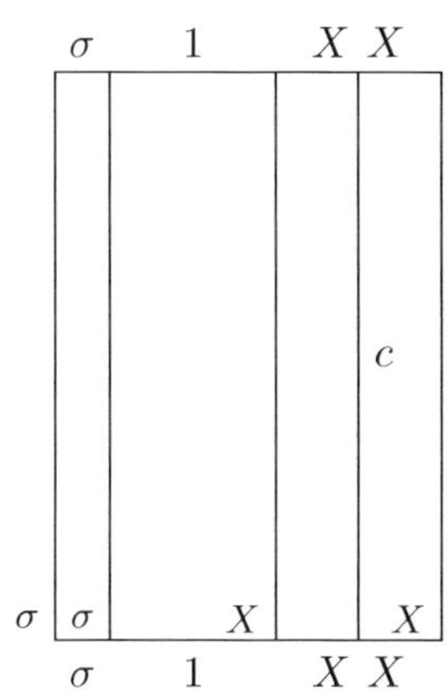

$$= d_{\sigma 1} d_X \sqrt{d_c}$$

Putting these steps together, we arrive at the formula

$$\frac{1}{d_X}\sum_c \sqrt{d_c}(R_c^{XX})^2\Big(\frac{S_{\sigma 1,c}}{S_{11,c}}\Big)^2 \left|[F_X^{XXX}]_{c,11}\right|^2 \frac{d_X^2}{d_c} d_{\sigma 1} d_X \sqrt{d_c} \tag{17}$$

$$= d_{\sigma 1} d_X^2 \sum_c (R_c^{XX})^2 \left|[F_X^{XXX}]^2_{c,11}\right|^2 \Big(\frac{S_{\sigma 1,c}}{S_{11,c}}\Big)^2 \tag{18}$$

C.3. Calculation of $\langle\psi|T|\psi\rangle$. The first step of resolving the defect braiding is identical to that of the previous subsection for $\langle 1|T|1\rangle$. However, when the fusion channel of the four monolayer Ising anyons $\sigma 1$ is given by $\psi 1$, the loops are not free to slide under the defect line as in Step 2. Instead, we perform an F-move in the middle of the diagram.

(19) 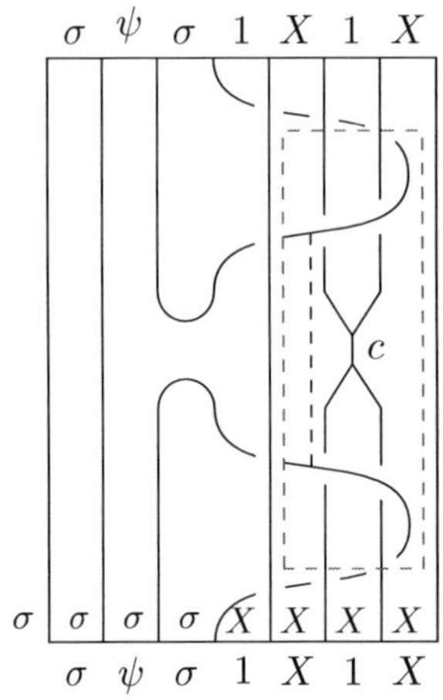 $= \sum_d [F_{\sigma 1}^{\sigma 1 \sigma 1 \sigma 1}]_{d,11}$ 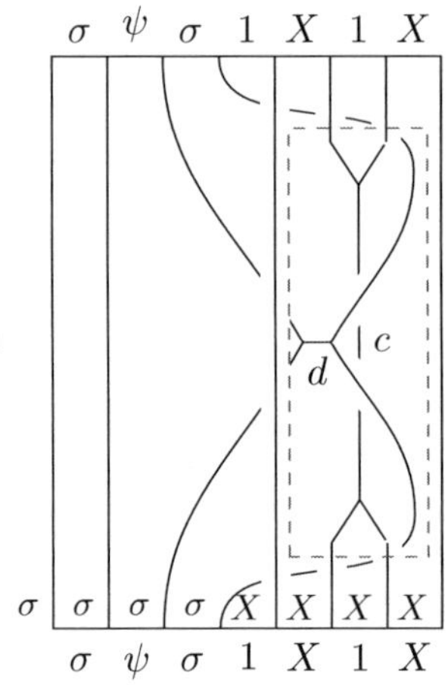

When d is the vacuum channel, $d = 11$, we can perform a sequence of sliding moves to resolve all crossings by the same arguments as in the previous subsection.

(20) 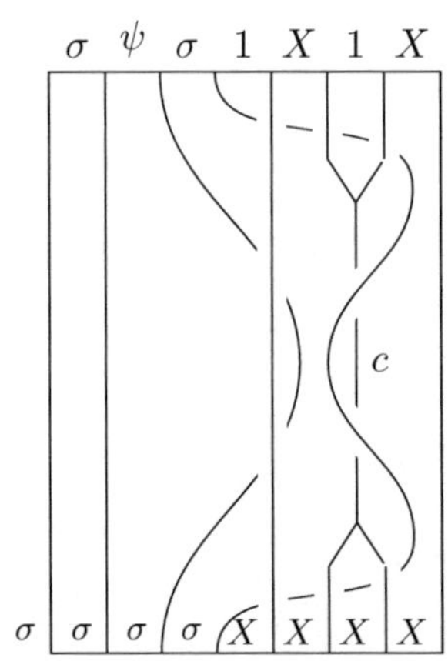 $=$ 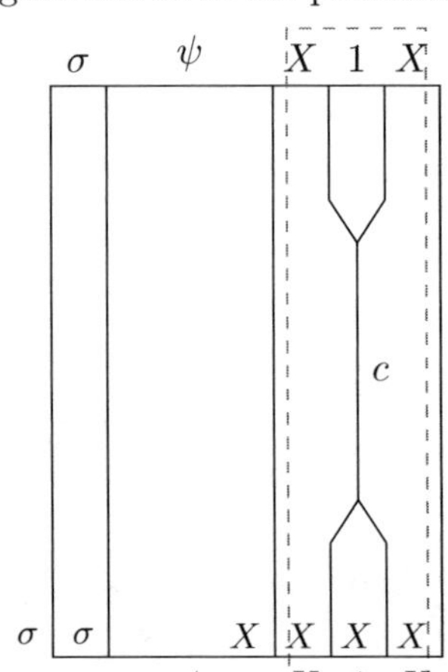

Now the righthand side can be resolved according as with steps 4 and 5 for $\langle 1|T|1\rangle$, with the minor difference that the fusion channel of the monolayer Ising anyons is given by $\psi 1$.

However, when $d = \psi 1$ we must use a more involved sequence of moves to resolve the diagram.

(21) 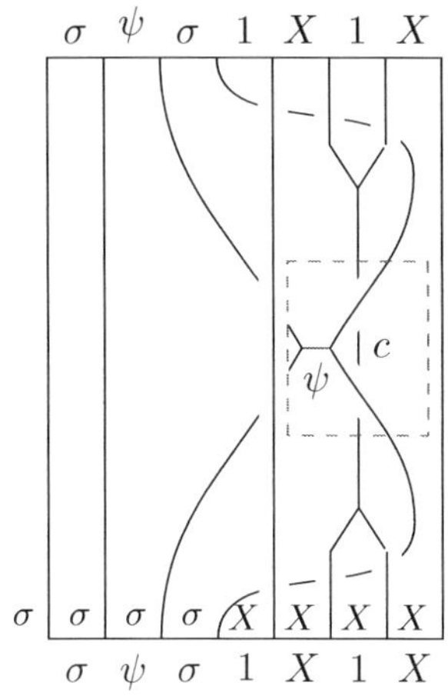

$=$ 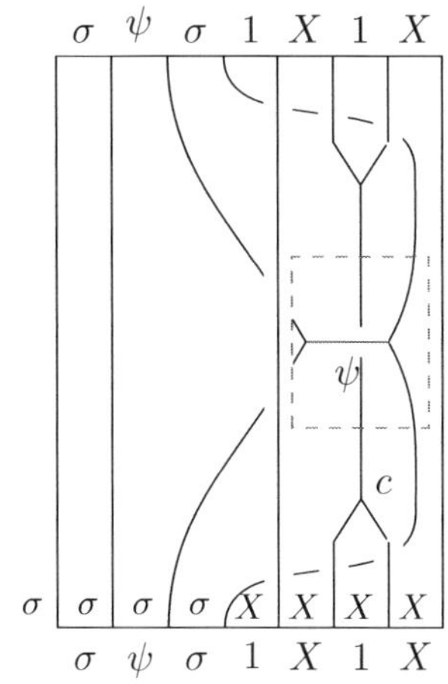

(22) $=$ 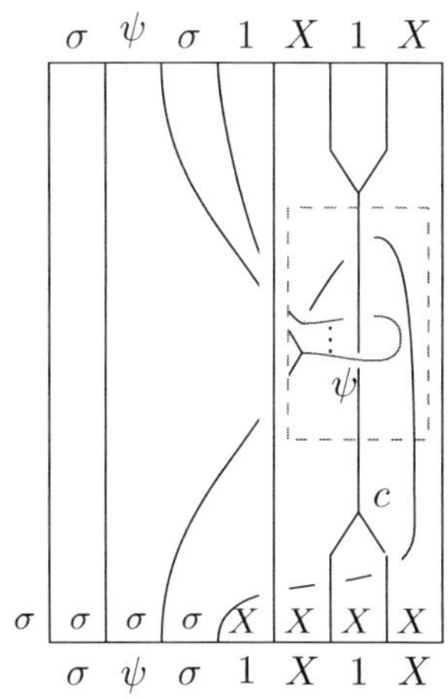

$=$ 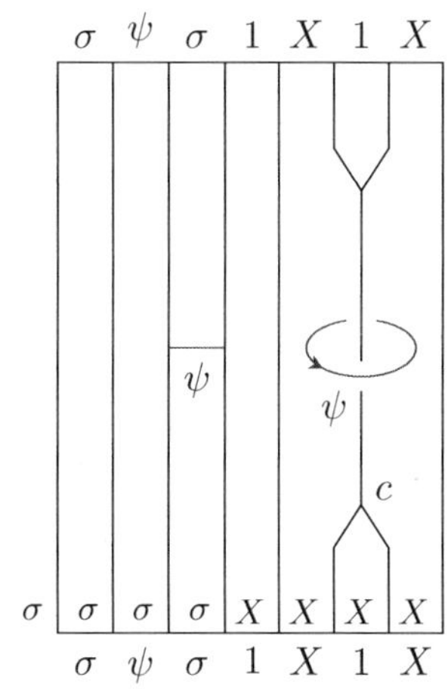

(23) $= \left(\frac{S_{\psi 1,c}}{S_{11,c}}\right)$ 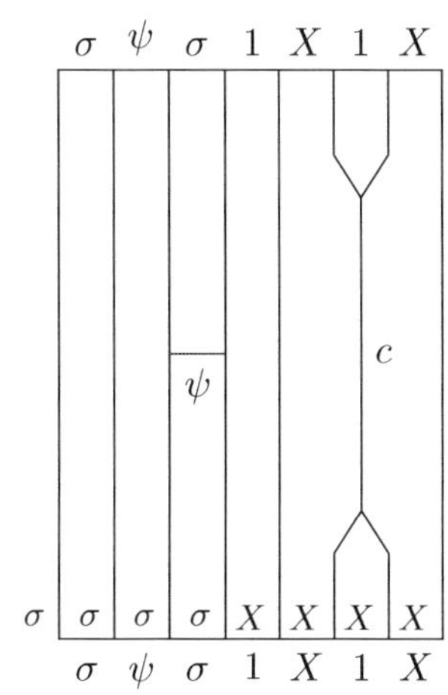

Although we have draw some horizontal lines here, the edges are understood to have the orientation they inherit from reading the diagram from the bottom up.

What remains can be resolved using a combination of F-moves and bubble-popping relations in a similar manner to the final steps of the calculation for $\langle 1|T|1\rangle$. The contributions from each step in the calculation result in the formula

(24)
$$\frac{1}{d_X}\sum_c \sqrt{d_c}(R_c^{XX})^2\left|[F_X^{XXX}]_{c,11}\right|^2\left(\sqrt{\frac{d_Xd_X}{d_c}}\right)^2\left([F_{\sigma 1}^{\sigma 1\sigma 1\sigma 1}]_{11,11}\left(d_{\sigma 1}^2\sqrt{d_c}d_X\right)\right.$$

(25)
$$\left.+\left(\frac{S_{\psi 1,c}}{S_{11,c}}\right)[F_{\sigma 1}^{\sigma 1\sigma 1\sigma 1}]_{\psi 1,11}[F_{\sigma 1}^{\sigma 1\sigma 1\sigma 1}]_{\psi 1,\psi 1}\left(\sqrt{\frac{d_{\sigma 1}d_{\sigma 1}}{d_{\psi 1}}}\right)^2\left(d_{\sigma 1}\sqrt{d_c}d_X\right)\right)$$

(26)
$$= d_{\sigma 1}d_{X_1}^2\sum_c\left(R_c^{X_1X_1}\right)^2\left|\left[F_{X_1}^{X_1X_1X_1}\right]_{c,11}\right|^2\left(1-\left(\frac{S_{\psi 1,c}}{S_{11,c}}\right)\right)$$

These formulas result in the ratio $\frac{\langle\psi|T|\psi\rangle}{\langle 1|T|1\rangle} = e^{\pi i/4}$.

We stress that the sequence of diagrammatic moves made here is constrained by what is known about the G-crossed data for the extension of the bilayer Ising category, and is only one sequence of many possible moves that calculates the same quantity. It is interesting to note that the calculation is independent of U- and η-symbols.

References

[1] R. Ainsworth and J. K. Slingerland. *Topological qubit design and leakage.* New Journal of Physics. Vol 13. (2011)

[2] M. Barkeshli, P. Bonderson, M. Cheng, Z. Wang. *Symmetry, defects, and gauging of topological phases.* arXiv preprint. arXiv:1410.4540v2. (2014).

[3] M. Barkeshli, P. Bonderson, C. M. Jian, M. Cheng, K. Walker. *Reflection and time reversal symmetry enriched topological phases of matter: path integrals, non-orientable manifolds and anomalies.* arXiv preprint. arXiv:1612.07792. (2016).

[4] Parsa Bonderson, Michael Freedman, and Chetan Nayak, *Measurement-only topological quantum computation*, Phys. Rev. Lett. **101** (2008), no. 1, 010501, 4, DOI 10.1103/PhysRevLett.101.010501. MR2429542

[5] P. Bruillard, P. Gustafson, J. Y. Plavnik, E. C. Rowell. *Dimensions as a quantum statistic and the classification of metaplectic categories.* arXiv:1710.10284 (2018).

[6] M. Barkeshli, C. M. Jian, X. L. Qi, *Genons, twist defects, and projective non-Abelian braiding statistics.* Phys. Rev. B, Vol. 87, No. 045130. (2013).

[7] A. Tran, unpublished work.

[8] Zhenghan Wang, *Topological quantum computation*, CBMS Regional Conference Series in Mathematics, vol. 112, Published for the Conference Board of the Mathematical Sciences, Washington, DC; by the American Mathematical Society, Providence, RI, 2010. MR2640343

[9] Shawn X. Cui, César Galindo, Julia Yael Plavnik, and Zhenghan Wang, *On gauging symmetry of modular categories*, Comm. Math. Phys. **348** (2016), no. 3, 1043–1064, DOI 10.1007/s00220-016-2633-8. MR3555361

[10] Shawn X. Cui, Seung-Moon Hong, and Zhenghan Wang, *Universal quantum computation with weakly integral anyons*, Quantum Inf. Process. **14** (2015), no. 8, 2687–2727, DOI 10.1007/s11128-015-1016-y. MR3370675

[11] Colleen Delaney, Eric C. Rowell, and Zhenghan Wang, *Local unitary representations of the braid group and their applications to quantum computing*, Rev. Colombiana Mat. **50** (2016), no. 2, 207–272, DOI 10.15446/recolma.v50n2.62211. MR3605648

[12] C. Delaney. *Topological quantum computation with symmetry defects.* Dissertation. (In preparation.)

[13] S. Das Sarma, M. Freedman, C. Nayak. *Majorana zero modes and topological quantum computation.* npj Quantum Information Vol. 1, No. 15001, (2015).

[14] Pavel Etingof, Dmitri Nikshych, and Victor Ostrik, *Fusion categories and homotopy theory*, Quantum Topol. **1** (2010), no. 3, 209–273, DOI 10.4171/QT/6. With an appendix by Ehud Meir. MR2677836

[15] Pavel Etingof, Shlomo Gelaki, Dmitri Nikshych, and Victor Ostrik, *Tensor categories*, Mathematical Surveys and Monographs, vol. 205, American Mathematical Society, Providence, RI, 2015. MR3242743

[16] T. Karzig. C. Knapp, R. M. Lutchyn, P. Bonderson, M. B. Hastings, C. Nayak, J. Alicea, K. Flensberg, S. Plugge, Y. Oreg, C. M. Marcus, M. H. Freedman. *Scalable designs for quasiparticle-poisoning-protected topological quantum computation with Majorana zero modes.* Phys. Rev. B 95, 235305. (2017).

[17] Alexei Kitaev, *Anyons in an exactly solved model and beyond*, Ann. Physics **321** (2006), no. 1, 2–111, DOI 10.1016/j.aop.2005.10.005. MR2200691

[18] Vaughan Jones, *Groupes de tresses, algèbres de Hecke et facteurs de type* II_1 (French, with English summary), C. R. Acad. Sci. Paris Sér. I Math. **298** (1984), no. 20, 505–508. MR753900

[19] M. Bischoff. *The rank of G-crossed braided extensions of modular tensor categories.* AMS Contemporary Mathematics Series, (2018).

[20] M. Müger. *On the structure of braided crossed G-categories (Appendix 5 to Homotopy quantum field theory) by V. Turaev.* European Mathematical Society. (2010).

[21] Deepak Naidu and Eric C. Rowell, *A finiteness property for braided fusion categories*, Algebr. Represent. Theory **14** (2011), no. 5, 837–855, DOI 10.1007/s10468-010-9219-5. MR2832261

[22] Eric C. Rowell and Zhenghan Wang, *Degeneracy and non-Abelian statistics*, Phys. Rev. A **93** (2016), no. 3, 030102(R), 4, DOI 10.1103/physreva.93.030102. MR3855013

[23] Daisuke Tambara and Shigeru Yamagami, *Tensor categories with fusion rules of self-duality for finite abelian groups*, J. Algebra **209** (1998), no. 2, 692–707, DOI 10.1006/jabr.1998.7558. MR1659954

Department of Mathematics, University of California, Santa Barbara, California
Email address: crdelane@iu.edu

Microsoft Research Station Q and Department of Mathematics, University of California, Santa Barbara, California
Email address: zhenghwa@microsoft.com

Contemporary Mathematics
Volume **747**, 2020
https://doi.org/10.1090/conm/747/15043

Topological quantum computation with gapped boundaries and boundary defects

Iris Cong and Zhenghan Wang

ABSTRACT. We survey some recent work on topological quantum computation with gapped boundaries and boundary defects: First, we review the Hamiltonians for gapped boundaries in quantum double models when the gapped boundaries correspond to subgroups with trivial cocycles. Then, the categorical formulation that excitations of gapped boundaries correspond to objects in dual categories, and defects between two gapped boundaries correspond to objects in functor categories is discussed. Finally, we list some open problems.

1. Introduction

A second quantum revolution in and around the construction of a large scale quantum computer is gaining momentum. A unique approach is topological quantum computation (TQC) based on topological phases of matter (TPMs) [**21**,**22**,**26**]. TPMs are new materials which exhibit many-body quantum entanglement, and their topological orders are patterns of long-range entanglement encoded algebraically by anyon models. TQC is maturing at the forefront of the second quantum revolution as a killer application of TPMs. The hardware of an anyonic quantum computer will be a TPM that harbors non-abelian objects. The low energy effective theory of a TPM is a unitary topological quantum field theory (TQFT) and the anyons are modeled by simple objects in the associated anyon model (mathematically unitary modular category). TPMs have degenerate ground states that are robust against any local perturbations. Those degenerate ground states are topological degrees of freedom, which can be used to construct qubits. Information that is encoded into topological degrees of freedom is automatically immune to local errors. It follows that protection of information from local interactions with the environment is conferred at a physical level, with no active error correction needed.

While universal anyons by braiding alone such as the Fibonacci anyon is theoretically a possibility, accessible anyons with current technology all belong to the class that is weakly integral (WI). If the property F conjecture [**32**] holds, then they cannot be universal for quantum computation by braiding alone. Since we are interested in using the non-abelian statistics for quantum computation, we will focus on gates obtained from non-abelian objects.

Recent studies in topological phases of matter revealed that certain topological phases of matter also support gapped boundaries [**7**, **9**, **25**]. Furthermore, if the

2010 *Mathematics Subject Classification.* Primary 81P68; Secondary 18D10.

©2020 American Mathematical Society

TPM supports multiple types of gapped boundaries, it will support *boundary defects* between different gapped boundary types (physical theories have been developed in [**4**–**6**, **13**, **15**, **24**, **29**, **33**]). Therefore, it is natural to ask if these cousins of anyons can be employed for quantum information processing. This is indeed the case [**13**, **16**].

In the UMC model of a 2D doubled topological order $\mathcal{B} = \mathcal{Z}(\mathcal{C})$, a stable gapped boundary is modeled by a Lagrangian algebra $\mathcal{A}$ in $\mathcal{B}$ [**25**, **27**]. The Lagrangian algebra $\mathcal{A}$ consists of a collection of bulk bosonic anyons that can be condensed to vacuum at the boundary, and the corresponding gapped boundary is a condensate of those anyons which behaves as a non-abelian anyon of quantum dimension $d_{\mathcal{A}}$. Lagrangian algebras in $\mathcal{B} = \mathcal{Z}(\mathcal{C})$ are in one-to-one correspondence with indecomposable module categories $\mathcal{M}$ over $\mathcal{C}$, which can also be used to label gapped boundaries [**18**].

A route to creating, manipulating, and measuring topological degeneracy for gapped boundaries in $\mathfrak{D}(\mathbb{Z}_3)$ in bilayer fractional quantum Hall states coupled to superconductors has been proposed [**2**]. Other experimentally reasonable designs are proposed to realize boundary defects in abelian fractional quantum Hall states [**2**, **11**, **12**, **23**, **30**].

The property F conjecture suggests that braidings alone for gapped boundaries would not be sufficient for universality. This leads us to consider more powerful computational primitives, such as the generalized topological charge measurement in [**13**]. We add our generalized topological charge measurement to braidings to obtain universal quantum computation from gapped boundaries in $\mathfrak{D}(\mathbb{Z}_3)$ [**16**].

2. Summary

Let G be a finite group. Ref. [**26**] provides a local commuting projector Hamiltonian to realize the Dijkgraaf-Witten gauge theory $\mathfrak{D}(G)$ from an arbitrary lattice of data qudits in a Hilbert space spanned by orthonormal basis $\{g \in G\}$. In these theories, gapped boundaries are classified by a subgroup $K \subseteq G$ up to conjugation, and a 2-cocycle $\omega \in H^2(K, \mathbb{C}^\times)$ [**13**, **14**, **25**]. In Refs. [**13**, **14**], we extend the Hamiltonians of [**7**, **8**, **26**], to realize gapped boundaries in $\mathfrak{D}(G)$ when the cocycle is trivial. The extension is as follows[1]:

Kitaev's famous toric code paper [**26**] presents a Hamiltonian to realize the Dijkgraaf-Witten theory based on any finite group G on a general directed lattice and perform topological quantum computation. For simplicity of illustration and calculation, we will assume throughout our paper that the lattice is the square lattice in the plane; however, it is clear that all of the developed theory here extends to arbitrary lattices. In this model, a data qudit is placed on each edge of the lattice, as shown in Fig. 2.1. The Hilbert space for each qudit has an orthonormal basis given by $\{|g\rangle : g \in G\}$, so the total Hilbert space is $\mathcal{L} = \otimes_e \mathbb{C}[G]$.

As discussed in Ref. [**26**], a Hamiltonian is used to transform the high-dimensional Hilbert space of all data qudits into a topological encoding. This Hamiltonian is built from several basic operators on a single data qudit:

$$L_+^{g_0}|g\rangle = |g_0 g\rangle \tag{2.1}$$

$$L_-^{g_0}|g\rangle = |g g_0^{-1}\rangle \tag{2.2}$$

[1]The Hamiltonians published in Ref. [**13**, **14**] contain redundancy and inaccuracy. The following is a corrected version.

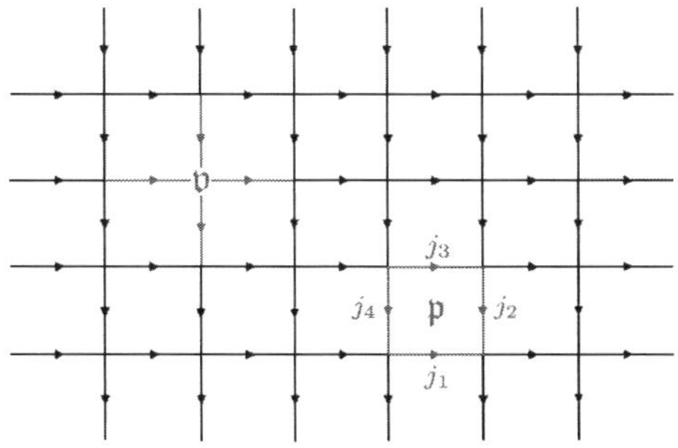

FIGURE 2.1. Lattice for the Kitaev model. For simplicity of illustration and calculation, we use a square lattice, but in general, one can use an arbitrary lattice. If the group G is non-abelian, it is necessary to define orientations on edges, as we have shown here. The edges j and $j_1, ...j_m$, used to obtain $A^g(v)$ and $B^h(p)$, are illustrated for this example of v, p.

$$T_+^{h_0}|h\rangle = \delta_{h_0,h}|h\rangle \tag{2.3}$$

$$T_-^{h_0}|h\rangle = \delta_{h_0^{-1},h}|h\rangle \tag{2.4}$$

where $\delta_{i,j}$ is the Kronecker delta function. These operators are defined for all elements $g_0, h_0 \in G$, and provide a faithful representation of the left/right multiplication and comultiplication in the Hopf algebra $\mathbb{C}[G]$. Using these operators, local gauge transformations and magnetic charge operators are defined as follows, on each vertex v and plaquette p [**26**]:

$$A^g(v,p) = A^g(v) = \prod_{j\in\text{star}(v)} L^g(j,v) \tag{2.5}$$

$$B^h(v,p) = \sum_{h_1\cdots h_k=h} \prod_{m=1}^{k} T^{h_m}(j_m, p) \tag{2.6}$$

Here, $j_1, ..., j_k$ are the boundary edges of the plaquette p in counterclockwise order originating from the vertex v (see Fig. 2.1), and L^g and T^h are defined as follows: if v is the origin of the directed edge j, $L^g(j,v) = L_-^g(j)$, otherwise $L^g(j,v) = L_+^g(j)$; if p is on the left (resp., right) of the directed edge j, $T^h(j,p) = T_-^h(j)$ $(T_+^h(j))$ [**26**].

Note that since the $A^g(v)$ satisfy $A^g(v)A^{g'}(v) = A^{gg'}(v)$, the set of all $A^g(v)$ (for fixed v) form a representation of G on the entire Hilbert space $\mathcal{L} = \otimes_e \mathbb{C}[G]$ of all data qudits [**8**].

Finally, two more linear combinations of these A^g and B^h operators are required to define the Hamiltonian:

$$A(v) = \frac{1}{|G|}\sum_{g\in G} A^g(v,p) \tag{2.7}$$

$$B(p) = B^1(v,p). \tag{2.8}$$

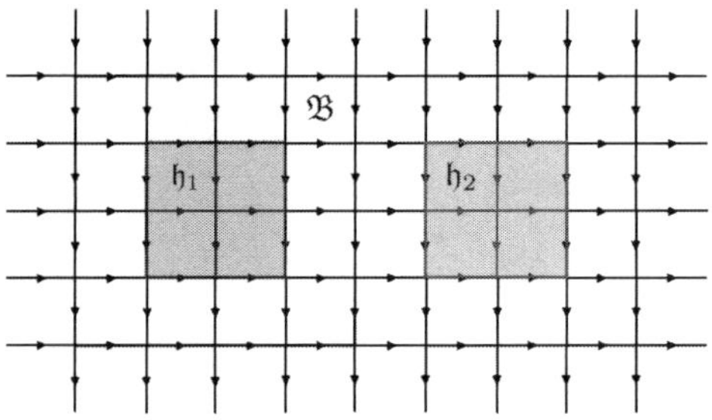

FIGURE 2.2. Square lattice with boundaries, used to define the gapped boundary Hamiltonian. The Hamiltonian $H_{\text{G.B.}}$ is a sum of two terms, where the boundary Hamiltonian $H_{(G,1)}^{(K,1)}$ acts on edges in the hole, and the bulk Hamiltonian $H_{(G,1)}$ acts on vertices and plaquettes in the bulk.

The Hamiltonian[2] for the Kitaev model is then defined as

$$H_{(G,1)} = \sum_v (1 - A(v)) + \sum_p (1 - B(p)) \tag{2.9}$$

We now extend this Hamiltonian to surfaces with boundaries (e.g. in Fig. 2.2). We will consider the model in which a gapped boundary is determined by a subgroup $K \subseteq G$ up to conjugation. In general, as shown in Ref. [**7**], a boundary is determined by both K and a 2-cocycle $\phi \in H^2(K, \mathbb{C}^\times)$, and it is straightforward to generalize our results.

Let us now define some new projector terms, which act on the edges of the lattice:

$$L^K(e) := \frac{1}{|K|} \sum_{k \in K} (L_+^k(e) + L_-^k(e)) \tag{2.10}$$

$$T^K(e) := \sum_{k \in K} T_+^k(e) \tag{2.11}$$

These projectors generalize the ones given in Ref. [**8**] to potentially non-normal subgroups K. The definitions of L^k and T^k for Eqs. (2.10-2.11) are based on Eqs. (2.1-2.4). The choice of using only T_+ in Eq. (2.11) is arbitrary, as using only T_- would yield the same operator.

Following Ref. [**8**], we can now define the following Hamiltonian[3]:

$$H_{(G,1)}^{(K,1)} = \sum_e ((1 - T^K(e)) + (1 - L^K(e)) \tag{2.12}$$

It is important to note that as in the Hamiltonian (2.9), all terms in this Hamiltonian commute with each other. Hence $H_{(G,1)}^{(K,1)}$ is also gapped. Using this,

[2]We call this Hamiltonian $H_{(G,1)}$, as this model is the Dijkgraaf-Witten theory with trivial cocycle (twist). In general, this Hamiltonian may be twisted by a 3-cocycle $\omega \in H^3(G, \mathbb{C}^\times)$, and may be written as $H_{(G,\omega)}$.

[3]As before, we write $H_{(G,1)}^{(K,1)}$ to leave room for the generalized version, where a boundary depends also on a 2-cocycle ϕ of K.

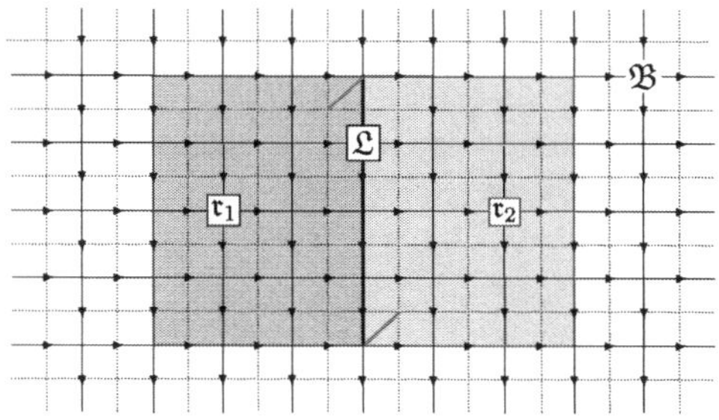

FIGURE 2.3. Boundary defects. The boundary defects are located at the junction of different boundary types.

a commuting Hamiltonian was defined to realize n gapped boundaries given by subgroups $K_1, ..., K_n$ (see Fig. 2.2 for example):

$$H_{\mathrm{G.B.}} = H_G(\mathfrak{B}) + \sum_{i=1}^{n} H_G^{K_i}(\mathfrak{h}_i). \tag{2.13}$$

In the bulk of the Dijkgraaf-Witten theory $\mathfrak{D}(G)$ based on input group G, elementary excitations are classified by pairs (C, π), where C is a conjugacy class of G, and π is an irreducible representation of the centralizer $E(C)$ of C. This can be shown directly from the Hamiltonian by introducing the notion of *ribbon operators* [**26**]. Refs. [**13, 14**] extend the definition of bulk ribbon operators to ribbon operators along the boundary, to show that the elementary excitations along a gapped boundary given by subgroup K are classified by pairs (T, R), where $T \in K\backslash G/K$ is a double coset, and R is an irreducible representation of the stabilizer $K^{r_T} = K \cap r_T K r_T^{-1}$ ($r_T \in T$ is any representative of the double coset). In particular, a correspondence is given between the bulk and boundary ribbon operators which characterizes the physical bulk-to-boundary condensation procedure [**13, 14**]. The correspondence describes how a bulk anyon (C, π) becomes a direct sum of boundary excitations (T, R) upon condensation to the boundary. These results are given in Sec. 2.5 for both Refs. [**13, 14**].

More generally, Sec. 2.6 of Ref. [**13**] and Sec. III of Ref. [**15**] also define a local commuting projector Hamiltonian to realize boundary defects, which occur at the junction points of two different boundary types (see Fig. 2.3). In the Dijkgraaf-Witten theory based on a finite group G, the defects between gapped boundaries given by K_1 and K_2 (and trivial cocycles) are classified by pairs (T, R), where $T \in K_1\backslash G/K_2$ is a double coset, and $R \in (K_1, K_2)_{\mathrm{ir}}^{r_T}$ is an irreducible representation of the stabilizer $(K_1, K_2)^{r_T} = K_1 \cap r_T K_2 r_T^{-1}$.

While these local commuting projector Hamiltonians provide a good way to understand gapped boundaries physically, a more elegant and abstract way of describing them is through the language of category theory. In particular, consider a Turaev-Viro TQFT which is a Drinfeld center $\mathcal{B} = \mathcal{Z}(\mathcal{C})$, for some unitary fusion category $\mathcal{C}$. The gapped boundaries in this theory are given by indecomposable module categories $\mathcal{M}$ of $\mathcal{C}$, or equivalently, Lagrangian algebras $\mathcal{A}$ in $\mathcal{B}$. Excitations on such a gapped boundary are given by the fusion category $\mathrm{Fun}_{\mathcal{C}}(\mathcal{M}, \mathcal{M})$ [**25**].

Similarly, while ribbon operators provide one way to describe bulk-to-boundary condensation, Sec. 3 of Refs. [**13, 14**] also present this condensation procedure as a quotient functor followed by an idempotent completion as in [**31**] (overall, a tensor functor):

$$F : \mathcal{Z}(\mathcal{C}) = \mathcal{B} \xrightarrow{\text{quotient}} \mathcal{B}/\mathcal{A} = \widetilde{\mathcal{Q}} \xrightarrow{\text{I.C.}} \mathcal{Q} = \text{Fun}_{\mathcal{C}}(\mathcal{M},\mathcal{M}). \tag{2.14}$$

(Here, I.C. denotes the idempotent completion.)

Let $\{M_i\}_{i=1}^n$ be a complete collection of gapped boundary types in the theory, corresponding to Lagrangian algebras $\{A_i\}_{i=1}^n$. Then one can form an $n \times n$ multi-fusion category which describes all possible boundary excitations and defects between gapped boundaries:

$$\mathfrak{C} = \{\mathcal{C}_{ij} := \text{Fun}_{\mathcal{C}}(\mathcal{M}_i,\mathcal{M}_j) : i,j = 1,2,...n\}. \tag{2.15}$$

This multi-fusion category can be used to compute important topological properties of boundary excitations and defects, such as their quantum dimension and fusion rules.

As discussed in Sec. V of Ref. [**15**], the bulk-to-boundary condensation procedure of Eq. (2.14) can also be generalized to boundary defects. In this case, the bulk category is not simply $\mathcal{B} = \mathcal{Z}(\mathcal{C})$, and some further setup is required. First, a global symmetry group of the category $\mathcal{B}$ is a group G with a homomorphism $\rho : G \to \text{Aut}_{\otimes}^{\text{br}}(\mathcal{B})$ from G to the braided tensor auto-equivalences of $\mathcal{B}$. Under ρ, each element of g is sent to a permutation ρ_g of the simple objects of $\mathcal{B}$. In particular, ρ_g can also act on a Lagrangian algebra (i.e. gapped boundary): $\rho_g(\mathcal{A}_i) = \mathcal{A}_{j_g}$.

Let G be a global symmetry group of our anyon theory $\mathcal{B} = \mathcal{Z}(\mathcal{C})$. Then, when all obstructions vanish, $\mathcal{B}$ can be extended to a G-graded fusion category $\mathcal{B}_G = \oplus_{g\in G}\mathcal{B}_g$ which describe all bulk anyons and bulk symmetry defects corresponding to the G-symmetry [**1**]. The generalized bulk-to-boundary condensation procedure, called *crossed condensation* [**15**], is then

$$F : \mathcal{B}_G \xrightarrow{\text{quotient}} \mathcal{B}_G/\mathcal{A}_i = \widetilde{\mathcal{Q}}(G,\mathcal{A}_i) \xrightarrow{\text{I.C.}} \mathcal{Q}(G,\mathcal{A}_i) = \oplus_{g\in G}\text{Fun}_{\mathcal{C}}(\mathcal{M}_i,\mathcal{M}_{j_g}). \tag{2.16}$$

Here, $\mathcal{M}_{j_g}$ is the indecomposable module category corresponding to the Lagrangian algebra $\mathcal{A}_{j_g} = \rho_g(\mathcal{A}_i)$, and as before, I.C. denotes the idempotent completion. Hence, bulk anyons condense to excitations on the boundary $\mathcal{A}_i$, while bulk symmetry defects in the flux sector $\mathcal{B}_g$ condense to boundary defects in the category $\mathcal{C}_{ij_g}$.

Using this understanding of crossed condensation, Ref. [**15**] provides an understanding for the projective braiding of boundary defects in $\mathcal{Q}(G,\mathcal{A}_i)$. Specifically, the tensor functor F corresponding to crossed condensation has an adjoint I, which sends boundary defects to their bulk symmetry defect counterparts. Because these bulk symmetry defects have a well-defined G-crossed braiding, this also gives a projective (G-crossed) braiding for boundary defects. As specific cases, Ref. [**15**] shows how this understanding explains the projective braid statistics of Majorana and parafermion zero modes when realized as boundary defects in topological phases [**30**], which is an important application to topological quantum computation.

Finally, Refs. [**13,16**] provide qudit encodings based on gapped boundaries. In particular, given two gapped boundaries (Lagrangian algebras) $\mathcal{A}_i$ and $\mathcal{A}_j$ in the

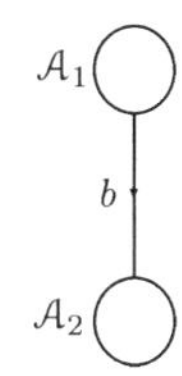

FIGURE 2.4. Qudit encoding. The ground state is given by the anyons that can tunnel from one boundary to another, without any energy cost.

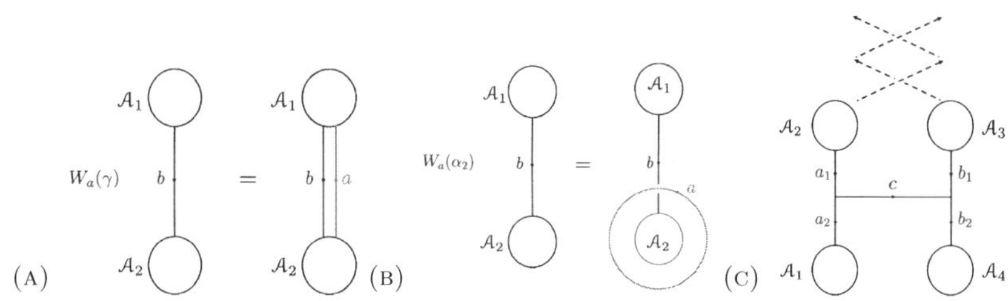

FIGURE 2.5. (A) Tunnel-a operator. (B) Loop-a operator. (C) Braiding gapped boundaries.

plane with total charge vacuum, the ground state is given by the Hilbert space

$$\mathrm{Hom}(\mathcal{A}_i, \mathcal{A}_j) \cong \mathbb{C}^d \tag{2.17}$$

for some d, where we identify $\mathrm{Hom}(\mathcal{A}_i, \mathcal{A}_j)$ as a matrix algebra and normalize the matrix units to an orthonormal basis. This Hilbert space may be used to encode a d-dimensional qudit, as shown in Fig. 2.4.

Given such topological qudits, there are several possible operations one can perform on the associated Hilbert space. First, one can tunnel an anyon a from the boundary $\mathcal{A}_i$ to the boundary $\mathcal{A}_j$, if the antiparticle $\bar{a}$ condenses on $\mathcal{A}_i$, and a condenses on $\mathcal{A}_j$. This is illustrated in Fig. 2.5a. Second, one can loop any bulk anyon around one of the gapped boundaries, as shown in Fig. 2.5b. Third, one can braid gapped boundaries around each other (Fig. 2.5c).

The final operation that can be performed is a new primitive introduced in Refs. [**13**, **16**] called *topological charge measurement*, which generalizes the topological charge projection discussed in [**3**]. This is a very general primitive, which measures arbitrary Hermitian operators along Wilson lines and loops; however, in the cases of interest, the required measurement is usually much more physical. For instance, Ref. [**14**] presents a universal quantum gate set, where the required topological charge measurement can be implemented in a symmetry-protected fashion. All of the topological qudit encodings and topological operations are discussed in the main text of Ref. [**16**] for the case of $\mathfrak{D}(\mathbb{Z}_3)$, and in the Supplementary Material of Ref. [**16**] or Sec. 5 of Ref. [**13**] in full generality.

Using these four topological operations, Refs. [**13**, **16**] show how universal quantum computation can be achieved using purely the gapped boundaries of the quantum double model $\mathfrak{D}(\mathbb{Z}_3)$. This is a very notable illustration of the additional computation power of gapped boundaries over anyons, as $\mathfrak{D}(\mathbb{Z}_3)$ is a purely abelian theory where anyon braidings are trivial.

3. Open questions

Gapped boundaries and boundary defects are non-abelian objects beyond anyons [**35**]. Most questions about non-abelian anyons have analogues for gapped boundaries and boundary defects plus some new ones. In this section, we compile some questions.

3.1. Beyond finite groups. In Refs. [**13**–**15**], we provided a Hamiltonian realization for gapped boundaries and boundary defects in $(2+1)$D Dijkgraaf-Witten theories for any finite group G, based on Kitaev's quantum double models [**26**].

The first problem is to generalize our theory to include nontrivial cocycles such as (K, ω) for the boundary of twisted DW theories. More generally, the Levin-Wen Hamiltonian [**28**] can realize a much larger class of $(2+1)$-D TQFTs. In this model, the representation category of G is generalized to any unitary fusion category $\mathcal{C}$. By [**10**], Kitaev's model can be made to realize the same TQFTs as the Levin-Wen model when G is replaced by a *weak Hopf algebra*. An interesting direction is to generalize the gapped boundary and boundary defect Hamiltonians of Refs. [**13**–**15**] to the Levin-Wen model by replacing the input group with a weak Hopf algebra.

3.2. Beyond two dimensions. Another interesting direction is to consider higher-dimensional analogs of topological phases and their gapped boundaries. One specific case is to examine the Walker-Wang model for the bulk of $(3+1)$-D TQFTs [**34**], and study this model in the presence of gapped boundaries and boundary defects. Even more generally, one could extend the models to the TQFTs and lattice model based on G-crossed braid categories [**17, 36**].

3.3. Beyond symmetry. There are general defects beyond those associated with topological symmetries. A notable example is the $A+C+D/A+B+2F$ boundary defect in the $\mathfrak{D}(S_3)$ theory, as discussed in Ref. [**15**]. An extension to such defects is an interesting direction.

3.4. Stability of gapped boundaries. The most interesting question that we have not touched on is the stability of the topological degeneracy in our model. Once our Hamiltonian moves off the fixed-point, finite-size splitting of the degeneracy would occur. It would be interesting to study the energy splitting of the ground state degeneracies of the Hamiltonians $H_{\text{G.B.}}$ and H_{dft} of Refs. [**13**–**15**] numerically under small perturbations.

3.5. Domain walls. Many physics papers have studied gapped domain walls between different topological phases. While gapped boundaries are often considered as a special case of gapped domain walls, by the folding trick [**25**], they also completely cover the domain wall theory mathematically. Physically, however, it is still interesting to analyze the general gapped domain walls following our work.

3.6. Applications to TQC. There are several open questions in our understanding of topological quantum computation with gapped boundaries and boundary defects.

(1) First, one important open question is to obtain a systematic method to perform TQC with boundary defects, including an efficient qudit encoding scheme and list of topologically protected operations. Furthermore, one should obtain the physical realizations of such quantum gates.
(2) Ref. [**15**] uses fusion channels to understand the degeneracy of many boundary defects on a ring. It would be useful to develop ribbon operator techniques to understand this degeneracy, as such a picture would give rise to a clearer picture of topological operations that can be performed on boundary defects.
(3) An interesting question is to determine computational power for a simple theory such as the $\mathbb{Z}_2$ toric code with gapped boundaries, symmetry defects, boundary defects all included.
(4) Ref. [**19**] analyzes the braid group representations resulting from braiding gapped boundaries. An interesting and more general question is to examine the gapped boundaries (and potentially also the boundary defects) of the Dijkgraaf-Witten theory $\mathfrak{D}(S_3)$ with entire braid group representation, tunneling operators (i.e. some non-unitary operators), and determine whether this set of operations is universal.
(5) Finally, another important implementation scheme for TQC with gapped boundaries is through quantum circuits and the surface code (e.g. Refs. [**13, 20**]). Ref. [**20**] discusses only the special case where the bulk is the $\mathbb{Z}_2$ toric code, and Ref. [**13**] provides an outline in Chapter 4. It would be interesting to find a more detailed implementation for general gapped boundaries.

3.7. Implementing universal gate set. Ref. [**15**] presents a symmetry-protected implementation of the topological charge measurement required for universality. One very interesting problem would be to obtain a purely topological implementation, as this would give a purely topological gate set using only an abelian theory.

References

[1] M. Barkeshli, P. Bonderson, M. Cheng, Z. Wang. *Symmetry, defects, and gauging of topological phases.* arXiv preprint arXiv:1410.4540 (2014).
[2] M. Barkeshli. *Charge 2e/3 superconductivity and topological degeneracies without localized zero modes in bilayer fractional quantum Hall states.* Phys. Rev. Lett. **117**, 096803 (2016).
[3] M. Barkeshli, M. Freedman. *Modular transformations through sequences of topological charge projections.* arXiv preprint arXiv:1602.01093 (2016).
[4] M. Barkeshli, C.-M. Jian, X.-L. Qi. *Twist defects and projective non-Abelian braiding statistics.* Physical Review B **87**(4), 045130 (2013).
[5] M. Barkeshli, C.-M. Jian, X.-L. Qi. *Theory of defects in Abelian topological states.* Physical Review B **88**(23), 235103 (2013).
[6] M. Barkeshli, C.-M. Jian, X.-L. Qi. *Classification of topological defects in Abelian topological states.* Physical Review B **88**(24), 241103 (2013).
[7] Salman Beigi, Peter W. Shor, and Daniel Whalen, *The quantum double model with boundary: condensations and symmetries*, Comm. Math. Phys. **306** (2011), no. 3, 663–694, DOI 10.1007/s00220-011-1294-x. MR2825505
[8] H. Bombin, M. A. Martin-Delgado. *Family of non-Abelian Kitaev models on a lattice: Topological condensation and confinement.* Phys. Rev. B **78**, 115421 (2008).
[9] S. B. Bravyi, A. Y. Kitaev. *Quantum codes on a lattice with boundary.* arXiv:quant-ph/9811052 (1998).

[10] Liang Chang, *Kitaev models based on unitary quantum groupoids*, J. Math. Phys. **55** (2014), no. 4, 041703, 20, DOI 10.1063/1.4869326. MR3390574

[11] M. Cheng. Phys. Rev. B, *Superconducting Proximity Effect on the Edge of Fractional Topological Insulators.* **86**, 195126 (2012)

[12] D. J. Clarke, J. Alicea and K. Shtengel. *Exotic non-Abelian anyons from conventional fractional quantum Hall states.* Nature Comm. **4**, 1348 (2013).

[13] Iris Cong, Meng Cheng, and Zhenghan Wang, *Universal quantum computation with gapped boundaries*, Phys. Rev. Lett. **119** (2017), no. 17, 170504, 5, DOI 10.1103/PhysRevLett.119.170504. MR3730719

[14] Iris Cong, Meng Cheng, and Zhenghan Wang, *Hamiltonian and algebraic theories of gapped boundaries in topological phases of matter*, Comm. Math. Phys. **355** (2017), no. 2, 645–689, DOI 10.1007/s00220-017-2960-4. MR3681387

[15] Iris Cong, Meng Cheng, and Zhenghan Wang, *Hamiltonian and algebraic theories of gapped boundaries in topological phases of matter*, Comm. Math. Phys. **355** (2017), no. 2, 645–689, DOI 10.1007/s00220-017-2960-4. MR3681387

[16] Iris Cong, Meng Cheng, and Zhenghan Wang, *Universal quantum computation with gapped boundaries*, Phys. Rev. Lett. **119** (2017), no. 17, 170504, 5, DOI 10.1103/PhysRevLett.119.170504. MR3730719

[17] Xingshan Cui, *Higher Categories and Topological Quantum Field Theories*, ProQuest LLC, Ann Arbor, MI, 2016. Thesis (Ph.D.)–University of California, Santa Barbara. MR3593287

[18] Alexei Davydov, Michael Müger, Dmitri Nikshych, and Victor Ostrik, *The Witt group of non-degenerate braided fusion categories*, J. Reine Angew. Math. **677** (2013), 135–177. MR3039775

[19] Nicolás Escobar-Velásquez, César Galindo, and Zhenghan Wang, *Braid group representations from braiding gapped boundaries of Dijkgraaf–Witten theories*, Pacific J. Math. **300** (2019), no. 1, 1–16, DOI 10.2140/pjm.2019.300.1. MR3985677

[20] A.G. Fowler, M. Mariantoni, J. M. Martinis, A. N. Cleland. *Surface codes: Towards practical large-scale quantum computation.* Phys. Rev. A **86**(3), 032324 (2012).

[21] Michael H. Freedman, *P/NP, and the quantum field computer*, Proc. Natl. Acad. Sci. USA **95** (1998), no. 1, 98–101, DOI 10.1073/pnas.95.1.98. MR1612425

[22] Michael H. Freedman, Alexei Kitaev, Michael J. Larsen, and Zhenghan Wang, *Topological quantum computation*, Bull. Amer. Math. Soc. (N.S.) **40** (2003), no. 1, 31–38, DOI 10.1090/S0273-0979-02-00964-3. Mathematical challenges of the 21st century (Los Angeles, CA, 2000). MR1943131

[23] S. Ganeshan, A. V. Gorshkov, V. Gurarie, V. M. Galitski. *Exactly soluble model of boundary degeneracy.* arXiv preprint arXiv:1604.02089.

[24] A. Kapustin, *Ground-state degeneracy for Abelian anyons in the presence of gapped boundaries*, Phys. Rev. B **89**, 125307 (2014)

[25] Alexei Kitaev and Liang Kong, *Models for gapped boundaries and domain walls*, Comm. Math. Phys. **313** (2012), no. 2, 351–373, DOI 10.1007/s00220-012-1500-5. MR2942952

[26] A. Yu. Kitaev, *Fault-tolerant quantum computation by anyons*, Ann. Physics **303** (2003), no. 1, 2–30, DOI 10.1016/S0003-4916(02)00018-0. MR1951039

[27] A. Kitaev. Bose-condensation and edges of topological quantum phases. Talk at modular categories and applications, Indiana University, March 19-22, 2009.

[28] M. A. Levin, X.-G. Wen. *String-net condensation: A physical mechanism for topological phases.* Phys. Rev. B **71**, 045110 (2005).

[29] M. Levin, *Protected edge modes without symmetry.* Phys. Rev. X **3**, 021009 (2013).

[30] N. H. Lindner, E. Berg, G. Refael and A. Stern. *Fractionalizing Majorana fermions: non-abelian statistics on the edges of abelian quantum Hall states.* Phys. Rev. X, **2**, 041002 (2012).

[31] Michael Müger, *Galois extensions of braided tensor categories and braided crossed G-categories*, J. Algebra **277** (2004), no. 1, 256–281, DOI 10.1016/j.jalgebra.2004.02.026. MR2059630

[32] Deepak Naidu and Eric C. Rowell, *A finiteness property for braided fusion categories*, Algebr. Represent. Theory **14** (2011), no. 5, 837–855, DOI 10.1007/s10468-010-9219-5. MR2832261

[33] J.C. Wang, X.-G. Wen. *Boundary degeneracy of topological order.* Phys. Rev. B **97**, 125124 (2015).

[34] K. Walker, Z. Wang. $(3+1)$-*TQFTs and Topological Insulators.* Frontiers of Physics, **7**(2), 150-159 (2012).

[35] Zhenghan Wang, *Beyond anyons*, Modern Phys. Lett. A **33** (2018), no. 28, 1830011, 8, DOI 10.1142/S0217732318300112. MR3853256

[36] Dominic J. Williamson and Zhenghan Wang, *Hamiltonian models for topological phases of matter in three spatial dimensions*, Ann. Physics **377** (2017), 311–344, DOI 10.1016/j.aop.2016.12.018. MR3606141

Dept of Physics, Harvard University, Cambridge, MA 02138, U.S.A.
Email address: cong@g.harvard.edu

Dept of Mathematics, University of California, Santa Barbara, CA 93106-6105, U.S.A.; and Microsoft Station Q, University of California, Santa Barbara, CA 93106-6105, U.S.A.
Email address: zhenghwa@microsoft.com

Contemporary Mathematics
Volume **747**, 2020
https://doi.org/10.1090/conm/747/15044

Classification of gapped quantum liquid phases of matter

Xiao-Gang Wen

Abstract. Quantum phases of matter correspond to patterns many-body entanglement in quantum many-body systems. This article summaries some recent results on classification of gapped quantum liquid phases of matter in low dimensions in terms of unitary braid fusion categories.

1. Introduction

Quantum condensed matter physics deals with local quantum systems (see Fig. 1). Similar quantum systems form a phase of matter. Quantum phases of matter are phases of matter at zero temperature. They can be divided into two classes: gapped quantum phases (see Fig. 2) and gapless quantum phases. For a long time, we believe that the gapped quantum phases of matter are described by Landau symmetry breaking theory. In 1989, it was realized that Landau symmetry breaking theory cannot describe all gapped quantum phases of matter. The new kind of gapped quantum phases were called topological orders [**36, 37, 41**].

At the beginning, topological order was defined by (a) the topology-dependent ground-state degeneracy [**36, 41**] and (b) the non-Abelian geometric phases of the degenerate ground states[**14, 37**], where both of them are *robust against any local perturbations* that can break any symmetry [**41**]. This is just like that superfluid order is defined by zero-viscosity and quantized vorticity, which are robust against any local perturbations that preserve the $U(1)$ symmetry. Chiral spin liquids [**13, 42**], integral/fractional quantum Hall states [**21, 33, 34**], $\mathbb{Z}_2$ spin liquids [**25, 29, 38**], non-Abelian fractional quantum Hall states [**26, 28, 39, 43**] etc, are examples of topologically ordered phases.

Microscopically, superfluid order is originated from boson or fermion-pair condensation. So it is natural for us to ask: what is the microscopic origin of topological order? What is the microscopic origin of robustness against *any* local perturbations? Recently, it was found that, microscopically, topological order is related to long-range entanglement [**16, 22**]. In fact, we can regard topological order as patterns of long-range entanglement [**5, 32, 45**] defined through generalized local unitary (LU) transformations.

In fact, the topologically ordered states are not arbitrary gapped states, but belong to a special kind of gapped quantum states, called *gapped quantum liquids* [**32, 45**]. 3D gapped states formed by stacking 2D fractional quantum Hall states

---

2010 *Mathematics Subject Classification.* 16D90.
This research is supported by NSF Grant No. DMR-1506475, and NSFC 11274192.

©2020 American Mathematical Society

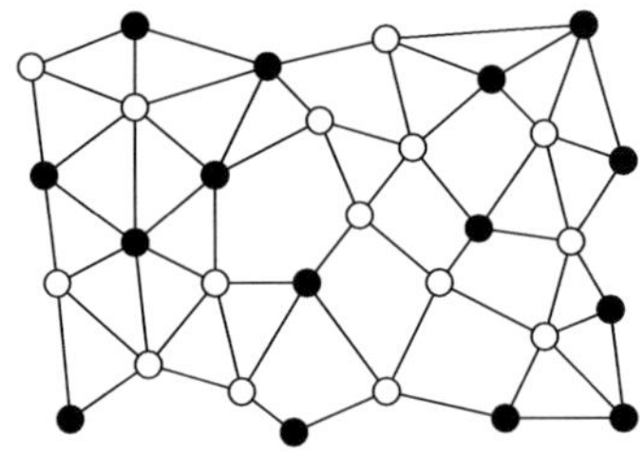

FIGURE 1. A local quantum system is given by a pair $\mathcal{V}_N, H_N$ plus a graph $\mathcal{G}$. There is a finite dimensional Hilbert space $\mathcal{V}^i_{\text{site}}$ associated with each of the N sites of the graph. The total Hilbert space $\mathcal{V}_N$ is given be the tensor product $\mathcal{V}_N = \otimes_{i=1}^N \mathcal{V}^i_{\text{site}}$. The hermitian operator H_N in the total Hilbert space $\mathcal{V}_N$ (called a Hamiltonian) have the following local form $H_N = \sum_i H_i + \sum_{ij} H_{ij} + \sum_{ijk} H_{ijk} + \cdots$, where, for example, H_{ijk} acts on the Hilbert space $\mathcal{V}^i_{\text{site}} \otimes \mathcal{V}^j_{\text{site}} \otimes \mathcal{V}^k_{\text{site}}$, and the distance between the sites i, j, k are bounded by the constant.

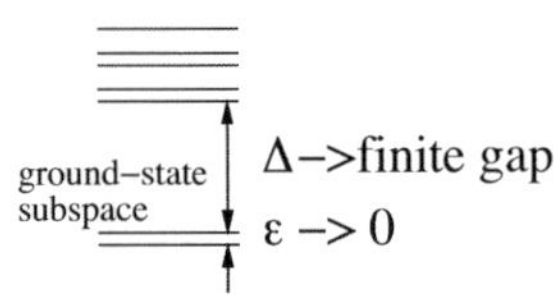

FIGURE 2. The eigenvalue spectrum of H_N in $N \to \infty$ limit for a gapped quantum system. It is also called energy spectrum in physics.

and Haah's cubic model are examples of gapped quantum states which is not a gapped quantum liquid [**10**]. Those gapped non-liquid states do not have a quantum field theory description (with a finite number of fields).

The mathematical foundation for many-body quantum physics is infinite dimensional linear algebra with an infinite tensor product decomposition. The mathematical foundation for topological order is the braided fusion category theory. This gives us a classification of gapped quantum phases in 1+1D and 2+1D, and a partial classification of gapped quantum liquid phases in 3+1D.

2. Classification of gapped quantum liquids

In history, we have completely classified some large classes of matter states only for a few times. The first time is the classification of all spontaneous symmetry breaking orders, which can be classified by a pair of groups:

$$(G_\Psi \subset G_H), \tag{2.1}$$

where G_H is the symmetry group of the system and G_Ψ, a subgroup of G_H, is the symmetry group of the ground state. This includes the classification of all 230 crystal orders in 3-dimensions.

TABLE 1. A braided fusion category BFC can be viewed as a set of point-like excitations, with fusion and braiding properties. The centralizer of a BFC is the set of point-like excitations in BFC that have trivial double-braiding with all the point-like excitations in BFC. A BFC becomes a MTC if its centralizer cantains only trivial point-like excitations. The following table lists the correspondences between point-like excitations and braided fusion category. Here, point-like excitations are created as the ground state subspace of Hamiltonian H_N with traps $\delta H(\boldsymbol{x}_i)$ (see Fig. 3).

Point-like excitations	Braided fusion category (BFC)
Ground state subspace of $H_N + \sum \delta H(\boldsymbol{x}_i)$	Fusion space of BFC: $\mathrm{Hom}(i \otimes j \otimes \cdots, \mathbf{1})$
Stable excitations	Simple objects
Excitations with accidental degeneracy	Composite objects
Deformation of excitations (i.e. $\delta H(\boldsymbol{x}_i)$)	Morphism of objects
Bound state of excitations	Fusion of objects (Grothendieck ring)
Statistics of excitations	Braiding of objects: $i \otimes j \to j \otimes i$
Mutual statistics of excitations	Double braiding of objects
Local excitations	Transparent objects (trivial double braiding with all objects)
Topological excitations	Non-transparent objects (non-trivial double braiding with some objects)
Internal degrees of freedom	Quantum dimension d_i
Angular momentum of an excitation	Topological spin s_i

The second time is the classification of all gapped 1-dimensional quantum states:

RESULT 2.1. *Gapped 1-dimensional quantum states with on-site symmetry G_H can be classified by a triple:*[**6, 31**]

$$[G_\Psi \subset G_H;\ \mathrm{pRep}(G_\Psi)], \tag{2.2}$$

where $\mathrm{pRep}(G_\Psi)$ *is a projective representation of* G_Ψ [**27**].

The third time is the classification of all gapped quantum phases in 2+1D. Since early on, it was conjectured that

RESULT 2.2. *All 2+1D bosonic topological orders (without symmetry) are classified by S, T modular matrices (plus other gauge connections)* [**37**], *or more precisely by a pair* [**15, 30, 35**]:

$$(\mathrm{MTC}, c), \tag{2.3}$$

where MTC is a unitary modular tensor category (see Table 1) and $c = c_{MTC}$ mod 8 is the chiral central charge c of the edge states.

Here, c_{MTC} is the central charge of the MTC given by $\mathrm{e}^{\mathrm{i}2\pi c_{MTC}/8} = \frac{\sum_i \mathrm{e}^{2\pi\,\mathrm{i}\,s_i} d_i^2}{\sqrt{\sum_i d_i^2}}$, and d_i, s_i are quantum dimensions and topological spins in MTC (see Table 1). We like to pointed out that the above classification include the invertible topological orders [**8**, **17**], which described by the central charge c.

For bosonic quantum systems with symmetry, we have the following result:

RESULT 2.3. *All 2+1D gapped bosonic phases with a finite unitary on-site symmetry G_H, are classified by* [**19**]

$$[G_\Psi \subset G_H;\ Rep(G_\Psi) \subset BFC \subset MTC;\ c]. \tag{2.4}$$

where $c = c_{MTC}$ mod 8.

Here $\mathrm{Rep}(G_\Psi)$ is the SFC formed by the representations of G_Ψ where all representations are assigned Bose statistics, In addition, $\mathrm{Rep}(G_\Psi)$ is the centralizer of BFC. Also MTC is a modular tensor category that contains BFC such that $\mathrm{Rep}(G_\Psi)$ is the centralizer of BFC in MTC, i.e. $\mathrm{Rep}(G_\Psi)$ is formed by the point-like excitations in MTC that that have trivial double braiding with all the point-like excitations in BFC. Such a MTC is called a minimal modular extension of the pair $\mathrm{Rep}(G_\Psi) \subset$ BFC.

Such a result is equivalent to an earlier result that all 2+1D gapped bosonic phases with a finite unitary on-site symmetry G_H, are classified by [**1**]

$$[G_\Psi \subset G_H;\ G_\Psi\text{-crossed MTC};\ c], \tag{2.5}$$

The above classification includes symmetry breaking orders, symmetry protected topological (SPT) orders [**4**, **7**, **9**], topological orders, and *symmetry-enriched topological orders* (SET) described by *projective symmetry group* [**40**]. SET orders of time-reversal/reflection symmetry are classified by Ref. [**2**]. Some more discussions on SET orders can be found in [**3**, **11**, **12**, **23**, **24**, **44**].

The above results for bosonic systems can be generalized to fermion systems:

RESULT 2.4. *2+1D fermionic topological orders are classified by a triple:*[**18**]

$$[sRep(Z_2^f) \subset \mathrm{BFC};\ c], \tag{2.6}$$

where $\mathrm{sRep}(Z_2^f)$ *is the symmetric fusion category (SFC) formed by the representations of the fermion-number-parity symmetry Z_2^f where the non-trivial representation is assigned Fermi statistics, and BFC is a unitary braided fusion category whose centralizer is* $\mathrm{sRep}(Z_2^f)$.

Here, $c = c_{MTC}$ mod $\frac{1}{2}$, and c_{MTC} is the central charge of a modular extension of the BFC.

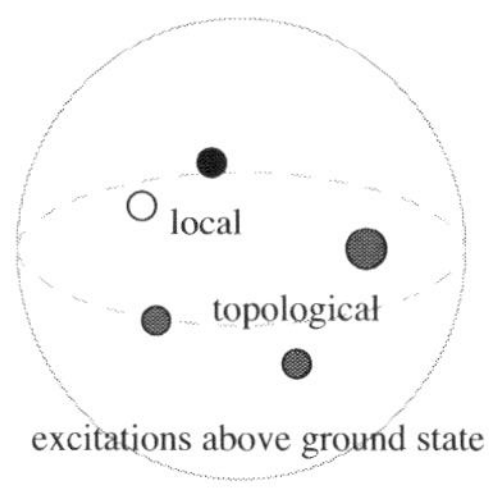

FIGURE 3. Point-like excitations at $\boldsymbol{x}_i$ correspond to the punctures in the space created the trap Hamiltonians $\delta H(\boldsymbol{x}_i)$ which is non-zero only near $\boldsymbol{x}_i$.

We also have a similar result for fermion systems:

RESULT 2.5. *All 2+1D gapped fermionic phases with unitary finite on-site symmetry G_H^f are classified by* [**19**]

$$[G_\Psi^f \subset G_H^f;\ sRep(G_\Psi^f) \subset BFC \subset MTC;\ c], \tag{2.7}$$

where $sRep(G_\Psi^f)$ is the SFC formed by the representations of G_Ψ^f where some representations are assigned Fermi statistics, and $c = c_{MTC}$ mod 8. Also $sRep(G_\Psi^f)$ is the centralizer of BFC in MTC.

In 3+1D, we only have a partial result [**20**]:

RESULT 2.6. *Consider 3+1D bosonic gapped quantum liquid phases without symmetry and with all the point-like excitations to be bosons. All such phases are classified by unitary pointed fusion 2-categories, which are one-to-one labeled by a finite group G and its group 4-cocycle $\omega_4 \in \mathcal{H}^4[G; U(1)]$ up to group automorphisms. Furthermore, all such 3+1D topological orders can be realized by Dijkgraaf-Witten gauge theories.*

The above classification results come from the realization that the point-like excitations in 2+1D topological orders are described by unitary braided fusion categories (see Table 1). Therefore, a classification of unitary braided fusion categories gives rise to a classification of possible point-like excitations, which in turn leads to a classification of 2+1D topological orders.

3. Summary

We have seen that to understand the symmetry breaking states, physicists need to learn group theory. From the above results, it appears that to understand patterns of many-body entanglement that correspond to topological order and SPT order, physicists will have to learn tensor category theory and group cohomology theory. In modern quantum many-body physics and in modern condensed matter physics, tensor category theory and group cohomology theory will be as useful as group theory. The days when physicists need to learn tensor category theory and group cohomology theory are coming, may be soon.

References

[1] M. Barkeshli, P. Bonderson, M. Cheng, and Z. Wang, *Symmetry, defects, and gauging of topological phases*, (2014).

[2] M. Barkeshli, P. Bonderson, C.-M. Jian, M. Cheng, and K. Walker, *Reflection and time reversal symmetry enriched topological phases of matter: path integrals, non-orientable manifolds, and anomalies*, (2016).

[3] Liang Chang, Meng Cheng, Shawn X. Cui, Yuting Hu, Wei Jin, Ramis Movassagh, Pieter Naaijkens, Zhenghan Wang, and Amanda Young, *On enriching the Levin-Wen model with symmetry*, J. Phys. A **48** (2015), no. 12, 12FT01, 19, DOI 10.1088/1751-8113/48/12/12FT01. MR3321463

[4] Xie Chen, Zheng-Cheng Gu, Zheng-Xin Liu, and Xiao-Gang Wen, *Symmetry protected topological orders and the group cohomology of their symmetry group*, Phys. Rev. B **87** (2013), 155114.

[5] Xie Chen, Zheng-Cheng Gu, and Xiao-Gang Wen, *Local unitary transformation, long-range quantum entanglement, wave function renormalization, and topological order*, Phys. Rev. B **82** (2010), 155138.

[6] Xie Chen, Zheng-Cheng Gu, and Xiao-Gang Wen, *Classification of gapped symmetric phases in 1D spin systems*, Phys. Rev. B **83** (2011), 035107.

[7] Xie Chen, Zheng-Xin Liu, and Xiao-Gang Wen, *Two-dimensional symmetry-protected topological orders and their protected gapless edge excitations*, Phys. Rev. B **84** (2011), 235141.

[8] Daniel S. Freed, *Anomalies and invertible field theories*, String-Math 2013, Proc. Sympos. Pure Math., vol. 88, Amer. Math. Soc., Providence, RI, 2014, pp. 25–45, DOI 10.1090/pspum/088/01462. MR3330283

[9] Zheng-Cheng Gu and Xiao-Gang Wen, *Tensor-entanglement-filtering renormalization approach and symmetry protected topological order*, Phys. Rev. B **80** (2009), 155131.

[10] J. Haah, *Local stabilizer codes in three dimensions without string logical operators*, Phys. Rev. A**83** (2011), no. 4, 042330.

[11] L.-Y. Hung and Y. Wan, *A K matrix Construction of Symmetry Enriched Phases of Matter*, (2013).

[12] L.-Y. Hung and X.-G. Wen, *Quantized topological terms in weak-coupling gauge theories with a global symmetry and their connection to symmetry enriched topological phases*, Phys. Rev. B **87** (2013), 165107.

[13] V. Kalmeyer and R. B. Laughlin, *Equivalence of the resonating-valence-bond and fractional quantum Hall states*, Phys. Rev. Lett. **59** (1987), 2095–2098.

[14] E. Keski-Vakkuri and Xiao-Gang Wen, *Ground state structure of hierarchical QH states on torus and modular transformation*, Int. J. Mod. Phys. B **7** (1993), 4227.

[15] Alexei Kitaev, *Anyons in an exactly solved model and beyond*, Ann. Physics **321** (2006), no. 1, 2–111, DOI 10.1016/j.aop.2005.10.005. MR2200691

[16] Alexei Kitaev and John Preskill, *Topological entanglement entropy*, Phys. Rev. Lett. **96** (2006), no. 11, 110404, 4, DOI 10.1103/PhysRevLett.96.110404. MR2214523

[17] L. Kong and X.-G. Wen, *Braided fusion categories, gravitational anomalies, and the mathematical framework for topological orders in any dimensions*, (2012), arXiv:1405.5858.

[18] T. Lan, L. Kong, and X.-G. Wen, *A theory of 2+1D fermionic topological orders and fermionic/bosonic topological orders with symmetries*, Phys. Rev. B**94** (2016), 155113.

[19] Tian Lan, Liang Kong, and Xiao-Gang Wen, *Classification of 2+1d topological orders and spt orders for bosonic and fermionic systems with on-site symmetries*, (2016).

[20] Tian Lan, Liang Kong, and Xiao-Gang Wen, *A classification of 3+1d bosonic topological orders (i): the case when point-like excitations are all bosons*, (2017).

[21] R. B. Laughlin, *Anomalous quantum Hall effect: An incompressible quantum fluid with fractionally charged excitations*, Phys. Rev. Lett. **50** (1983), 1395–1398.

[22] Michael Levin and Xiao-Gang Wen, *Detecting topological order in a ground state wave function*, Phys. Rev. Lett. **96** (2006), 110405.

[23] Y.-M. Lu and A. Vishwanath, *Classification and Properties of Symmetry Enriched Topological Phases: A Chern-Simons approach with applications to Z2 spin liquids*, (2013).

[24] A. Mesaros and Y. Ran, *A classification of symmetry enriched topological phases with exactly solvable models*, Phys. Rev. B **87** (2013), 155115.

[25] R. Moessner and S. L. Sondhi, *Resonating valence bond phase in the triangular lattice quantum dimer model*, Phys. Rev. Lett. **86** (2001), 1881.

[26] Gregory Moore and Nicholas Read, *Nonabelions in the fractional quantum Hall effect*, Nuclear Phys. B **360** (1991), no. 2-3, 362–396, DOI 10.1016/0550-3213(91)90407-O. MR1118790

[27] F. Pollmann, E. Berg, A. M. Turner, and M. Oshikawa, *Entanglement spectrum of a topological phase in one dimension*, Phys. Rev. B **81** (2010), 064439.

[28] Iuliana P. Radu, J. B. Miller, C. M. Marcus, M. A. Kastner, L. N. Pfeiffer, and K. W. West, *Quasiparticle tunneling in the fractional quantum Hall state at* $\nu = 5/2$, Science **320** (2008), 899.

[29] N. Read and Subir Sachdev, *Large-N expansion for frustrated quantum antiferromagnets*, Phys. Rev. Lett. **66** (1991), 1773.

[30] Eric Rowell, Richard Stong, and Zhenghan Wang, *On classification of modular tensor categories*, Comm. Math. Phys. **292** (2009), no. 2, 343–389, DOI 10.1007/s00220-009-0908-z. MR2544735

[31] N. Schuch, D. Perez-Garcia, and I. Cirac, *Classifying quantum phases using Matrix Product States and PEPS*, Phys. Rev. B **84** (2011), 165139.

[32] B. Swingle and J. McGreevy, *Renormalization group constructions of topological quantum liquids and beyond*, Phys. Rev. B **93** (2016), 045127.

[33] D. C. Tsui, H. L. Stormer, and A. C. Gossard, *Two-dimensional magnetotransport in the extreme quantum limit*, Phys. Rev. Lett. **48** (1982), 1559–1562.

[34] K. von Klitzing, G. Dorda, and M. Pepper, *New method for high-accuracy determination of the fine-structure constant based on quantized Hall resistance*, Phys. Rev. Lett. **45** (1980), 494–497.

[35] Xiao-Gang Wen, *A theory of 2+1d bosonic topological orders*, Natl. Sci. Rev. **3** (2016), 68–106.

[36] Xiao-Gang Wen, *Vacuum degeneracy of chiral spin state in compactified spaces*, Phys. Rev. B **40** (1989), 7387.

[37] X. G. Wen, *Topological orders in rigid states*, Internat. J. Modern Phys. B **4** (1990), no. 2, 239–271, DOI 10.1142/S0217979290000139. MR1043302

[38] X. G. Wen, *Mean-field theory of spin-liquid states with finite energy gap and topological orders*, Phys. Rev. B **44** (1991), 2664.

[39] X. G. Wen, *Nonabelian statistics in the fractional quantum Hall states*, Phys. Rev. Lett. **66** (1991), no. 6, 802–805, DOI 10.1103/PhysRevLett.66.802. MR1089602

[40] X. G. Wen, *Quantum orders and symmetric spin liquids (extended version at http://arxiv.org/abs/cond-mat/0107071)*, Phys. Rev. B **65** (2002), 165113.

[41] Xiao-Gang Wen and Q. Niu, *Ground state degeneracy of the FQH states in presence of random potentials and on high genus riemann surfaces*, Phys. Rev. B **41** (1990), 9377.

[42] Xiao-Gang Wen, F. Wilczek, and A. Zee, *Chiral spin states and superconductivity*, Phys. Rev. B **39** (1989), 11413.

[43] R. Willett, J. P. Eisenstein, H. L. Strörmer, D. C. Tsui, A. C. Gossard, and J. H. English, *Observation of an even-denominator quantum number in the fractional quantum hall effect*, Phys. Rev. Lett. **59** (1987), 1776.

[44] C. Xu, *Three-dimensional* Z_2 *topological phases enriched by time-reversal symmetry*, Phys. Rev. B**88** (2013), no. 20, 205137.

[45] Bei Zeng and Xiao-Gang Wen, *Gapped quantum liquids and topological order, stochastic local transformations and emergence of unitarity*, Phys. Rev. B **91** (2015), 125121.

Department of Physics, Massachusetts Institute of Technology, Cambridge, Massachusetts 02139, USA

Email address: xgwen@mit.edu

Contemporary Mathematics
Volume **747**, 2020
https://doi.org/10.1090/conm/747/15045

Schur-type invariants of branched G-covers of surfaces

Eric Samperton

Abstract. Fix a finite group G and a conjugacy invariant subset $C \subseteq G$. Let Σ be an oriented surface, possibly with punctures. We consider the question of when two homomorphisms $\pi_1(\Sigma) \to G$ taking punctures into C are equivalent up to an orientation preserving diffeomorphism of Σ. We provide an answer to this question in a stable range, meaning that Σ has enough genus and enough punctures of every conjugacy type in C. If C generates G, then we can assume Σ has genus 0 (or any other constant). The main tool is a classifying space for (framed) C-branched G-covers, and related homology classes we call branched Schur invariants, since they take values in a torsor over a quotient of the Schur multiplier $H_2(G)$. We conclude with a brief discussion of applications to (2+1)-dimensional G-equivariant TQFT and symmetry-enriched topological phases.

1. Introduction

Let $\Sigma_{g,n}$ denote an oriented genus g surface with n distinct marked points $p_1, \ldots, p_n \in \Sigma_g$, thought of as punctures. Throughout, we assume $\Sigma_{g,n}$ has a fixed basepoint distinct from the punctures. Recall that the pointed mapping class group $\mathrm{MCG}_*(\Sigma_{g,n})$ consists of isotopy classes of orientation-preserving diffeomorphisms of $\Sigma_{g,n}$ that fix the basepoint. In particular, representatives of elements of $\mathrm{MCG}_*(\Sigma_{g,n})$ are allowed to permute the punctures. Since the basepoint of $\Sigma_{g,n}$ is fixed, $\mathrm{MCG}_*(\Sigma_{g,n})$ acts on $\pi_1(\Sigma_{g,n})$.

Fix a finite group G. The action of $\mathrm{MCG}_*(\Sigma_{g,n})$ on $\pi_1(\Sigma_{g,n})$ induces an action of $\mathrm{MCG}_*(\Sigma_{g,n})$ on the finite set of G-representations

$$\hat{R}_{g,n} \stackrel{\text{def}}{=} \{\pi_1(\Sigma_{g,n}) \to G\}.$$

The goal of the present paper is to understand the orbits of this action.

We make significant progress by deploying three different $\mathrm{MCG}_*(\Sigma_{g,n})$-invariants of $\hat{R}_{g,n}$. When combined, these invariants are powerful enough that we can specify orbits uniquely in a certain "stable range." The precise results are provided by Theorem 1.1. For now we simply remark that as g and n grow, the stable range includes almost all homomorphisms in $\hat{R}_{g,n}$ with respect to the uniform counting measure. After first accounting for two elementary invariants, our approach is to interpret the homomorphisms in $\hat{R}_{g,n}$ as branched G-covers of Σ_g, where the punctures form the branch locus. We then use some algebraic topology to construct invariant homology classes called branched Schur invariants.

2010 *Mathematics Subject Classification.* Primary 57M12; Secondary 57R85, 55R40, 81T45, 14H30.

©2020 American Mathematical Society

The first invariant is the image of a homomorphism: if $\phi \in \hat{R}_{n,g}$ has image $H \leq G$, then so does $\tau \cdot \phi$ for all $\tau \in \mathrm{MCG}_*(\Sigma_{n,g})$. Accordingly, we only need to consider the subset

$$R_{g,n} \stackrel{\text{def}}{=} \{\pi_1(\Sigma_{g,n}) \twoheadrightarrow G\} \subseteq \hat{R}_{g,n}$$

consisting of all *surjective* homomorphisms. In terms of branched covers, this will mean we only consider *connected* covers of $\Sigma_{g,n}$. This reduces us to the narrower question: what are the orbits of the action of $\mathrm{MCG}_*(\Sigma_{g,n})$ on $R_{g,n}$?

The second invariant is the branch type of a homomorphism, defined as follows. We begin with some notation: given a conjugacy invariant subset $C \subseteq G$, we let $C /\!/ G$ denote the set of conjugacy classes intersecting C. If $c \in G$, we denote the conjugacy class of c by $\overline{c} \in G /\!/ G$.

For each puncture p_i, pick a simple closed loop $\gamma_i \in \pi_1(\Sigma_{g,n})$ such that γ_i winds once counterclockwise around p_i, and γ_i does not wind around any of the other punctures. It is not quite correct to say that $\mathrm{MCG}_*(\Sigma_{g,n})$ preserves the conjugacy class of $f(\gamma_i)$, since $\mathrm{MCG}_*(\Sigma_{g,n})$ can permute the punctures. However, if we form a vector

$$v_\phi \in \mathbb{Z}_{\geq 0}^{G/\!/G}$$

by letting the component of $\overline{c} \in G /\!/ G$ in v_ϕ be

$$(v_\phi)(\overline{c}) \stackrel{\text{def}}{=} |\{1 \leq i \leq n \mid \phi(\gamma_i) \in \overline{c}\}|,$$

then elementary algebraic topology shows $\mathrm{MCG}_*(\Sigma_{g,n})$ preserves v_ϕ. We call v_ϕ the *branch type*, or *branching data*, of ϕ.

We can interpret the branch type v_ϕ as a multiset of cardinality n, meaning the sum of the entries of the vector v_ϕ is n. Given any branching data $v \in \mathbb{Z}_{\geq 0}^{G/\!/G}$ of cardinality n, we define the $\mathrm{MCG}_*(\Sigma_{g,n})$-invariant subset

$$R_{g,v} \stackrel{\text{def}}{=} \{\phi \in R_{g,n} \mid v_\phi = v\} \subseteq R_{g,n}.$$

In fact, we will refine our approach by fixing a conjugacy invariant subset $C \subseteq G$ and only considering

$$v \in \mathbb{Z}_{\geq 0}^{C/\!/G} \subseteq \mathbb{Z}_{\geq 0}^{G/\!/G}.$$

We emphasize that C need not be closed under inversion. In everything that follows, we allow $C = \emptyset$, although then we must set $n = 0$.

In Section 2.1, we define the notion of *(framed) C-branched G-cover* of a smooth manifold. Then, given $\phi \in R_{g,v}$, we construct a C-branched G-cover of Σ_g with branch locus consisting of the punctures $p_1, \ldots, p_n$. The details of this are provided in Section 3.1. By a small abuse of notation, we continue to denote this cover by ϕ.

In Section 2.2, we describe a classifying space for (concordance classes of) C-branched G-covers of smooth manifolds, denoted BG_C. The C-branched G-cover ϕ yields a (homotopy class) of a map $\phi_\# : \Sigma_g \to BG_C$. The *$C$-branched Schur invariant* of ϕ is the integral homology class

$$\mathrm{sch}_C(\phi) \stackrel{\text{def}}{=} \phi_*[\Sigma_g] \in H_2(BG_C)$$

where $[\Sigma_g] \in H_2(\Sigma)$ is the orientation of Σ_g. (All of the homology groups in this paper have integral coefficients.) Since we only consider orientation-preserving mapping classes, $\mathrm{sch}_C(\phi)$ is $\mathrm{MCG}_*(\Sigma_{g,n})$-invariant.

Our main theorem shows that the branched Schur invariant completely determines the orbits of $\mathrm{MCG}_*(\Sigma_{g,n})$ acting on $R_{g,v}$ whenever the genus g and branching data v are "large enough." When we say g is *large enough*, we mean this is in the usual sense for integers. However, when we say v is *large enough*, we mean that every conjugacy class $\bar{c} \in C \mathbin{/\!/} G$ occurs in v with enough multiplicity, *i.e.* that all of the integers $v(\bar{c})$ are large enough. In particular, when we say v is large enough, we do not simply mean that the cardinality of v is large enough.

Before stating the main theorem, we introduce one more definition. Recall that the group homology $H_*(G)$ is equivalent to the singular homology $H_*(BG)$, where BG is the classifying space for G.

We say that a homology class in $H_2(G)$ is a *C-torus* if it can be represented by a pointed map from the torus $(S^1 \times S^1, *)$ to $(BG, *)$ such that the induced map $\pi_1(S^1 \times S^1) \to G$ sends the loop winding once counterclockwise around the first factor of S^1 to an element of C. Define the *C-reduced Schur multiplier* of G as the quotient

$$M(G)_C \stackrel{\text{def}}{=} H_2(G)/\langle C\text{-tori}\rangle.$$

In Section 2.3, we compute the low-dimensional homology of BG_C, and learn that $M(G)_C$ plays an important role.

THEOREM 1.1. *Fix a finite group G and a conjugacy invariant subset $C \subseteq G$. Let $v \in \mathbb{Z}_{\geq 0}^{C/\!/G}$ be a branch type of cardinality n.*

1. *If v and g are large enough, then the C-branched Schur invariant is a complete invariant for the orbits of the action of* $\mathrm{MCG}_*(\Sigma_{g,n})$ *on $R_{g,v}$.*
2. *If v is large enough and C generates G, then the C-branched Schur invariant is a complete invariant for the orbits of the action of* $\mathrm{MCG}_*(\Sigma_{g,n})$ *on $R_{g,v}$* for all $g \geq 0$.

In both cases, the set of orbits

$$R_{g,v}/\,\mathrm{MCG}_*(\Sigma_{g,n})$$

is a torsor for $M(G)_C$.

In principle, Theorem 1.1 gives a practical way to compute the orbits of $R_{g,v}$ in the *stable range*, meaning when v and g are large enough. One caveat, however, is that we do not have any upper bounds on when the stable range begins. In other words, how large v and g must be in order to guarantee the conclusions of Theorem 1.1 is unclear to us. One thing is clear: the answer depends intimately on G and C. See Section 4.1 for more discussion.

Theorem 1.1 is not entirely new. On one hand, Dunfield and Thurston solved the unbranched case where $C = \emptyset$ and $n = 0$ [**6**]. In this case, $BG_C = BG$ and $M(G)_C = M(G) = H_2(G)$, and we recover their results. On the other hand, Ellenberg, Venkatesh and Westerland solved the braid case where $g = 0$ and $\langle C\rangle = G$ [**9**]. Their techniques build on work of Fried and Volklein [**11**], who used unpublished ideas of Conway and Parker to solve the braid case with the additional assumptions that $C = G$ and $M(G)_C = 0$. Thus, our Theorem 1.1 can be understood as an interpolation between the two extremal cases of [**6**] and [**9**]. We note that our approach via the classifying space BG_C is briefly mentioned, but left undeveloped, in [**9**].

Section 3.4 contains the proof of Theorem 1.1. The first statement of the theorem follows by applying Proposition 3.1 in a manner similar to how Dunfield and Thurston applied Livingston's stable equivalence theorem [**15**]. The second statement follows from the Hopf-Whitney classification, a kind of generalization of the Hurewicz isomorphism. The final part of the theorem follows, in part, from a computation of the second homology of BG_C that we carry out in Section 2.3. Our proof exploits the well-known fact that homology and oriented bordism are the same in dimension 2. Section 4 contains more remarks on the stable range, and a brief discussion of potential applications to symmetry-enriched topological phases.

Acknowledgments

I want to thank Greg Kuperberg for helpful feedback concerning earlier drafts, in particular for bringing my attention to a gap in the first version's Lemma 3.2, as well as helping to streamline the proof of the second part of Theorem 1.1 by using the Hopf-Whitney classification. I also want to thank Greg for encouraging me to learn TikZ; if nothing else, the figures in this version should make it look better than the first version. Finally, my thanks to the anonymous referee for patient and helpful comments.

2. C-branched G-covers

Throughout this section, let G be a discrete group, and let $C \subseteq G$ be a conjugacy invariant subset, possibly empty and not necessarily inverse-closed.

2.1. Definitions and examples. Let M be a smooth, connected manifold, possibly with boundary. A *C-branched G-cover of M* consists of the following data:

1. A smooth map $f : \tilde{M} \to M$, where $\tilde{M}$ is a smooth manifold.
2. A codimension 2 properly embedded submanifold $K \subseteq M$, possibly empty, called the *branch locus*.
3. A framing of K, *i.e.* a trivialization of the unit disk bundle
$$N(K) \cong K \times D^2,$$
where D^2 has the standard orientation it inherits as a submanifold of $\mathbb{R}^2$. We will conflate $N(K)$ with a closed regular neighborhood of K.

This data must satisfy the following conditions:

1. $f \mid f^{-1}(M \smallsetminus K)$ is a regular G-cover.
2. (C-branched condition) The monodromy homomorphism $\pi_1(M \smallsetminus K) \to G$ associated to the regular G-cover $f \mid f^{-1}(M \smallsetminus K)$ sends a counterclockwise loop around the boundary of each fiber D^2 of $N(K)$ into C. ("Counterclockwise" is determined by the framing of K.)
3. $f \mid f^{-1}(K)$ is a cover over each component of K.

If M is not connected, then a C-branched G-cover of M is simply a C-branched G-cover of each component of M. We will variously abuse notation by referring to $\tilde{M}$ or M as a C-branched G-cover of M, and taking the other structures for granted. Also, C and G will be fixed throughout, so if we say "branched cover," we always mean C-branched G-cover.

We make several remarks regarding this definition. First, if either $C = \emptyset$ or $K = \emptyset$, then a C-branched G-cover of M is just a regular, unbranched G-cover of M. Second, a branched cover of M is connected if and only if the regular cover over $M \smallsetminus K$ is connected if and only if the monodromy homomorphism $\pi_1(M \smallsetminus K) \to G$ is surjective.

Finally, note that if we pick a component of K, then for any two fibers of $N(K)$ over that component, their counterclockwise boundary loops map to conjugate elements of G. Thus, to every component of K we associate a conjugacy class in C, called the *branch type* of the component. Relatedly, if $K \subseteq M$ does not admit a framing, then, by our definition, M does not have any C-branched G-covers with branch locus K. Thus, it is arguably more precise to call our covers "*framed* C-branched G-covers." There are alternative reasonable definitions for what it means to be C-branched in the absence of a framing. For example, one may suppose the weaker condition that $N(K)$ admit a consistent orientation of its fibers. However, the classifying space BG_C we exploit later classifies framed covers, so we will absorb the adjective "framed" into the adjective "C-branched." See the end of Section 2.2 for more discussion.

There are three equivalence relations on C-branched G-covers that we will need: equivalence, concordance, and cobordism. Each of these is coarser than the preceding one. As we shall see, Theorem 1.1 can be interpreted as saying that cobordism is sometimes enough to guarantee equivalence anyway.

Two C-branched G-covers $\tilde{M}_0$ and $\tilde{M}_1$ of M_0 and M_1, respectively, are *equivalent* if there is a diffeomorphism from $\tilde{M}_0$ to $\tilde{M}_1$ that takes $f_0^{-1}(K_0)$ diffeomorphically to $f_1^{-1}(K_1)$ so that the framings are identified, and that is an equivalence of G-covers on the complement of the branch loci. If M_0 and M_1 are oriented, we require equivalences to preserve orientations.

Two C-branched G-covers $\tilde{M}_0$ and $\tilde{M}_1$ are *cobordant* if there is a manifold W such that $\partial W = M_0 \sqcup M_1$, and a C-branched G-cover $\tilde{W}$ that is equivalent to $\tilde{M}_0$ when restricted to M_0 and equivalent to $\tilde{M}_1$ when restricted to M_1. In particular, the framing of the branch locus of W has to extend the framings of K_0 and K_1. We can also talk about oriented cobordism. In the sequel, when we say cobordism we will always mean oriented cobordism.

If $W = M \times I$ is a cobordism, then we say $\tilde{M}_0 = M \times \{0\}$ and $\tilde{M}_1 = M \times \{1\}$ are *concordant*. Every equivalence yields a concordance by taking the mapping cylinder. Not all concordances are cylinders because there can be births and deaths of components of the branch loci.

The next lemma shows that a C-branched G-cover of M is uniquely specified by K, a framing of K, and a homomorphism $\pi_1(M \setminus K) \to G$ satisfying the C-branched condition. The proof explains the requirement that $N(K)$ be trivializable (or, at least, that the fibers of $N(K)$ be coherently orientable).

LEMMA 2.1. *Let K be a codimension 2 properly embedded submanifold of M such that $N(K)$ is trivializable. Then for every choice of framing of K and for every regular G-cover of $M \setminus K$ that satisfies the C-branched condition with respect to the chosen framing, there is a unique (up to equivalence) C-branched G-cover of M with branch locus K with the given framing and G-cover of $M \setminus K$.*

PROOF. Let $\bar{c} \in C \,/\!/\, G$. There is a standard way to construct a branched G-cover of the oriented disk D^2, with a single branch point of type $\bar{c}$. Each component is modeled on the the pth power map $z \mapsto z^p$ on the unit disk in the complex plane, where p is the order of the group element c; there are $|G|/p$ components. Call this branched cover $D^2_{\bar{c}}$.

Given a trivialization of $N(K)$, let $K_{\bar{c}}$ be a component of K with branch type $\bar{c}$. Then $K_{\bar{c}} \times D^2_{\bar{c}}$ is a C-branched G-cover of $N(K_{\bar{c}})$ with branch locus $K_{\bar{c}}$. We form a C-branched G-cover of $N(K)$ by taking the union over components of K.

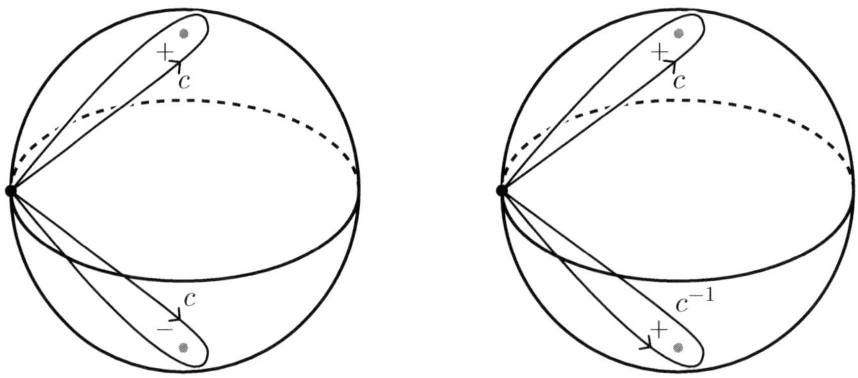

FIGURE 1. Two examples of C-branched G-covers of S^2. The branch locus K of both covers (indicated in green) consists of two points. The framing of K is indicated by signs on the branch points. The left cover can be extended into the 3-ball by adding an oriented arc that connects the two branch points and using the blackboard framing. Thus, the left cover is null-cobordant. The right cover is not null-cobordant, because the two branch points have the same framing. Note that this is true even if $c = c^{-1}$, and even though the unbranched G-covers of $S^2 \setminus K$ are the same. Thus we see the choice of framing can affect the equivalence class of a C-branched G-cover whenever C is nonempty.

The C-branched condition implies the restriction of this branched cover to $\partial N(K)$ is an unbranched regular G-cover that is equivalent to the cover of $\partial N(K)$ induced by the given regular G-cover of $M \setminus K$. So we can glue the regular G-cover of $M \setminus N(K)^\circ$ to the C-branched G-cover of $N(K)$ along $\partial N(K)$. Note that, before gluing, it may be necessary to homotope one of the covers in a neighborhood of $\partial N(K)$ to guarantee the result is smooth. This yields a C-branched G-cover of M with branch locus K with the given trivialization of K and G-cover of $M \setminus K$. It is straightforward to verify that any other branched cover of M with this data is equivalent. □

We conclude this subsection with some examples of C-branched G-covers. We first explain how we shall describe the covers. Lemma 2.1 says we can specify a well-defined C-branched G-cover of M by specifying K, a trivialization of $N(K)$, and a G-cover of $M \setminus K$ satisfying the C-branched condition. We describe the G-cover of $M \setminus K$ by picking a basepoint and fixing a homomorphism $\pi_1(M \setminus K) \to G$.

When M is 2-dimensional, K is simply a collection of points, which we will call *branch points*. The disk bundle $N(K)$ is the union of the unit tangent disks at K. Up to oriented equivalence, a trivialization of a 2-disk bundle over a point is just an orientation of the disk. We conclude that if M is oriented, we can specify a framing of K by labelling each branch point in K with a sign $+$ or $-$ to indicate whether the trivialization of the normal disk agrees with the orientation of M or not.

Similar remarks apply when M is oriented and 3-dimensional. In this case, K is a link in M, and we can specify a framing of K by assigning an integer to each

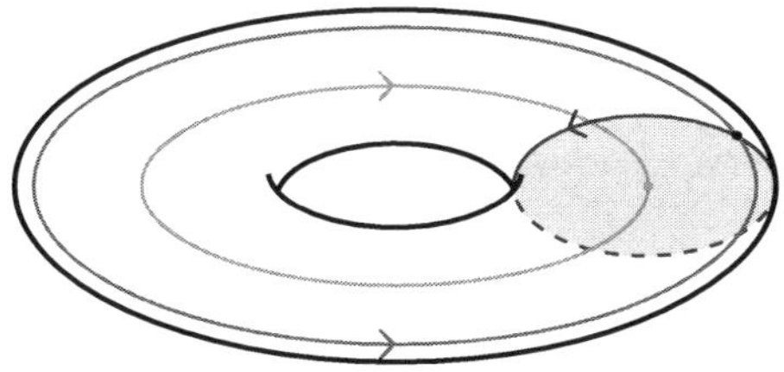

FIGURE 2. This figure shows that every C-torus is null-cobordant as a C-branched G-cover. The red and blue curves represent the standard generators of $\pi_1(S^1 \times S^1)$. Suppose the blue curve is mapped to $a \in C$ and the red curve is mapped to some element $b \in G$ that commutes with a. We can extend this (unbranched) G-cover of $S^1 \times S^1$ to a C-branched G-cover of the solid torus $D^2 \times S^1$. The branch locus K is the core, indicated by the green curve, equipped with the blackboard framing.

component. In practice, when drawing figures, we will always use the blackboard framing, which is determined by an orientation.

Figure 1 provides two different C-branched G-covers of S^2 with the same unbranched G-covers over $S^2 \setminus K$. These examples show how a C-branched G-cover depends on the choice of framing.

Figure 2 shows that every C-torus is null-cobordant when considered as a C-branched G-cover of $S^1 \times S^1$. In fact, we can generalize this example.

LEMMA 2.2. *If C generates G, then every unbranched G-cover of a closed, oriented surface is cobordant to a branched cover of S^2.*

PROOF. We begin by showing that every unbranched G-cover of a torus with boundary is cobordant to a C-branched G-cover of the disk D^2. Consider Figure 3. Suppose the blue α curve is mapped to $a \in G$ and the red β curve is mapped to $b \in G$. This describes an unbranched G-cover of the torus with boundary. We glue on a C-branched G-cover of a 2-handle as indicated in the figure. The branch locus K is given by the green curves, and the framing of K is understood to be the blackboard framing determined by the orientations of the curves. Send the loops γ_i to any $x_i \in C$ so that

$$x_k \cdots x_1 = a,$$

and send the δ_i to y_i in C so that

$$y_l \cdots y_1 = b.$$

This construction can now be used to construct the desired cobordism.

To prove the lemma, just do the above on every handle. Precisely, let $\alpha_1, \ldots, \alpha_g$ be a maximal collection of non-isotopic, disjoint, essential simple closed curves on Σ_g, and glue a branched cover of a 2-handle along each α_i as in the previous paragraph. □

2.2. Classifying space. We now construct a classifying space BG_C such that homotopy classes of maps from M to BG_C are naturally bijective with concordance classes of C-branched G-covers of M. The first such constructions were provided by Brand [**3**], who considered the case of irregular n-sheeted branched covers with various restrictions on the branch types. The covers he considers can be identified

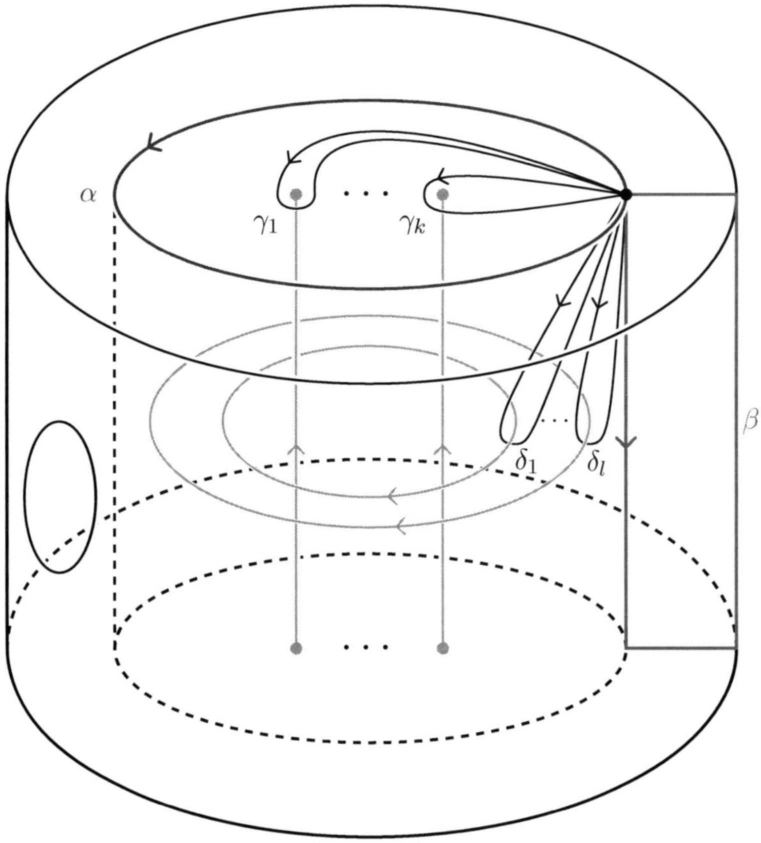

FIGURE 3. Surgering a G-handle. The black circle on the left of the figure is a boundary component.

with regular branched G-covers where $G = S_n$, and BG_C recovers Brand's constructions in this case. The more general construction we use was first described, to our knowledge, by Ellenberg, Venkatesh and Westerland in [**9**], where they use the notation $A(G, C)$ for BG_C. Since [**9**] was never published, and they do not provide the proof that BG_C is a classifying space, we do so here.

Let $LBG = \text{Maps}(S^1, BG)$ be the loop space of BG, which is the space of all maps from the circle S^1 into BG with the compact-open topology. The set of components of LBG is the space of free homotopy classes of maps $S^1 \to BG$, which after orienting S^1 can be naturally identified with the set of conjugacy classes $G /\!/ G$. Let $L^C BG$ be the union of components of LBG corresponding to $C /\!/ G$. Then

$$BG_C \stackrel{\text{def}}{=} BG \bigsqcup_{ev} L^C BG \times D^2,$$

where

$$ev : L^C BG \times S^1 \to BG$$

is the evaluation map. Intuitively, for every loop in BG representing a free homotopy class in $C /\!/ G$, we glue a disk to BG so that the boundary of the disk is sent to that loop. Our next theorem formalizes what we mean when we say that BG_C is a classifying space.

THEOREM 2.3. *Fix a discrete group G and a conjugacy invariant subset $C \subseteq G$. Let M be a smooth manifold. Then homotopy classes of maps from M to BG_C are in natural bijection with concordance classes of C-branched G-covers of M.*

PROOF. Let $\phi : M \to BG_C$ be a continuous map. Fix $0 < \epsilon < 1$ and let $D_\epsilon(0) \subseteq D^2$ be a small disk of radius ϵ centered at 0. Let $\pi : L^C BG \times D^2 \to D^2$ be the projection map. The Whitney approximation theorem (see [**14**, Thm. 6.19]) implies that after applying some homotopy to ϕ, we can guarantee that

$$\pi \circ \phi : (\pi \circ \phi)^{-1}(D_\epsilon(0)) \to D_\epsilon(0)$$

is smooth. By Sard's theorem, $\pi \circ \phi$ has a regular value $z \in D_\epsilon(0)$. The preimage $\phi^{-1}(L^C BG \times \{z\})$ is a codimension two submanifold $K \subseteq M$. Since $N(K)$ is the pullback of the normal bundle of z inside D^2, which is trivial, we get a trivialization of $N(K)$. Note that

$$BG_C \setminus L^C BG \times \{z\}$$

is homotopy equivalent to BG. Thus

$$\phi : M \setminus K \to BG_C \setminus L^C BG \times \{z\}$$

classifies some G-cover of $M \setminus K$. Clearly this cover satisfies the C-branched condition. Applying Lemma 2.1, this G-cover of $M \setminus K$ and the trivialization of $N(K)$ assemble to form a C-branched G-cover over M with branch locus K.

Let ϕ and ψ be continuous maps $M \to BG_C$, and let

$$\Phi : M \times [0,1] \to BG_C$$

be a homotopy from ϕ to ψ. Without loss of generality, we assume that there is an $0 < \epsilon < 1$ so that

$$\pi \circ \phi : (\pi \circ \phi)^{-1}(D_\epsilon(0)) \to D_\epsilon(0)$$

and

$$\pi \circ \psi : (\pi \circ \psi)^{-1}(D_\epsilon(0)) \to D_\epsilon(0)$$

are both smooth. Homotope Φ rel $M \times \{0,1\}$ so that

$$\pi \circ \Phi : (\pi \circ \Phi)^{-1}(D_\epsilon(0)) \to D_\epsilon(0)$$

is smooth, and construct a C-branched G-cover of $M \times [0,1]$ as in the previous paragraph. The resulting cover is a concordance between the covers associated to ϕ and ψ. Thus the homotopy class of ϕ yields a well-defined concordance class of C-branched G-covers of M.

Conversely, let $\tilde{M}$ be a C-branched G-cover of M with branch locus K. Then there is a map $\phi : M \setminus N(K)^\circ \to BG$ classifying the G-cover of $M \setminus N(K)^\circ$. The trivialization of $N(K)$ is equivalent to a map $N(K) \to D^2$ that takes K to 0 and is a diffeomorphism on the fibers of $N(K)$. Define a map $N(K) \to L^C BG$ by sending a point in $N(K)$ to the loop in BG determined by ϕ and the boundary of the fiber that point lives in (the parametrization of the loop is determined by the trivialization). The product of these maps from $N(K)$ is a map $N(K) \to L^C BG \times D^2$ that agrees with ϕ on $\partial N(K)$ when composed with the evaluation map $L^C BG \times S^1 \to BG$. Thus $\tilde{M}$ determines a map $M \to BG_C$. Applying this construction to a concordance between two branched covers of M yields a homotopy between the corresponding maps to BG_C. □

It is interesting to consider how to modify the construction of BG_C so that framings are no longer induced by maps to BG_C. We expect that instead of gluing a copy of D^2 to BG for every map $S^1 \to BG$ in the conjugacy class of $c \in G$, one should glue a copy of a cone over the classifying space $B\langle c\rangle$ of the cyclic group generated by c for every map $B\langle c\rangle \to BG$ in the appropriate homotopy class. One might call this new space BG_C^{unfr}. Since, *a priori*, the statement of Theorem 1.1 is not concerned with framings, it is reasonable to ask for a proof that does not unnecessarily introduce framings, and, hence, a more thorough development of BG_C^{unfr}. However, this would require more algebraic topology than we want to invoke, and so, for expediency's sake, we will not pursue such a direction in this paper; instead, we employ a geometric technique to circumvent framings in Section 3.3.

2.3. Invariants of BG_C. Let us compute some basic topological invariants of BG_C.

First we consider $L^C BG$. Clearly

$$|\pi_0(L^C BG)| = |C \mathbin{/\!/} G|.$$

A point p in $L^{\bar{c}}BG$ is a loop in the free homotopy class $\bar{c}$. A loop in $L^{\bar{c}}BG$ with basepoint p is a map from a torus $S^1 \times S^1$ to BG which restricts to p on the meridian $S^1 \times \{1\}$. If we fix a basepoint of BG, then we can pick p so that its *pointed* homotopy class represents $c \in \bar{c}$. A little more work shows

LEMMA 2.4. *Let G be a discrete group and let $\bar{c} \in G \mathbin{/\!/} G$ be a conjugacy class in G with representative c. Then $\pi_1(L^{\bar{c}}BG) \cong Z_G(c)$, the centralizer of c in G. In particular, $H_1(L^{\bar{c}}BG) \cong Z_G(c)_{ab}$.* □

In fact, $L^{\bar{c}}BG$ is a $K(Z_G(c), 1)$ (see, for instance, [**16**]). Thus the homology of $L^{\bar{c}}BG$ is the same as the group homology of $Z_G(c)$. We will not use this fact, rather we simply state it so that the following lemma may be interpreted as a satisfying calculation of $H_n(BG_C)$.

LEMMA 2.5. *Let G be a discrete group and let $C \subseteq G$ be a conjugacy invariant subset. Then the homology of BG_C fits into an exact sequence*

$$\cdots \to H_{n-1}(L^C BG) \to H_n(BG) \to$$
$$H_n(BG_C) \to H_{n-2}(L^C BG) \to \cdots.$$

PROOF. Decompose BG_C into the two subspaces

$$BG \bigsqcup_{ev} L^C BG \times (D^2 \setminus \{0\})$$

and

$$L^C BG \times D_\epsilon(0)$$

where $D_\epsilon(0) \subseteq D^2$ is a small disk of radius $0 < \epsilon < 1$ centered at 0. The former subspace deformation retracts onto BG, while the latter subspace deformation retracts to $L^C BG \times \{0\} \simeq L^C BG$. The intersection of the two subspaces is homotopy equivalent to $L^C BG \times S^1$, with the inclusion

$$i : L^C BG \times S^1 \to BG$$

the evaluation map, and the inclusion

$$j : L^C BG \times S^1 \to L^C BG$$

the projection onto the first factor.

Consider the Mayer-Vietoris sequence

$$\cdots \to H_n(L^C BG \times S^1) \to H_n(L^C BG) \oplus H_n(BG) \to H_n(BG_C) \to H_{n-1}(L^C BG \times S^1) \to \cdots .$$

The Künneth formula shows

$$H_n(L^C BG \times S^1) \cong H_{n-1}(L^C BG) \oplus H_n(L^C BG).$$

The induced map

$$j_* : H_n(L^C BG \times S^1) \to H_n(L^C BG)$$

simply projects out $H_{n-1}(L^C BG)$. Thus, by exactness, we can modify the last term of the sequence, yielding

$$\cdots \to H_{n-1}(L^C BG) \oplus H_n(L^C BG) \to H_n(L^C BG) \oplus H_n(BG) \to H_n(BG_C) \to H_{n-2}(L^C BG) \to \cdots .$$

The image of $H_n(L^C BG)$ inside $H_n(L^C BG) \oplus H_n(BG)$ is the graph of the induced map $H_n(L^C BG) \to H_n(BG)$. The quotient of a direct sum by the graph of a linear map is the codomain, so we get the new exact sequence

$$\cdots \to H_{n-1}(L^C BG) \to H_n(BG) \to H_n(BG_C) \to H_{n-2}(L^C BG) \to \cdots .$$

□

Let us analyze this exact sequence in low dimensions. When $n = 1$, we have

$$\mathbb{Z}^{C/\!/G} \to H_1(BG) \to H_1(BG_C) \to 0.$$

Of course, $H_1(BG) = H_1(G) = G_{ab}$, the abelianization of G. The map $\mathbb{Z}^{C/\!/G} \to G_{ab}$ is the evaluation map defined by taking an element of $C /\!/ G$ to its image in G_{ab}. We deduce

$$H_1(BG_C) \cong (G/\langle C \rangle)_{ab}.$$

Note that because C is conjugacy invariant, $\langle C \rangle$ is automatically normal. In fact, it is a straightforward exercise to compute $\pi_1(BG_C)$.

LEMMA 2.6. *For any discrete group G and conjugacy invariant subset $C \subset G$, $\pi_1(BG_C) \cong G/\langle C \rangle$.* □

Consider the exact sequence of Lemma 2.5 when $n = 2$, and substitute the result of Lemma 2.4:

$$\bigoplus_{\overline{c} \in C/\!/G} Z_G(c)_{ab} \to H_2(G) \to H_2(BG_C) \to \mathbb{Z}^{C/\!/G}.$$

Let N be the kernel of the evaluation map

$$\mathbb{Z}^{C/\!/G} \to G_{ab}.$$

Then by the $n = 1$ discussion, we can extend the sequence to

$$\bigoplus_{\overline{c} \in C/\!/G} Z_G(c)_{ab} \to H_2(G) \to H_2(BG_C) \to N \to 0.$$

To interpret this sequence, we use the fact that oriented bordism is the same as homology in dimension 2. Combining this with Theorem 2.3, every element of $H_2(BG_C)$ can be represented by a C-branched G-cover of a closed, oriented surface Σ. The map $H_2(BG_C) \to N$ takes this cover of Σ to its *homological branch type*, which algebraically counts the different conjugacy types of the branch points, with a sign that depends on whether the trivialization of the normal plane at a branch point agrees with the orientation of Σ or not. In particular, the homological branch type can easily be computed from the usual branch type. (See Section 3.1 for more details on branch types.)

The image of $Z_G(c)_{ab}$ inside $H_2(G)$ can be understood as those homology classes in $H_2(G)$ that are represented by sums of tori where the meridian maps to the conjugacy class $\overline{c}$, that is, $\overline{c}$-tori. Thus, $H_2(BG_C)$ fits into the exact sequence

$$0 \to M(G)_C \to H_2(BG_C) \to N \to 0,$$

where, recall from the introduction, the C-branched Schur multiplier of G is defined as

$$M(G)_C = H_2(G)/\langle C\text{-tori}\rangle.$$

Since $N \subseteq \mathbb{Z}^{C/\!/G}$ is free abelian, the sequence splits (but not naturally). If G is finite, then we can identify $M(G)_C$ with the torsion subgroup of $H_2(BG_C)$:

$$M(G)_C \cong H_2(BG_C)_{tor}.$$

3. Stable orbits

In this section we prove Theorem 1.1.

3.1. Framed branching data and branched Schur invariants. Let ϕ be a homomorphism in $R_{g,v}$ and let $K = \{p_1, \dots, p_n\} \subseteq \Sigma_g$ be the set of punctures. To associate a C-branched G-cover of Σ_g to ϕ, we must first choose a framing of K. As discussed at the end of Section 2.1, we can specify a framing simply by decorating each point p_i with a sign $o_i \in \{+1, -1\}$.

The simplest choice of framing is the *positive framing* with $o_i = +1$ for all $i = 1, \dots, n$. If $v \in \mathbb{Z}_{\geq 0}^{C/\!/G}$ has cardinality n, it follows immediately from the definition of $R_{g,v}$ in Section 1 that the homomorphism ϕ satisfies the C-branched condition with respect to the positive framing. Thus, by Lemma 2.1, we can form a C-branched G-cover of Σ_g with branch locus K.

Theorem 1.1 is only concerned with homomorphisms in $R_{g,v}$, and, hence, branched covers with positive trivializations. However, since our proof uses BG_C, it requires us to study covers with trivializations that are not positive. The definition of branching data we provided in Section 1 is only sufficient for positive framings, so we now define and introduce notation for more general framings.

Let

$$T : K \to \{+1, -1\}$$
$$p_i \mapsto o_i$$

denote a framing of K and pick a homomorphism

$$\phi : \pi_1(\Sigma_{g,n}) \to G$$

that satisfies the C-branched condition with respect to T. Lemma 2.1 shows the pair (T, ϕ) determines a C-branched G-cover of Σ_g with branch locus K, which we will often refer to simply by (T, ϕ). We define the *branching data* of this cover to be the vector

$$v_{(T,\phi)} \in \mathbb{Z}_{\geq 0}^{C /\!/ G \times \{+1,-1\}}$$

such that

$$(v_{(T,\phi)})(\overline{c}, o) \stackrel{\text{def}}{=} |\{1 \leq i \leq n \mid o_i = o, \phi(\gamma_i)^o \in \overline{c}\}|,$$

where $\overline{c}$ is a conjugacy class in $C /\!/ G$ and $o \in \{+1, -1\}$. Thus, $(v_{(T,\phi)})(\overline{c}, o)$ counts the number of branch points such that winding around them in the direction o yields monodromy in $\overline{c}$.

In Section 1 we defined the branching data of a homomorphism in $R_{g,n}$. Implicit in that definition was the use of the positive framing of K. From now on, we make the identification

$$\mathbb{Z}_{\geq 0}^{C /\!/ G} = \mathbb{Z}_{\geq 0}^{C /\!/ G \times \{+1\}} \subseteq \mathbb{Z}_{\geq 0}^{C /\!/ G \times \{+1,-1\}}.$$

So our new definition generalizes the previous definition: $v_{(T,\phi)} = v_\phi$ if T is the constant function $+1$, *i.e.* a positive framing.

As before, we say that a vector in v in $\mathbb{Z}_{\geq 0}^{C /\!/ G \times \{+1,-1\}}$ has cardinality n if the sum of its entries is n. Given any v of cardinality n, we define

$$R_{g,v} \stackrel{\text{def}}{=} \{(T, \phi) \mid \phi \text{ onto and } C\text{-branched w.r.t. } T,\ v_{(\phi,T)} = v\}.$$

The pointed mapping class group $\mathrm{MCG}_*(\Sigma_{g,n})$ acts on $R_{g,v}$. The action on the ϕ component is the usual action of $\mathrm{MCG}_*(\Sigma_{g,n})$ on a homomorphism, and the action on the T component is induced by the permutation action on K. If

$$v \in \mathbb{Z}_{\geq 0}^{C /\!/ G} = \mathbb{Z}_{\geq 0}^{C /\!/ G \times \{+1\}} \subseteq \mathbb{Z}_{\geq 0}^{C /\!/ G \times \{+1,-1\}}$$

then T must be the positive trivialization, and the definition of $R_{g,v}$ given is the same as the definition of $R_{g,v}$ given in Section 1.

It is useful to interpret $R_{g,v}$ as a parametrization of the set of equivalence classes of connected C-branched G-covers of Σ_g with branching data v. Indeed, every connected C-branched G-cover of Σ_g is equivalent to a cover specfied by an element of $R_{g,v}$ for some unique v. Of course, there could be many such elements of $R_{g,v}$; the goal of this paper is to understand them.

By Theorem 2.3, the cover $(T, \phi) \in R_{g,v}$ induces a homotopy class of a map

$$(T, \phi)_\# : \Sigma_g \to BG_C$$

which in turn induces a homomorphism

$$(T, \phi)_* : H_2(\Sigma_g) \to H_2(BG_C).$$

The *C-branched Schur invariant of* (T, ϕ) is the homology class

$$\mathrm{sch}_C(T, \phi) \stackrel{\text{def}}{=} (T, \phi)_* ([\Sigma_g]) \in H_2(BG_C),$$

where $[\Sigma_g]$ is the orientation of Σ_g.

3.2. Stable equivalence lemmas. We now explain the role of the Schur invariant.

Let (T, ϕ) be a C-branched G-cover of Σ_g. A *handle stabilization* of (T, ϕ) is any C-branched G-cover of a surface that is equivalent to the connect sum of (T, ϕ) with the trivial C-branched G-cover over the torus $S^1 \times S^1$.

For any $\overline{c} \in C /\!/ G$, let $S^2_{\overline{c}}$ denote any C-branched G-cover of the oriented sphere S^2 such that the branch locus consists of one point with branch type $(\overline{c}, +1)$ and one point with branch type $(\overline{c}, -1)$. For example, see the cover on the left side of Figure 1. If the conjugacy class $\overline{c}$ has more than one element, then there are inequivalent covers which we denote $S^2_{\overline{c}}$. However, there will never be any ambiguity because of how we use these covers, which we now explain.

A *$\overline{c}$-stabilization* of (T,ϕ) is any C-branched G-cover of a surface that is equivalent to the connect sum of (T,ϕ) with a cover of the form $S^2_{\overline{c}}$. Note that if (T,ϕ) is connected (*i.e.* ϕ is surjective), it does not matter which $S^2_{\overline{c}}$ we use, since stabilizing by any of them yields equivalent covers.

A *puncture stabilization* of (T,ϕ) is some sequence of $\overline{c}$-stabilizations of (T,ϕ) with various $\overline{c}\in C /\!/ G$. We call a connect sum of copies of $S^2_{\overline{c}}$, for possibly varying c, a *puncture stabilizing sphere.*

Note that if $(T,\phi)\in R_{g,v}$, then a $\overline{c}$-stabilization of (T,ϕ) has branch type

$$v+\delta_{(\overline{c},+1)}+\delta_{(\overline{c},-1)}\in \mathbb{Z}_{\geq 0}^{C/\!/G\times\{+1,-1\}}$$

where $\delta_{(\overline{c},\pm 1)}$ is the delta function on $(\overline{c},\pm 1)$. In particular, a puncture stabilization never has positive trivialization. (This fact is responsible for the "dilation" map we introduce in the next subsection.)

Finally, we say two C-branched G-covers are *stably equivalent* if they are equivalent after applying some sequence of handle and puncture stabilizations to each of them.

PROPOSITION 3.1. *Suppose $(T,\phi)\in R_{g,v}$ and $(S,\psi)\in R_{g',w}$ are connected C-branched G-covers of oriented surfaces. Then $\mathrm{sch}_C(T,\phi)=\mathrm{sch}_C(S,\psi)$ if and only if (T,ϕ) and (S,ψ) are stably equivalent.*

Suppose, moreover, that $\langle C\rangle = G$ and $g=g'$. Then handle stabilization is unnecessary. That is, $\mathrm{sch}_C(T,\phi)=\mathrm{sch}_C(S,\psi)$ if and only if (T,ϕ) and (S,ψ) are puncture-stabilization equivalent.

Livingston proved this result in the unbranched case for which C is empty [**15**]. Moreover, he considers the question of extending his results to the branched case, but without employing puncture stabilization. He even gives an example of two unbranched covers that are bordant as branched covers, but not as unbranched covers. Our proposition seems to be the appropriate generalization of Livingston's result to the (framed) branched case.

PROOF. As in Livingston's proof, the main idea is to use the fact that integral homology and oriented cobordism are the same in dimension 2. Thus, if the homology classes $\mathrm{sch}_C(T,\phi)$ and $\mathrm{sch}_C(S,\psi)$ in $H_2(BG_C)$ are equal, then there is a branched cover of some oriented 3-manifold M with boundary the union of (T,ϕ) and (S,ψ). Let $K\subseteq M$ be the branch locus of this cover.

Construct a relative handle decomposition of M and arrange so that all 1-handles are attached before 2-handles, and so that the attaching maps of the handles avoid the branch loci on Σ_g and $\Sigma_{g'}$. Because (T,ϕ) is connected, we can slide the attaching maps for the 1-handles in the complement of the branch locus to guarantee that the G-cover over each of these new handles is trivial. We can do the same thing for the 2-handles because (S,ψ) is connected. See [**15**] for details. Let Σ be the resulting relative Heegaard surface, which is the intersection of the compression bodies

$$H_1 \stackrel{\text{def}}{=} \Sigma_g\cup\{\text{1-handles}\},$$

$$H_2 \stackrel{\text{def}}{=} \Sigma_h\cup\{\text{2-handles}\}.$$

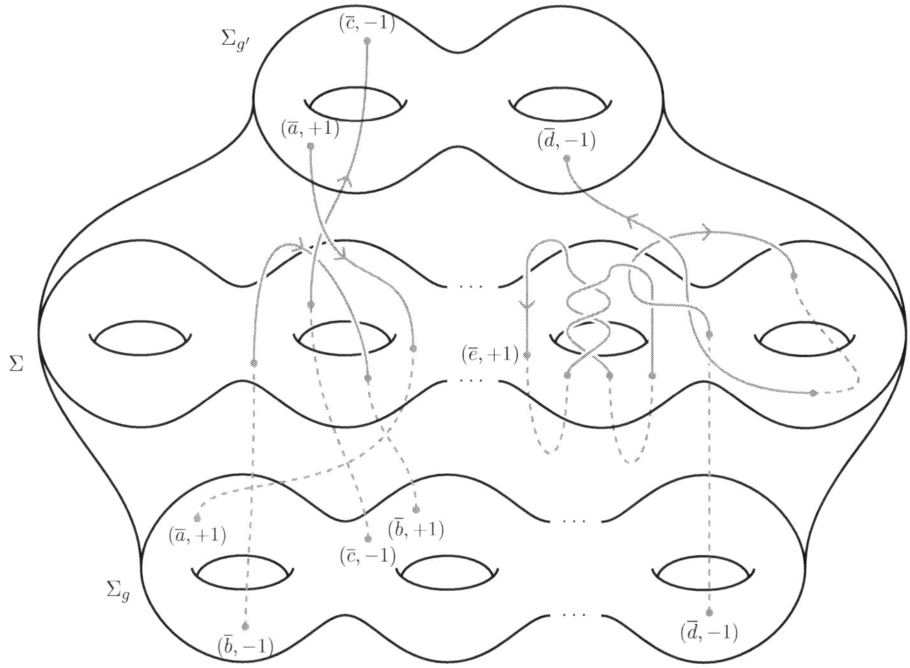

FIGURE 4. A cobordism M between the branched cover (T, ϕ) of Σ_g and the branched cover (S, ψ) of $\Sigma_{g'}$ provides a mutual stabilization of the two covers. The branch locus $K \subseteq M$ is drawn in green. The dashed green arcs denote K_1, which is the part of K between Σ_g and the relative Heegaard splitting Σ. The solid green arcs denote K_2, which is the part of K between Σ and $\Sigma_{g'}$. The labels a, b, c, d, e denote elements in C. To avoid overcrowding this schematic, we have not drawn a Heegaard diagram on Σ.

Isotope K so that it is transverse to Σ, all maxima of K are on the top side of Σ, and all minima are on the bottom side. The latter two conditions can be understood as saying that Σ is a bridge surface for the tangle $K \subseteq M$. The C-branched G-cover of M restricts to a C-branched G-cover over Σ. See Figure 4.

We claim that this cover of Σ is a stabilization of both (T, ϕ) and (S, ψ). Let

$$K_1 \stackrel{\text{def}}{=} K \cap H_1,$$
$$K_2 \stackrel{\text{def}}{=} K \cap H_2.$$

So K_1 is the portion of K that lies between Σ_g and Σ, while K_2 is the portion of K that lies between Σ and $\Sigma_{g'}$. Let $\alpha_1, \dots, \alpha_k \subseteq H_1$ be the components of K_1 without endpoints on Σ_g. The conditions on the maxima and minima of K guarantee that K_1 can be isotoped inside H_1 so that the arcs $\alpha_1, \dots, \alpha_k$ lie inside Σ simultaneously. Equivalently, there is a collection of disjoint disks $D_1, \dots, D_k$ in H_1 such that $\alpha_i \subseteq \partial D_i$ and $\partial D_i \setminus \alpha_i \subseteq \Sigma$. If we restrict the cover over Σ to a regular neighborhood of $\partial D_i \setminus \alpha_i^\circ$ in Σ, we see a summand equivalent to $S^2_{\overline{c}}$ for some $\overline{c} \in C \mathbin{/\!/} G$. We already slid the 1-handles to ensure the cover over them is trivial, so we conclude that Σ is a stabilization of (T, ϕ). This same argument with

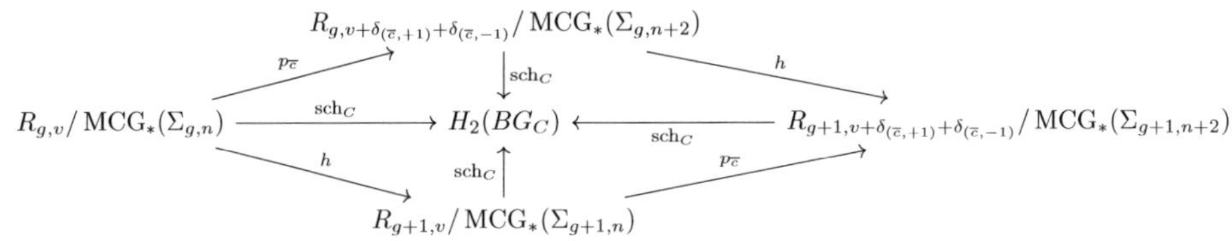

FIGURE 5. The diagram commutes.

$\Sigma_{g'}$, H_2 and K_2 used in place of Σ_g, H_1 and K_1 shows that the cover over Σ is a stabilization of (S, ψ). Thus (T, ϕ) and (S, ψ) are stably equivalent.

Conversely, suppose (T, ϕ) and (S, ψ) are stably equivalent. It follows from Figures 1 and 2 that $S^2_{\overline{c}}$ and the trivial cover of $S^1 \times S^1$ are both null-cobordant. Since homology and oriented bordism are the same in dimension 2, the branched Schur invariants of these covers are both trivial. The Schur invariant of a connect-sum is the sum of the Schur invariants. Hence puncture stabilization and handle stabilization both preserve the Schur invariant, and so $\mathrm{sch}_C(T, \phi) = \mathrm{sch}_C(S, \psi)$.

Finally, suppose $\langle C \rangle = G$, $g = g'$, and $\mathrm{sch}_C(T, \phi) = \mathrm{sch}_C(S, \psi)$. Lemma 2.6 says $\pi_1(BG_C)$ is trivial, so the Hopf-Whitney classification [**21**, Corollary V.6.19] and the universal coefficient theorem imply that homotopy classes of maps $\Sigma_g \to BG_C$ are determined by their induced maps $H_2(\Sigma_g) \to H_2(BG_C)$, which of course are determined by the values taken on $[\Sigma_g]$, *i.e.* the Schur invariant. Thus, (T, ϕ) and (S, ψ) induce homotopic maps to BG_C. As in the proof of Theorem 2.3, this implies (T, ϕ) and (S, ψ) are cobordant via a cylinder $M = \Sigma_g \times [0, 1]$. As before, isotope the branch locus $K \subset M$ so that it is in bridge position with respect to the surface $\Sigma = \Sigma_g \times \{1/2\}$. The restriction of the cover over M to Σ is a mutual puncture stabilization of (T, ϕ) and (S, ψ). $\square$

We remark that Proposition 3.1 is true even if G is infinite. However, all of our other results in this section use finiteness of G in some way.

Any two $\overline{c}$-stabilizations of a given connected cover are equivalent. Thus, $\overline{c}$-stabilization yields a well-defined map

$$p_{\overline{c}} : R_{g,v}/\,\mathrm{MCG}_*(\Sigma_{g,n}) \to R_{g,v+\delta_{(\overline{c},+1)}+\delta_{(\overline{c},-1)}}/\,\mathrm{MCG}_*(\Sigma_{g,n+2}).$$

Similarly, any two handle stabilizations are equivalent, so handle stabilization yields a map

$$h : R_{g,v}/\,\mathrm{MCG}_*(\Sigma_{g,n}) \to R_{g+1,v}/\,\mathrm{MCG}_*(\Sigma_{g+1,n}).$$

Since two elements of $R_{g,v}$ in the same $\mathrm{MCG}_*(\Sigma_{g,n})$ orbit are equivalent as C-branched G-covers, they must have the same Schur invariant. Thus, the Schur invariant yields a map

$$\mathrm{sch}_C : R_{g,v}/\,\mathrm{MCG}_*(\Sigma_{g,n}) \to H_2(BG_C).$$

All of these maps commute. More precisely, see the commuting diagram in Figure 5. A similar diagram commutes where, instead of involving a puncture stabilization and a handle stabilization, there are two puncture stabilizations of different types.

The next two lemmas show that whenever g and v are large enough, everything is a stabilization.

LEMMA 3.2. *If $g > |G|$, then h is surjective.*

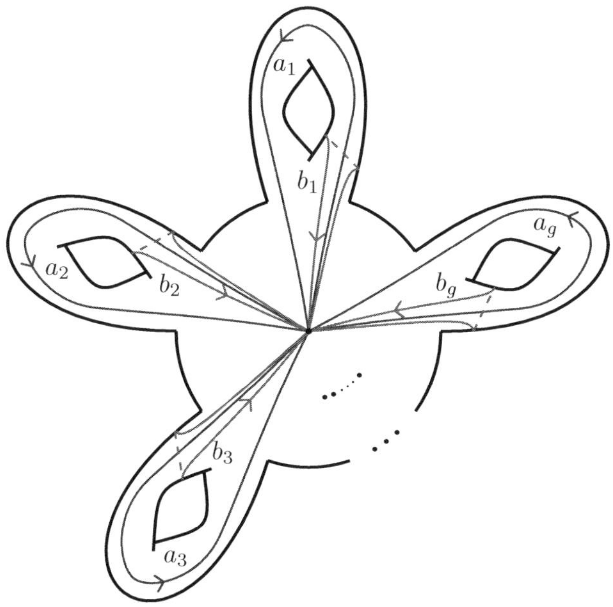

FIGURE 6. The elements of $\pi_1(\Sigma_{g,n})$ used in the proof of Lemma 3.2.

PROOF. The proof is exactly the same as the proof of Proposition 6.16 in Dunfield and Thurston's work [**6**]. In particular, the branch locus plays no role.

Let $(T, \phi) \in R_{g,v}$. We want to find a torus with one boundary component inside $\Sigma_{g,n}$ where the cover (T, ϕ) is trivial over it. Let $a_1, b_1, \dots, a_g, b_g$ be the elements of $\pi_1(\Sigma_{g,n})$ indicated in Figure 6, and let $w_i = a_i \cdots a_2 \cdot a_1$ for each $i = 1, \dots, g$. Since $g > |G|$, the pigeonhole principle guarantees $\phi(w_i) = \phi(w_j)$ for some $i < j$, hence $\phi(a_j \cdots a_{i+1}) = 1$. Thus, $a_j \cdots a_{i+1}$ is a non-separating simple closed curve on Σ_g in the kernel of ϕ. Let $c_1, \dots, c_k$ be a maximal disjoint collection of such curves. After applying some equivalence, we can make $c_1 = b_1, \dots, c_k = b_k$. The same argument as before shows that some $w = a_j \cdots a_{i+1}$ is in the kernel of ϕ. Since the collection $c_1, \dots, c_k$ is maximal, w must intersect one of the curves c_l. Since w and c_l intersect in exactly one point, a regular neighborhood of the two curves is a torus with boundary. By construction, ϕ is trivial on this torus, which shows (T, ϕ) is a handle stabilization. □

We say v is *larger* than w if

$$v(\overline{c}, o) > w(\overline{c}, o)$$

for all $(\overline{c}, o) \in C /\!\!/ G \times \{+1, -1\}$. For each conjugacy class $\overline{c}$ in $C /\!\!/ G$, define a special branch type

$$U^{\overline{c}} \in \mathbb{Z}_{\geq 0}^{C /\!\!/ G \times \{+1,-1\}}$$

that consists of the single element $(\overline{c}, +1)$ with multiplicity $|\overline{c}| \cdot |\operatorname{Inn}_{\overline{c}}(c)|$, where $|\operatorname{Inn}_{\overline{c}}(c)|$ is the order of c acting as an inner automorphism on the conjugacy class $\overline{c}$. That is,

$$U^{\overline{c}} = |\overline{c}| \cdot |\operatorname{Inn}_{\overline{c}}(c)| \cdot \delta_{(\overline{c},1)}.$$

LEMMA 3.3. *If v is larger than $U^{\overline{c}}$, then the puncture stabilization map $p_{\overline{c}}$ is surjective.*

PROOF. This lemma and its proof extend Lemma 4 of the appendix in Fried and Volklein's paper [**11**].

Let (T, ϕ) be an element of $R_{g,v+\delta_{(\overline{c},+1)}+\delta_{(\overline{c},-1)}}$. Pick a disk D in $\Sigma_{g,n+2}$ containing the branch set K. Pick a basepoint on the disk, and consider a generating set for $\pi_1(D \setminus K)$ consisting of small loops around the branch points, oriented according to T. Apply a braid to arrange so that a branch point of type $(\overline{c}, -1)$ is to the right of all the other branch points. The condition on v and the pigeonhole principle guarantee that there is some $c' \in \overline{c}$ that appears as the monodromy around $|\operatorname{Inn}_{\overline{c}}(c)|$ many of the positively-oriented branch points. Move all of these branch points to the left of the disk, being sure to braid them over the rest of the branch points, not under. This conjugates the monodromies of the branch points that pass underneath, but preserves the monodromies of the $|\operatorname{Inn}_{\overline{c}}(c)|$ points passing over. The resulting monodromies, arranged from left to right, look as follows:

$$c_1,\ c_1,\ c_1, \ldots\ c_1, *,\ *, \cdots *, c_2^{-1},$$

where there are $|\operatorname{Inn}_{\overline{c}}(c)|$ copies of c_1 on the left, $c_2 \in \overline{c}$, and the asterisks $*$ denote arbitrary elements of C. Choose a smaller disk D' that contains only the $|\operatorname{Inn}_{\overline{c}}(c)|$ branch points on the left. Since (T, ϕ) is connected, there is some loop on $\Sigma_{g,n+2}$ around which we can slide D' so that the result is

$$c_2,\ c_2,\ c_2, \ldots\ c_2, *,\ *, \cdots *, c_2^{-1}.$$

In particular, this sliding of D' does not change c_2^{-1} because any time c_2^{-1} gets conjugated by a loop around a puncture in D', it gets conjugated by all $|\operatorname{Inn}_{\overline{c}}(c)|$ of them. Now apply another braid to arrange so that a point of type c_2 is adjacent to the point of type c_2^{-1}. This shows that ϕ is equivalent to a $\overline{c}$-stabilization. ☐

3.3. Dilation replacement. In this subsection, we introduce a method called *dilation* for converting a branched cover with negatively trivialized branch points into a cover with a positive trivialization. The dilation of branching data is a linear map

$$\begin{aligned} \operatorname{dil} : \mathbb{Z}_{\geq 0}^{C/\!/G \times \{+1,-1\}} &\to \mathbb{Z}_{\geq 0}^{C/\!/G} \\ w &\mapsto w_{\text{dil}} \end{aligned}$$

where w_{dil} is defined by

$$w_{\text{dil}}(\overline{c}) \stackrel{\text{def}}{=} w(\overline{c}, +1) + (\operatorname{ord}(c) - 1) \cdot w(\overline{c}, -1).$$

Here $\operatorname{ord}(c)$ is the order of the group element c. If w has cardinality n, then w_{dil} has cardinality $n + N$ where

$$N = \sum_{\overline{c} \in C/\!/G} (\operatorname{ord}(c) - 2) w(\overline{c}, -1).$$

Extend this notion of dilation to elements of $R_{g,w}$ via the map

$$\begin{aligned} \operatorname{dil} : R_{g,w} &\to R_{g,w_{\text{dil}}} \\ (T, \phi) &\mapsto (+1, \phi_{\text{dil}}) = \phi_{\text{dil}} \end{aligned}$$

where $+1$ denotes the positive trivialization and ϕ_{dil} is defined by replacing each puncture of type $(\overline{c}, -1)$ with $\operatorname{ord}(c) - 1$ punctures of type $(\overline{c}, +1)$ as in Figure 7. Note that this map dil depends on a choice of generating set of $\pi_1(\Sigma_{g,n})$; we use

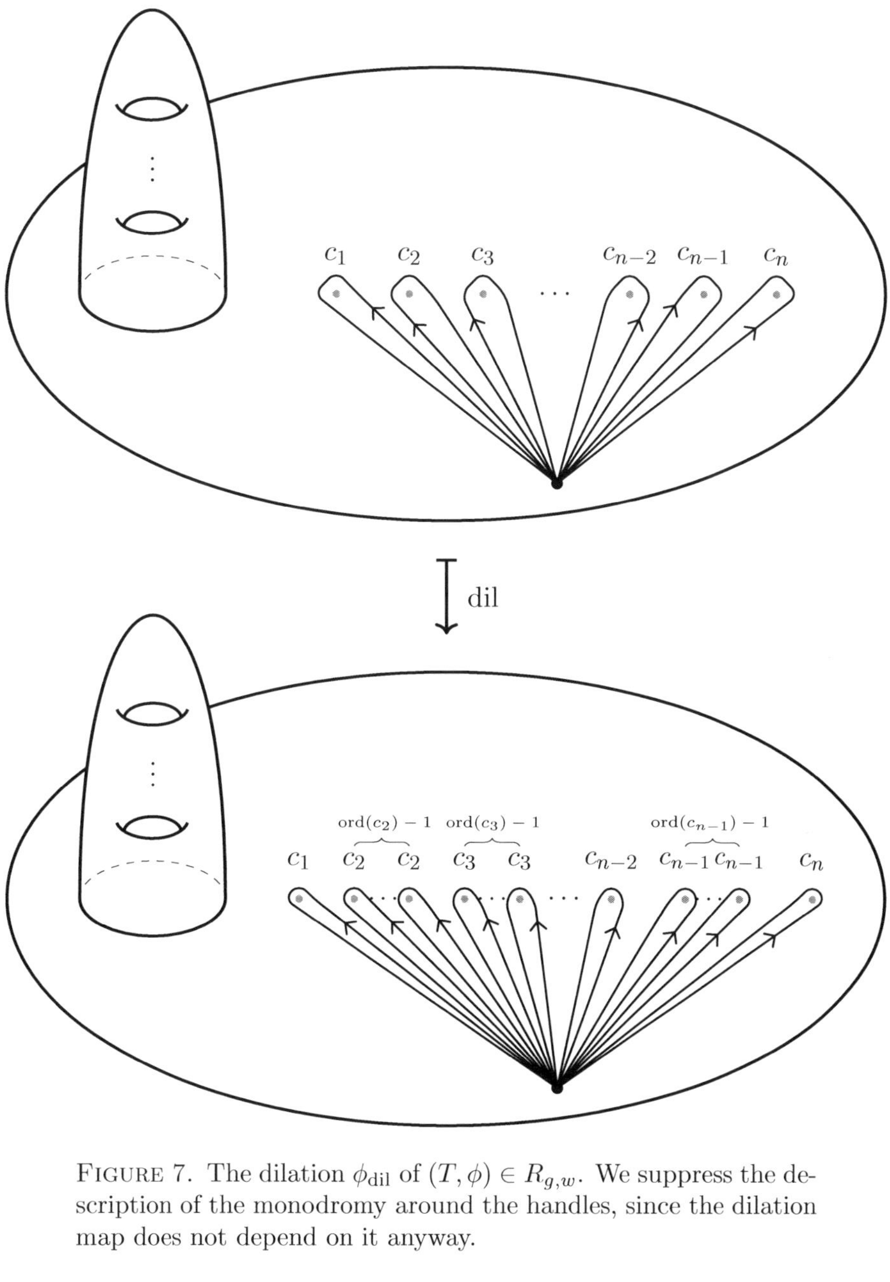

FIGURE 7. The dilation ϕ_{dil} of $(T, \phi) \in R_{g,w}$. We suppress the description of the monodromy around the handles, since the dilation map does not depend on it anyway.

any standard set of generators that contains those indicated in the figure. However, this choice is irrelevant for our purposes, because of our next observation.

When a mapping class in $\mathrm{MCG}_*(\Sigma_{g,w})$ acts on an element (T, ϕ) in $R_{g,w}$, there is an induced action on ϕ_{dil} by banding together the dilated points. Thus, dil descends to a well defined map

$$\mathrm{dil} : R_{g,w}/\,\mathrm{MCG}_*(\Sigma_{g,n}) \to R_{g,w_{\mathrm{dil}}}/\,\mathrm{MCG}_*\, i(\Sigma_{g,n+N}).$$

LEMMA 3.4. *If w satisfies the inequality*

$$w_{\mathrm{dil}}(\overline{c}) > |\overline{c}| \cdot (\mathrm{ord}(c) - 1) \cdot w(\overline{c}, -1)$$

for all $\overline{c} \in C /\!/ G$, then

$$\mathrm{dil} : R_{g,w}/\,\mathrm{MCG}_*(\Sigma_{g,n}) \to R_{g,w_{\mathrm{dil}}}/\,\mathrm{MCG}_*(\Sigma_{g,n+N})$$

is surjective.

In particular, the condition of the lemma is satisfied if

We reiterate that $|\overline{c}|$ is the cardinality of the conjugacy class $\overline{c}$, while $\mathrm{ord}(c)$ is the order of the group element c. Moreover, since

$$w_{\mathrm{dil}}(\overline{c}) = w(\overline{c}, +1) + (\mathrm{ord}(c) - 1) \cdot w(\overline{c}, -1),$$

the inequality in the lemma is equivalent to

$$w(\overline{c}, +1) > (|\overline{c}| - 1) \cdot (\mathrm{ord}(c) - 1) \cdot w(\overline{c}, -1).$$

So the conditions of the lemma are stronger than saying w is large enough. Fortunately, we will later use this lemma in a "backwards" way, *i.e.* by picking w_{dil} first and then choosing w so that it satisfies the conditions of the lemma. See Lemma 3.7.

PROOF OF LEMMA 3.4. Let $\psi \in R_{g,w_{\mathrm{dil}}}$. We must show ψ is equivalent to ϕ_{dil} for some $(T, \phi) \in R_{g,w}$.

For each $\overline{c}$, pick a disk $D_{\overline{c}} \subseteq \Sigma_g$ containing the basepoint and all of the branch points of type $\overline{c}$. Arranged from left to right, the monodromies around these points reads

$$c_1, \; c_2, \; c_3, \; \ldots, \; c_{w_{\mathrm{dil}}(\overline{c})},$$

where $c_1, c_2, \ldots, c_{w_{\mathrm{dil}}(\overline{c})} \in \overline{c}$. By iteratively applying the pigeonhole principle and a similar argument as in the proof of Lemma 3.3, we can find a braid that takes this to

$$\overbrace{d_1 \cdots d_1}^{\mathrm{ord}(c)-1} \quad \overbrace{d_2 \cdots d_2}^{\mathrm{ord}(c)-1} \quad \cdots \quad \overbrace{d_k \cdots d_k}^{\mathrm{ord}(c)-1} \qquad * \; \cdots \; *$$

where $k = w(\overline{c}, -1)$ and the asterisks $*$ denote various leftover elements in $\overline{c}$. If we pick all of our disks $D_{\overline{c}}$ to be disjoint (except for the basepoint), then this shows ψ is equivalent to a dilation. Indeed, since the braids for each $\overline{c}$ are supported in their respective disks, they do not interfere with each other. □

Dilation does not preserve the Schur invariant. However, we can compute the Schur invariant of a dilation in certain cases. We require a definition: let w and v' be branching data in $\mathbb{Z}_{\geq 0}^{C/\!/G \times \{+1,-1\}}$. We say w is a *stabilization* of v' if

$$(w - v')(\overline{c}, +1) = (w - v')(\overline{c}, -1)$$

for all $\overline{c} \in C /\!/ G$.

LEMMA 3.5. *Let w and v' be branching data in $\mathbb{Z}_{\geq 0}^{C/\!/G\times\{+1,-1\}}$, and suppose w is a stabilization of v'. If v' is larger than*

$$\sum_{\overline{c}\in C/\!/G} U^{\overline{c}},$$

then there exists a homology class $\chi_{\mathrm{dil}(w-v')}\in H_2(BG_C)$ such that

$$\mathrm{sch}_C(\mathrm{dil}(T,\phi)) = \mathrm{sch}_C(T,\phi) + \chi_{\mathrm{dil}(w-v')}$$

for all $(T,\phi)\in R_{g,w}$.

PROOF. By Lemma 3.3, every element of $R_{g,w}$ is a stabilization of an element in $R_{g,v'}$. That is, there exists some puncture stabilizing sphere, which we shall denote by $S^2_{w-v'}$, such that for all $(T,\phi)\in R_{g,w}$, (T,ϕ) is equivalent to the connect sum of $S^2_{w-v'}$ and some element $(S,\psi)\in R_{g,v'}$. There exists another branched cover of the sphere, which we denote by $S^2_{\mathrm{dil}(w-v')}$, such that $\mathrm{dil}(T,\phi)$ is equivalent to the connect sum of (S,ψ) and $S^2_{\mathrm{dil}(w-v')}$. Let $\chi_{\mathrm{dil}(w-v')} = \mathrm{sch}_C(S^2_{\mathrm{dil}(w-v')})$. Then

$$\begin{aligned}\mathrm{sch}_C(\mathrm{dil}(T,\phi)) &= \mathrm{sch}_C(S,\psi) + \mathrm{sch}_C(S^2_{\mathrm{dil}(w-v')})\\ &= \mathrm{sch}_C(T,\phi) - \mathrm{sch}_C(S^2_{w-v'}) + \mathrm{sch}_C(S^2_{\mathrm{dil}(w-v')})\\ &= \mathrm{sch}_C(T,\phi) - 0 + \chi_{\mathrm{dil}(w-v')}\\ &= \mathrm{sch}_C(T,\phi) + \chi_{\mathrm{dil}(w-v')}.\end{aligned}$$

□

3.4. Proof of Theorem 1.1. Proposition 3.1 implies that any two elements of $R_{g,v}$ with the same Schur invariant are equivalent in some $R_{g+M,w}$. In other words, two orbits in $R_{g,v}/\operatorname{MCG}_*(\Sigma_{g,n})$ that have the same Schur invariant eventually merge into one orbit inside some $R_{g+M,w}/\operatorname{MCG}_*(\Sigma_{g+M,n+2N})$. Because G is finite, all of the sets of orbits are finite, and so Lemmas 3.3 and 3.2 imply that all of the handle and puncture stabilization maps are bijections when g and v are large enough. We conclude that only a finite amount of merging must occur before the Schur invariant maps

$$\mathrm{sch}_C : R_{g+M,w}/\operatorname{MCG}_*(\Sigma_{g+M,n+2N}) \to H_2(BG_C)$$

are injections. This proves

PROPOSITION 3.6. *For all $w\in\mathbb{Z}_{\geq 0}^{C/\!/G\times\{+1,-1\}}$ and g large enough, the Schur invariant yields an injective map*

$$\mathrm{sch}_C : R_{g,w}/\operatorname{MCG}_*(\Sigma_{g,n}) \to H_2(BG_C).$$

□

At first glance, this proposition is very close to the statement of Theorem 1.1(1); however, the proposition allows negatively trivialized branch points, while Theorem 1.1 does not. We now use dilation to address this. By combining the proposition and Lemma 3.4, it is straightforward to prove the following

LEMMA 3.7. *For all $v\in\mathbb{Z}_{\geq 0}^{C/\!/G}$ large enough, we can write $v = w_{dil}$ where $w\in\mathbb{Z}_{\geq 0}^{C/\!/G\times\{+1,-1\}}$ and the following conditions hold:*

1. $\mathrm{sch}_C : R_{g,w}/\operatorname{MCG}_*(\Sigma_{g,n}) \to H_2(BG_C)$ *is injective,*
2. $\mathrm{dil} : R_{g,w}/\operatorname{MCG}_*(\Sigma_{g,n}) \to R_{g,v}/\operatorname{MCG}_*(\Sigma_{g,n+N})$ *is surjective,*

3. *w is a stabilization of positive branching data $v' \in \mathbb{Z}_{\geq 0}^{C/\!/G \times \{+1\}}$, and*
4. *v' is larger than $\sum_{\overline{c} \in C/\!/G} U^{\overline{c}}$.* □

Now suppose $v \in \mathbb{Z}_{\geq 0}^{C/\!/G}$ is large enough for the lemma to hold. Let $\phi, \psi \in R_{g,v}$ and suppose $\mathrm{sch}_C\, \phi = \mathrm{sch}_C\, \psi$. By condition 2, there are (T', ϕ') and (S', ψ') in $R_{g,w}$ such that $\phi = \mathrm{dil}(T', \phi')$ and $\psi = \mathrm{dil}(S', \psi')$. Lemma 3.5 shows

$$\begin{aligned} \mathrm{sch}_C(T', \phi') &= \mathrm{sch}_C(\phi) - \chi_{w-v'} \\ &= \mathrm{sch}_C(\psi) - \chi_{w-v'} = \mathrm{sch}_C(S', \psi') \end{aligned}$$

By condition 1, (T', ϕ) and (S', ψ') are in the same $\mathrm{MCG}_{g,n}$. Condition 2 now implies that ϕ and ψ represent the same orbit in $R_{g,v}/\,\mathrm{MCG}_*(\Sigma_{g,n+N})$. This proves part 1 of Theorem 1.1.

To prove part 2, observe that when C generates G, the second part of Proposition 3.1 says that we never need to introduce handle stabilizations in the above argument. That is, two orbits in $R_{g,v}/\,\mathrm{MCG}_*(\Sigma_{g,n})$ that have the same Schur invariant merge into one orbit inside some $R_{g,w}/\,\mathrm{MCG}_*(\Sigma_{g,n+2N})$. Now continue with the same argument to finish the proof of part 2.

Finally, we show that $R_{g,v}/\,\mathrm{MCG}_*(\Sigma_{g,n})$ is a torsor for $M(G)_C$ in the cases where g and v are large enough, and where v is large enough and C generates G. At the end of Section 2.3 we described an exact sequence

$$0 \to M(G)_C \to H_2(BG_C) \to N \to 0,$$

where the map $H_2(BG_C) \to N$ is the homological branch type. All of the elements of $R_{g,v}$ have the same branch type, so, in particular, they have the same homological branch type $[v] \in N$. Thus, for any $(T, \phi), (S, \psi) \in R_{g,v}$,

$$\mathrm{sch}_C(T, \phi) - \mathrm{sch}_C(S, \psi) \in M(G)_C.$$

To conclude, we must show that every element of $M(G)_C$ can be represented by such a difference.

We first consider the case where g and v are large enough. Fix a branched cover of the oriented disk with branch type v. Since N is the kernel of the evaluation map $\mathbb{Z}^{C/\!/G} \to G_{ab}$, the boundary monodromy of this cover is in the commutator subgroup $[G, G]$. Let g_1 be the commutator length of G, *i.e.*

$$g_1 = \max_{x \in [G,G]} cl(x),$$

where

$$cl(x) = \min\{l \mid \exists a_1, b_1, \ldots, a_l, b_l \in G, [a_1, b_1] \cdot [a_l, b_l] = x\}.$$

We can extend the chosen branched cover of the disk to a C-branched G-cover (T, ϕ) of a closed oriented surface with the same branch type v and genus g_1; if necessary, we can pick g_1 even larger so that $(T, \phi_1) \in R_{g_1,v}$ is connected. Let g_2 be large enough so that every element $\chi \in M(G)_C$ can be represented by an unbranched G-cover ϕ_χ of a closed surface of genus g_2. Let $g_3 = g - g_1 - g_2$ and let ϕ_3 be the trivial G-cover over a surface of genus g_3. Let (T, ψ_χ) be any element of $R_{g,v}$ equivalent to the connect sum

$$(T, \phi) \# \phi_\chi \# \phi_3.$$

The Schur invariant of this cover is

$$\mathrm{sch}(T, \psi_\chi) = \mathrm{sch}_C(T_1, \phi_1) + \chi.$$

In particular, for every $\chi \in M(G)_C$,

$$\mathrm{sch}(T, \psi_\chi) - \mathrm{sch}(T, \psi_0) = \chi.$$

Finally, suppose v is large enough and C generates G. Lemma 2.2 shows that every element of $M(G)_C$ can be represented by an unbranched cover of S^2. Better yet, there exists a branch type $v_1 \in \mathbb{Z}_{\geq 0}^{C/\!/G\times\{+1,-1\}}$ such that every element $\chi \in M(G)_C$ can be represented by an element (T_χ, ϕ_χ) of R_{0,v_1}. Fix any element (T_1, ϕ_1) in $R_{0,v-v_1}$. For every $\chi \in M(G)_C$, let (T, ψ_χ) be an element of $R_{0,v}$ that is equivalent to the connect sum

$$(T_\chi, \phi_\chi)\#(T_1, \phi_1)$$

As before,

$$\mathrm{sch}(T, \psi_\chi) = \mathrm{sch}_C(T_1, \phi_1) + \chi,$$

and in particular,

$$\mathrm{sch}(T, \psi_\chi) - \mathrm{sch}(T, \psi_0) = \chi$$

for every $\chi \in M(G)_C$. □

4. Outlook

4.1. Understanding the stable range. We remark that while Lemmas 3.2, 3.3 and 3.4 provide specific lower bounds, the merging argument in the proof of Theorem 1.1 does not give any bounds whatsoever on when merging stops. Thus, we do not have an upper bound for when the stable range begins. The problem of computing the stable range is closely related to the *quantitative Steenrod problem* in $H_2(G)$, *i.e.* the question of finding a smallest genus representative of a given homology class in $H_2(G)$. It could be interesting to try to bound the stable range for specific families of finite groups, such as solvable groups or the non-abelian simple groups.

4.2. Applications to G-equivariant TQFT. The subject of this paper first received attention as early as Nielsen in 1937 [**17**]. Despite its age, it takes on a new life in the context of topological quantum field theory.

The linearization of the permutation action of $\mathrm{MCG}_*(\Sigma_{g,n})$ on $\hat{R}_{g,n}$ is closely related to one of the simplest examples of a topological quantum field theory, the untwisted Dijkgraaf-Witten theory with gauge group G [**5, 10**]. In fact, this linearized action is precisely the representation of $\mathrm{MCG}_*(\Sigma_{g,n})$ afforded by the G-crossed modular tensor category of G-graded vector spaces. Every G-crossed modular tensor category gives rise to an extended, pointed $(2+1)$-dimensional homotopy quantum field theory with target $K(G,1)$, and, hence, a (projective) representation of $\mathrm{MCG}_*(\Sigma_{g,n})$ on a vector space

$$V^\times(\Sigma_{g,n}) = \bigoplus_{\phi\in \hat{R}_{g,n}} V^\times(\Sigma_{g,n}, \phi)$$

consisting of blocks indexed by the G-representation set $\hat{R}_{g,n}$. [**19, 20**]. The linear action of $\mathrm{MCG}_*(\Sigma_{g,n})$ on $V^\times(\Sigma_{g,n})$ refines the permutation action of $\mathrm{MCG}_*(\Sigma_{g,n})$ on the index set $\hat{R}_{g,n}$.

In recent works, the author and Kuperberg study the action of $\mathrm{MCG}_*(\Sigma_{g,n})$ on $\hat{R}_{g,n}$ under the assumption that G is a nonabelian simple group [**12, 13**]. In this case, we understand more than just the orbits of the action: building on

[**6**] and [**18**], the results of [**12**] and [**13**] establish a precise version of *classical* topological computing via this action. Complexity-theoretic hardness results for combinatorial 3-manifold invariants ensue. As we now explain, we advertise this result here because we believe it may be of interest in the study of topological quantum computing with symmetry enriched topological phases (see [**4**] for an introduction).

Recall that the algebraic model of a $(2+1)$-dimensional G-symmetry enriched topological phase of matter (G-SET phase) is believed to be a $(2+1)$-dimensional G-crossed unitary modular tensor category [**1**]. The physical interpretation says that a G-representation $\phi \in \hat{R}_{g,n}$ is a background field that gauges some internal G-symmetry of a topological phase residing on $\Sigma_{g,n}$. The subspaces $V^{\times}(\Sigma_{g,n}, \phi)$ are called twisted sectors. In the language of condensed matter physics, this paper seeks to understand when two gauge fields on $\Sigma_{g,n}$ are equivalent under a modular transformation. This question recently received some attention in the physics literature, where the case of $n = 0$ and G abelian was solved [**1, 2**]. (We note that the cyclic case was already known to Nielsen [**17**], and the abelian and metacyclic cases were solved by Edmonds in the early 1980s [**7, 8**].)

When every twisted sector is 1-dimensional, a G-SET phase is instead called a G-symmetry protected topological phase, or G-SPT phase. It is well-known that topological quantum computing with a G-SPT phase is never quantum universal. Nevertheless, the results of [**12, 13**] imply that, at least for nonabelian simple G, a kind of topological computing with a G-SPT phase is #P-complete via parsimonious reduction, a precise notion of classical (*i.e.* non-quantum) universality. More generally, it follows that for such G, every G-SET phase can model classical reversible circuits.

Accordingly, when G is nonabelian simple, one might hope that for the topological operations available from a G-SET phase to be quantum universal, the only thing left to find is a single entangling gate between two states in different twisted sectors. Unfortunately, it is expected that no such entangling gate exists using topological operations. Moreover, unless one is willing to believe that quantum computers can efficiently solve problems in #P, the results of [**12, 13**] can be understood as evidence that preparing, measuring and topologically manipulating arbitrary gauge fields on a surface is too hard for a quantum computer to do efficiently. Nevertheless, knowledge of the action of $\mathrm{MCG}_*(\Sigma_{g,n})$ on $\hat{R}_{g,n}$ could help when designing protocols for universal gate sets augmented with *non-topological* operations, as in [**4**].

References

[1] Maissam Barkeshli, Parsa Bonderson, Meng Cheng, and Zhenghan Wang, *Symmetry, defects, and gauging of topological phases*, 2014, arXiv:1410,4540.

[2] Maissam Barkeshli and Xiao-Gang Wen, *$U(1) \times U(1) \rtimes Z_2$ Chern-Simons theory and Z_4 parafermion fractional quantum Hall states*, Phys. Rev. B **81** (2010), no. 4, arXiv:0909.4882.

[3] Neal Brand, *Classifying spaces for branched coverings*, Indiana Univ. Math. J. **29** (1980), no. 2, 229–248, DOI 10.1512/iumj.1980.29.29015. MR563208

[4] Colleen Delaney and Zhenghan Wang, *Symmetry defects and their application to topological quantum computing*, to appear in AMS Contemporary Math Series.

[5] Robbert Dijkgraaf and Edward Witten, *Topological gauge theories and group cohomology*, Comm. Math. Phys. **129** (1990), no. 2, 393–429. MR1048699

[6] Nathan M. Dunfield and William P. Thurston, *Finite covers of random 3-manifolds*, Invent. Math. **166** (2006), no. 3, 457–521, DOI 10.1007/s00222-006-0001-6. MR2257389

[7] Allan L. Edmonds, *Surface symmetry. I*, Michigan Math. J. **29** (1982), no. 2, 171–183. MR654478

[8] Allan L. Edmonds, *Surface symmetry. II*, Michigan Math. J. **30** (1983), no. 2, 143–154, DOI 10.1307/mmj/1029002844. MR718259

[9] Jordan S. Ellenberg, Akshay Venkatesh, and Craig Westerland, *Homological stability for Hurwitz spaces and the Cohen-Lenstra conjecture over function fields, II*, 2012, arXiv:1212.0923v1.

[10] Daniel S. Freed and Frank Quinn, *Chern-Simons theory with finite gauge group*, Comm. Math. Phys. **156** (1993), no. 3, 435–472. MR1240583

[11] Michael D. Fried and Helmut Völklein, *The inverse Galois problem and rational points on moduli spaces*, Math. Ann. **290** (1991), no. 4, 771–800, DOI 10.1007/BF01459271. MR1119950

[12] Greg Kuperberg and Eric Samperton, *Coloring invariants of knots and links are often intractable*, in preparation.

[13] Greg Kuperberg and Eric Samperton, *Computational complexity and 3-manifolds and zombies*, Geo. Top. **12** (2018), no. 6, 3623–3670, arXiv:1707.03811.

[14] John M. Lee, *Introduction to smooth manifolds*, 2nd ed., Graduate Texts in Mathematics, vol. 218, Springer, New York, 2013. MR2954043

[15] Charles Livingston, *Stabilizing surface symmetries*, Michigan Math. J. **32** (1985), no. 2, 249–255, DOI 10.1307/mmj/1029003192. MR783579

[16] MathOverflow, *What is the equivariant cohomology of a group acting on itself by conjugation?*, https://mathoverflow.net/q/20671.

[17] Jakob Nielsen, *Die Strucktur periodisher Transformationen von Flachen*, Danske Vid Selsk., Math.-Fys. Medd. **15** (1937), 1–77.

[18] David P. Roberts and Akshay Venkatesh, *Hurwitz monodromy and full number fields*, Algebra Number Theory **9** (2015), no. 3, 511–545, DOI 10.2140/ant.2015.9.511. MR3340543

[19] Vladimir Turaev and Alexis Virelizier, *On 3-dimensional homotopy quantum field theory II: The surgery approach*, Internat. J. Math. **25** (2014), no. 4, 1450027, 66, DOI 10.1142/S0129167X1450027X. MR3195545

[20] Vladimir Turaev, *Homotopy quantum field theory*, EMS Tracts in Mathematics, vol. 10, European Mathematical Society (EMS), Zürich, 2010. Appendix 5 by Michael Müger and Appendices 6 and 7 by Alexis Virelizier. MR2674592

[21] George W. Whitehead, *Elements of homotopy theory*, Graduate Texts in Mathematics, vol. 61, Springer-Verlag, New York-Berlin, 1978. MR516508

UNIVERSITY OF CALIFORNIA, SANTA BARBARA
Email address: eric@math.ucsb.edu

Contemporary Mathematics
Volume **747**, 2020
https://doi.org/10.1090/conm/747/15046

Quantum error-correcting codes over finite Frobenius rings

Andreas Klappenecker, Sangjun Lee, and Andrew Nemec

Abstract. The idealized model of quantum computation promises for certain problems a considerable speed-up over classical computation. However, protecting quantum information requires a considerable effort. We give a brief survey of the construction of quantum error-correcting codes over finite Frobenius rings. We illustrate the construction of families of quantum error-correcting codes that are obtained from quadratic residue codes over finite commutative chain rings.

1. Introduction

Quantum computation offers a superpolynomial speedup over classical computation for many computational problems in algebra and number theory, cf. [**14**]. However, building a reliable quantum computer capable of performing these computations is a formidable task. There are two main problems.

(a) Keeping the information safely stored in quantum states is a challenge, since the quantum bits tend to interact with their environment.
(b) Performing the computational steps without error is another challenge, since it requires a level of quantum control that is difficult to engineer.

We can improve the reliability of stored quantum information by using quantum error-correcting codes, so this addresses problem (a). We can reduce errors in the computation using fault-tolerant operations, so this addresses problem (b).

The modest goal of this paper is to introduce quantum error-correcting codes in a way that is accessible to mathematicians. In the conclusions, we will very briefly sketch the idea behind fault-tolerant operations and give pointers to the literature.

Most of the research literature is focused on binary quantum error-correcting codes. We will emphasize nonbinary quantum error-correcting codes, focusing on quantum codes over *nice rings*. These rings occur naturally in certain constructions of quantum error-correcting codes with *nice error bases*.

In Section 2, we discuss the discretization of quantum errors, the concept of nice error bases, and the related notion of nice rings. We include a proof that a ring is nice if and only if it is a finite Frobenius ring, following [**12**]. In Section 3, we recall the basic constructions of stabilizer codes over finite Frobenius rings, and

2010 *Mathematics Subject Classification.* Primary 81P70; Secondary 94B60.

©2020 American Mathematical Society

their relation to classical codes. In Section 4, we show how to construct quantum codes from quadratic residue codes and develop more generally the basic theory of duadic codes over finite Frobenius rings. Section 5 concludes the paper with a brief discussion of fault-tolerant operations that are available for the quantum codes constructed in the previous section.

2. Discretization and Nice Rings

Let q be a positive integer. Suppose that we have n q-level quantum systems. The state space is given by the q^n-dimensional complex vector space

$$H = \underbrace{\mathbb{C}^q \otimes \mathbb{C}^q \otimes \cdots \otimes \mathbb{C}^q}_{n \text{ times}} \cong \mathbb{C}^{q^n}$$

equipped with the usual Hermitian inner product (taken to be antilinear in the first argument and linear in the second), so it forms a finite-dimensional Hilbert space.

A *quantum error-correcting code* is simply a subspace Q of the state space H. If P is an orthogonal projector from H onto Q, and E is a bounded operator acting on H, then Q is said to *detect* E if and only if there exists a complex number λ_E such that

$$PEP = \lambda_E P.$$

A K-dimensional quantum error-correcting code Q of the state space H that can detect all errors affecting less than d quantum systems is called an $((n, K, d))_q$ *quantum error-correcting code.* We call n the *length* and K the *dimension* of the quantum error-correcting code. We say that d is the minimum distance of Q if and only if Q is an $((n, K, d))_q$ but not an $((n, K, d+1))_q$ quantum code. If there exists an integer k such that $K = q^k$, then an $((n, K, d))_q$ quantum code is also called an $[[n, k, d]]_q$ quantum code.

A striking difference between quantum error-correcting codes and classical error-correcting codes concerns the set of possible errors. Quantum bits storing the information are affected not just by a finite or discrete set of errors. The state of a quantum bit can be altered by any unitary operator acting on the state space H. The state of a single quantum bit can be altered for example by applying

$$\begin{pmatrix} \cos\theta & \sin\theta \\ -\sin\theta & \cos\theta \end{pmatrix}, \quad \theta \in \mathbb{R},$$

or some diagonal unitary matrix or products thereof. Fortunately, quantum mechanics is a linear theory. This means that a quantum error-correcting code capable of correcting a set of errors can also correct any linear combination of these errors. Thus, the solution is to expand errors in a basis that eases the recovery from errors. We use a special class of nice error bases that have some additional structure that is beneficial in the construction of quantum error-correcting codes.

A *nice error basis* $\mathcal{E} = \{D(g) \mid g \in G\}$ is a set consisting of unitary $q \times q$ matrices $D(g)$ that are parametrized by a finite group G, called the index group, such that

(a) D is a faithful irreducible unitary projective representation of the group G,
(b) the trace $\mathrm{Tr}(D(g)^* D(h))$ is zero for all g, h of G with $g \neq h$.

Here * denotes the Hermitian transpose of a matrix. Condition (b) ensures that there are at most q^2 matrices in $\mathcal{E} = \{D(g) \mid g \in G\}$, since they are orthogonal with respect to the Hilbert-Schmidt inner product

$$\langle A \mid B\rangle = \mathrm{Tr}(A^* B).$$

Since the representation is irreducible, there are exactly q^2 matrices in $\mathcal{E}$.

Since the matrices in $\mathcal{E}$ form a projective representation, the product of two matrices in $\mathcal{E}$ is contained in $\mathcal{E}$ up to multiplication with a scalar factor, that is,

$$D(g)D(h) = w(g,h)D(gh)$$

holds for all g, h in G, and some complex number $w(g,h)$ of absolute value 1. It follows from the associativity of the matrix product that w is a representative of a 2-cocycle of G, so it satisfies the cocycle equation

$$w(h,k)w(g,hk) = w(g,h)w(gh,k)$$

for all g, h, k in G.

EXAMPLE 2.1. A well-known example of a nice error basis is the Pauli basis

$$\{X^a Z^b \mid a, b \in \mathbb{F}_2\}$$

where $\mathbb{F}_2$ is the finite field with two elements, and X and Z are the Pauli matrices given by

$$X = \begin{pmatrix} 0 & 1 \\ 1 & 0 \end{pmatrix} \quad \text{and} \quad Z = \begin{pmatrix} 1 & 0 \\ 0 & -1 \end{pmatrix}.$$

In this case, the index group G is given by the additive group $G = \mathbb{F}_2 \times \mathbb{F}_2$.

We can find a higher-dimensional generalization as follows. Let R be a finite ring with q elements. We assume throughout that all rings have a unit element 1. We fix an orthonormal basis of $\mathbb{C}^q$ and denote it by

$$\{|x\rangle | x \in R\}.$$

Let χ be an additive character of R. For all $a, b \in R$, we define a *shift operator* $X(a)$ and a *multiplication operator* $Z(b)$ by

$$X(a)|x\rangle = |x+a\rangle, \qquad Z(b)|x\rangle = \chi(bx)|x\rangle,$$

for all $x \in R$. Let us form the set $\mathcal{E}_\chi = \{X(a)Z(b) \mid a, b \in R\}$.

The set $\mathcal{E}_\chi$ is always a projective representation of the additive group $G = R \times R$, but it is not necessarily a nice error basis. For example, if we choose the trivial character for the character χ, then $\mathcal{E}_\chi$ cannot be an nice error basis when $q > 1$.

It turns out that some finite rings do not have an irreducible character χ such that $\mathcal{E}_\chi$ is a nice error basis. We will call a finite ring R *nice* if and only if there exists a character χ such that the set $\mathcal{E}_\chi$ generated by the shift and multiplication errors forms a nice error basis. In other words, a nice ring supports the construction of a nice error basis that generalizes the Pauli X and Z operators.

Our next goal is to characterize nice rings using ring-theoretic properties. We begin with a simple characterization of a character χ that can form a nice error basis $\mathcal{E}_\chi$. Given a character χ and a ring element b, we will use the notation χ_b to denote the character $\chi_b(x) = \chi(bx)$ obtained from χ by multiplying the argument with b.

LEMMA 2.2. *Let R be a finite ring. The matrices in $\{X(a)Z(b) \mid a, b \in R\}$ are pairwise orthogonal with respect to the Hilbert-Schmidt inner product if and only if the characters $\{\chi_b \mid b \in R\}$ are pairwise orthogonal.*

PROOF. For a, b, a', b' in R, the trace $\mathrm{Tr}((X(a')Z(b'))^* X(a)Z(b))$ is equal to

$$\begin{cases} 0 & \text{if } a \neq a' \\ \sum_{x \in R} \overline{\chi(b'x)}\chi(bx) = \langle \chi_{b'} \mid \chi_b \rangle & \text{if } a = a' \ . \end{cases}$$

Indeed, we have

$$\mathrm{Tr}((X(a')Z(b'))^* X(a)Z(b)) = \mathrm{Tr}(X(-a'+a)Z(b)Z(b')^*),$$

and the entries on the diagonal of $X(-a'+a)Z(b)Z(b')^*$ are 0 when $-a'+a \neq 0$. If $-a'+a = 0$, then the entries on the diagonal are determined by $Z(b)Z^*(b')|x\rangle = \overline{\chi(b'x)}\chi(bx)|x\rangle$ for $x \in G$. The claim follows immediately. $\square$

The lemma allows us to characterize nice rings a bit better. A finite ring is *nice* if and only if it has an irreducible generating character χ, meaning that the set of all irreducible characters can be obtained by "scaling" the character χ, so

$$\{\chi_b \mid b \in R\} = \mathrm{Irr}((R, +)).$$

Now the question arises which finite rings have such a generating character. This question has been answered by Hirano [**5**] and also by Wood [**16**]. They proved that a finite ring is nice if and only if it is a finite Frobenius ring (naturally, the concept of a nice ring had not been introduced back then, but their papers contain the result). Since the methods are quite instructive, we include a proof for this result, following Sarma and Klappenecker [**12**].

Before we give the definition of a finite Frobenius ring, we recall some notions from ring theory. The interested reader can find more background for example in Anderson and Fuller [**1**] and Lam [**8**].

If we are given a ring R and a module M over R, then we write M_R to indicate that M is a right R-module and ${}_RM$ to indicate that M is a left module. In particular, we write R_R when we view the ring R as a right R-module over itself, and ${}_RR$ when we view R as a left R-module over itself.

The *Jacobson radical* $\mathrm{rad}(R)$ of a ring R is defined as the intersection of the maximal left-ideals of the R. Despite the apparent lack of left-right symmetry, the Jacobson radical is an ideal of R.

The *left socle* $\mathrm{soc}({}_RR)$ of a ring R is the sum of the minimal left ideals of R. Similarly, the *right socle* $\mathrm{soc}(R_R)$ of a ring R is the sum of the minimal right ideals of R. The left socle and the right socle of a ring are both ideals, but they are not necessarily the same ideal. More generally, the *socle* $\mathrm{soc}(M)$ of a module over a ring R is defined as the sum of the simple submodules of M. A module is semisimple if and only if $\mathrm{soc}(M) = M$.

Let us write $\overline{R} = R/\mathrm{rad}(R)$. An Artinian ring R is called a *Frobenius ring* if and only if $\mathrm{soc}({}_RR) \cong {}_R\overline{R}$ and $\mathrm{soc}(R_R) \cong \overline{R}_R$. If R is not just Artinian but finite, then it suffices to check one of the two isomorphism. Indeed, Honold [**6**] has shown that a finite ring R is a Frobenius ring if and only if $\mathrm{soc}(R_R) \cong \overline{R}_R$ holds.

EXAMPLE 2.3. (1) The residue class ring $R = \mathbb{Z}/12\mathbb{Z}$ has the maximal (left) ideals $2R$ and $3R$, so the Jacobson radical $\mathrm{rad}(R) = 2R \cap 3R = 6R \cong$

$\mathbb{Z}/6\mathbb{Z}$. The minimal ideals of R are $4R$ and $6R$, so $\mathrm{soc}(R) = 4R + 6R = 2R \cong \mathbb{Z}/6\mathbb{Z} \cong R/\mathrm{rad}(R)$. Therefore, $R = \mathbb{Z}/12\mathbb{R}$ is a finite Frobenius ring.

(2) More generally, the residue class ring $\mathbb{Z}/n\mathbb{Z}$ is a finite Frobenius ring for all integers $n \geq 2$.

(3) A finite commutative ring whose ideals form a chain is called a *finite chain ring*. A finite chain ring is a finite Frobenius ring.

Frobenius rings have many interesting properties. In a Frobenius ring, the left socle and the right socle coincide, as is apparent from the definition of a Frobenius ring. It is perhaps less obvious that if R is a Frobenius ring, then R_R is an injective right R-module and ${}_RR$ is an injective left R-module. This injective module structure is the starting point for our exploration of finite Frobenius rings. We will need the concept of an injective hull that we briefly review here. A thorough discussion of essential extensions and injective hulls can be found in Lam [**8**, Chapter §3D].

If M and N are R-modules such that $M \subseteq N$, then N is called an *essential extension* of M if and only if every nontrivial submodule of N intersects M nontrivially. The module N is called a *maximal essential extension* of M if and only if no R-module properly containing N can be an essential extension of M. Every module M has a maximal essential extension. A maximal essential extension is necessarily an injective module. Furthermore, any two maximal essential extensions of a module M are isomorphic. We write $E(M)$ for "the" maximal essential extension of the module M and call it the *injective hull* or *injective envelope* of M.

PROPOSITION 2.4 (See [**12**]). *Let R be a finite ring and let $\overline{R} = R/\mathrm{rad}(R)$. Then R is a finite Frobenius ring if and only if $R_R \cong E(\overline{R}_R)$.*

PROOF. $\Longrightarrow$: Suppose that R is a finite Frobenius ring, so in particular $\overline{R}_R \cong \mathrm{soc}(R_R)$. Since a Frobenius ring R is right Artinian, every right ideal contains a minimal right ideal, which implies that the right module R_R is an essential extension of its socle $\mathrm{soc}(R_R)$. As R is a right-self-injective ring and an essential extension of its socle, we must have that R_R is the injective hull of $\mathrm{soc}(R_R)$, so

$$R_R = E(\mathrm{soc}(R_R)) \cong E(\overline{R}_R).$$

$\Longleftarrow$: Conversely, let us suppose that R is a finite ring satisfying $R_R \cong E(\overline{R}_R)$. It follows that R is a right-self-injective ring, hence a quasi-Frobenius ring. For any R-module M, the socle of its injective hull is given by $\mathrm{soc}(E(M)) \cong \mathrm{soc}(M)$, see Sharpe [**13**, p. 70]. It follows that

$$\mathrm{soc}(R_R) \cong \mathrm{soc}(E(\overline{R}_R)) \cong \mathrm{soc}(\overline{R}_R) = \overline{R}_R,$$

where the last equality holds because $\overline{R}_R = (R/\mathrm{rad}(R))_R$ is a semisimple R-module. Therefore, R is a finite Frobenius ring. □

PROPOSITION 2.5 (See [**12**]). *Let R be a finite ring with identity. Then the character module* $\mathrm{Hom}_{\mathbb{Z}}({}_RR, \mathbb{Q}/\mathbb{Z})$ *of ${}_RR$ is isomorphic to an injective hull $E(\overline{R}_R)$ of the right R-module $\overline{R}_R = (R/\mathrm{rad}\, R)_R$.*

PROOF. Let ${}_R\mathrm{Mod}$ (resp. Mod_R) denote the category of finitely generated left (right) R-modules. Let F denote the contravariant functor $\mathrm{Hom}_{\mathbb{Z}}(-, \mathbb{Q}/\mathbb{Z})$ from ${}_R\mathrm{Mod}$ to Mod_R, and G the contravariant functor $\mathrm{Hom}_{\mathbb{Z}}(-, \mathbb{Q}/\mathbb{Z})$ from Mod_R to ${}_R\mathrm{Mod}$.

By Pontrajagin duality, the functor GF is naturally isomorphic to the identity functor on ${}_R\mathrm{Mod}$ using the natural isomorphism $x \mapsto (\chi \mapsto \chi(x))$, and FG is naturally isomorphic to the identity functor on Mod_R. Thus, F and G form a duality between the categories ${}_R\mathrm{Mod}$ and Mod_R.

A simple left R-module is mapped by F to a simple right R-module of the same cardinality. Since the additive functor F respects direct sums, it follows that $F({}_R(R/\operatorname{rad} R) \cong (R/\operatorname{rad} R)_R$.

Since ${}_RR$ is a projective cover of ${}_R(R/\operatorname{rad} R)$, we can conclude that $F({}_RR) = \mathrm{Hom}({}_RR, \mathbb{Q}/\mathbb{Z})$ is the injective hull of $F({}_R(R/\operatorname{rad} R)) = (R/\operatorname{rad} R)_R$, because projective covers are mapped to injective hulls by a duality functor, see Anderson & Fuller [**1**, p. 276]. □

THEOREM 2.6 (Hirano [**5**], Wood [**16**]). *Let R be a finite ring. Then R is a nice ring if and only if it is a Frobenius ring.*

PROOF. If R is a nice ring, then $R_R \cong \mathrm{Hom}_{\mathbb{Z}}({}_RR, \mathbb{Q}/\mathbb{Z})$. By Proposition 2.5, we have $R_R \cong \mathrm{Hom}_{\mathbb{Z}}({}_RR, \mathbb{Q}/\mathbb{Z}) \cong E(\overline{R}_R)$. Therefore, Proposition 2.4 implies that R_R is a Frobenius ring.

If R is a Frobenius ring, then $R_R \cong E(\overline{R}_R)$ by Proposition 2.4. It follows from Proposition 2.5 that $R_R \cong E(\overline{R}_R) \cong \mathrm{Hom}_{\mathbb{Z}}({}_RR, \mathbb{Q}/\mathbb{Z})$. Therefore, we can conclude that R is a nice ring. □

3. Quantum Error-Correcting Codes

Let R be a nice ring with generating character χ and the corresponding nice error basis $\mathcal{E} = \{X(a)Z(b) \mid a, b \in R\}$. We will use this nice error basis for a single quantum system. Let n be a positive integer. We can form a nice error basis $\mathcal{E}_n$ for a system of n quantum systems as follows.

For $a = (a_1, ..., a_n)$ and $b = (b_1, ..., b_n) \in R^n$, we abbreviate the tensor product of X and Z operators as follows:

$$X(a) = X(a_1) \otimes \cdots \otimes X(a_n),$$
$$Z(b) = Z(b_1) \otimes \cdots \otimes Z(b_n).$$

Then $\mathcal{E}_n = \{X(a)Z(b) \mid a, b \in R^n\}$ is a nice error basis. We will use this nice error basis when constructing quantum codes that encode the quantum information into n quantum systems with $q = |R|$ levels each. Note that we used the same nice error basis in each tensor component. For simplicity, we will confine ourselves to this homogeneous situation.

Suppose that the multiplicative group $\langle X(a)Z(b) \,|\, a, b \in R^n \rangle$ generated by the matrices $X(a)Z(b)$ has exponent m. Let ω denote a primitive m-th root of unity, $\omega = \exp(2\pi i/m)$. We define the *error group* G_n generated by the operators $X(a)Z(b)$, $a, b \in R^n$, as

$$G_n = \langle \omega^c X(a)Z(b) \,|\, a, b \in R^n, c \in \mathbb{Z} \rangle.$$

The center, $Z(G_n)$, of the group G_n is given by $Z(G_n) = \{\omega^c 1 \,|\, c \in \mathbb{Z}\}$, and is a group of order m.

We define the *weight*, $\mathrm{wt}(g)$, of an element g in the error group G_n to be the number of tensor components of g that are not multiples of the identity. For $(a|b)$ in R^{2n}, we define its symplectic weight $\mathrm{swt}((a|b))$ to be the number of indices i

such that $a_i \neq 0$ or $b_i \neq 0$. It follows that

$$\mathrm{wt}(\omega^c X(a)Z(b)) = \mathrm{swt}((a|b)).$$

Suppose that the ring R has q elements. Let $H = (\mathbb{C}^q)^{\otimes n}$ and S be a subgroup of G_n. The *stabilizer code* $\mathrm{Fix}(S)$ associated with S is given by

$$\mathrm{Fix}(S) = \{v \in H \mid Ev = v \text{ for all } E \in S\}.$$

In general, a quantum code, Q, is a subspace of H. We call Q a *stabilizer code* if and only if there exists a subgroup S of G_n such that $\mathrm{Fix}(S) = Q$.

The methodology for constructing stabilizer code is quite interesting. Suppose that we have n quantum systems into which we encode quantum information. We choose a K-dimensional subspace of $H = (\mathbb{C}^q)^{\otimes n}$. Let's say our ultimate goal is to construct a quantum error-correcting code that can detect errors of weight less than d. Specifying a K-dimensional subspace Q directly that can detect all errors of weight less than d is challenging. In the stabilizer code construction, we use the error operators also to define such an $((n, K, d))$ quantum error-correcting code Q. We simply need to find an abelian subgroup S of the error group such that $\mathrm{Fix}(S) = Q$. The best part is that we can determine all parameters $((n, K, d))$ from the group S.

In fact, we can simplify matters even more. Instead of choosing S directly, it suffices to find its image modulo the center $Z(G_n)$. We have to find an additive subgroup $C = S/Z(G_n)$ of $R^{2n} = G_n/Z(G_n)$. We call C a *classical code.* There are certain conditions that the classical code $C = S/Z(G_n)$ needs to satisfy such that its preimage S is an abelian group, namely the elements of C need to be orthogonal to each other for a certain bilinear form (see below).

The next lemma gives the reason why we can decide whether two operators $\omega^c X(a)Z(b)$ and $\omega^{c'} X(a')Z(b')$ in G_n commute by looking at a bilinear relation of $(a|b) \in R^{2n}$ and $(a'|b') \in R^{2n}$.

LEMMA 3.1. *Let R be a nice ring with generating character χ. Let a, b and a', b' be elements of R^n. Then the operators $X(a)Z(b)$ and $X(a')Z(b')$ of the error group commute if and only if $\chi(b \cdot a' - b' \cdot a) = 1$ (where '.' represents the regular dot inner product) holds.*

PROOF. For any $x \in R^n$, we have

$$X(a)Z(b)|x\rangle = X(a)\chi(b \cdot x)|x\rangle = \chi(b \cdot x)|x + a\rangle,$$
$$Z(b)X(a)|x\rangle = Z(b)|x + a\rangle = \chi(b \cdot x)\chi(b \cdot a)|x + a\rangle,$$

hence $\chi(b \cdot a)X(a)Z(b) = Z(b)X(a)$. It follows that

$$\begin{aligned} X(a)Z(b)X(a')Z(b') &= \chi(b \cdot a')X(a + a')Z(b + b') \\ X(a')Z(b')X(a)Z(b) &= \chi(b' \cdot a)X(a + a')Z(b + b'). \end{aligned}$$

Therefore, we can conclude that $X(a)Z(b)$ and $X(a')Z(b')$ commute if and only if $\chi(b \cdot a' - b' \cdot a) = 1$, as claimed. □

Consider the additive group of the ring R. For each character χ in $\mathrm{Hom}(R, \mathbb{C}^\times)$ there exists a unique function ψ in $\mathrm{Hom}(R, \mathbb{Q}/\mathbb{Z})$ such that

$$\chi(x) = \exp(2\pi i \psi(x)).$$

We can define a form $\langle \cdot \mid \cdot \rangle_\chi \colon R^{2n} \times R^{2n} \to \mathbb{Q}/\mathbb{Z}$ such that

$$\langle (a|b) \mid (a'|b') \rangle_\chi = \psi(b \cdot a' - b' \cdot a)$$

for all $(a|b)$ and $(a'|b')$ in R^{2n}. It is easy to see that $\langle\cdot\,|\,\cdot\rangle_\chi$ is a nondegenerate bilinear form, see [**11**]. We write $u \perp v$ if and only if $\langle u\,|\,v\rangle_\chi = 0$.

Stabilizer codes over finite Frobenius rings can be constructed from classical error-correcting codes, as the next theorem shows. We call a classical code C *additive* if and only if it is an additive subgroup of R^{2n}.

THEOREM 3.2. *An $((n,K,d))_R$ stabilizer code exists if and only if there exists an additive code $C \leq R^{2n}$ of size $|C| = |R|^n/K$ such that $C \leq C^\perp$ and $\mathrm{swt}(C^\perp \setminus C) = d$ for $K > 1$, and $\mathrm{swt}(\mathbb{C}^\perp - \{0\}) = d$ for $K = 1$.*

PROOF. The proof is a straightforward generalization of [**3**, Theorem 1] that prove this for the binary field with two elements. A detailed proof for the generalization to finite Frobenius rings can be found in [**11**]. □

A special case is the so-called CSS construction that derives stabilizer codes of length n from a pair of classical codes of length n. A quantum error-correcting code Q is called *pure to* p if and only if $\langle u \mid E \mid v\rangle = 0$ holds for all u, v in Q and all non-scalar errors E of weight p or less. A quantum code is called *pure* if and only if it is pure to the minimum distance. In the next theorem, the orthogonality is understood with respect to the dot inner product on R^n.

A classical code C is called *linear* if and only if it is an R-module. A classical linear code C has parameters $[n,k,d]_R$ if and only if it is a submodule of R^n, has $|R|^k$ elements, and any two codewords in C differ in at least d coordinates. We call n the *length*, k the *dimension*, and d the *minimum distance* of the classical code C.

THEOREM 3.3. [**10**, Theorem 9] *Let C_1 and C_2 denote two classical linear codes with parameters $[n,k_1,d_1]_R$ and $[n,k_2,d_2]_R$ such that $C_2^\perp \leq C_1$. Then there exists a $[[n, k_1+k_2-n, d]]_R$ stabilizer code with minimum distance*

$$d = \min\left\{\mathrm{wt}\,(c)\,|c \in \left(C_1 \setminus C_2^\perp\right) \cup \left(C_2 \setminus C_1^\perp\right)\right\}$$

that is pure to $\min\{d_1, d_2\}$.

4. Quadratic Residue Codes

The goal of this section is to construct some quantum codes over finite commutative chain rings. Finite commutative chain rings are arguably the most accessible examples of finite Frobenius rings. We will use Theorem 3.3 to construct the quantum codes, so the starting point is a pair of classical error-correcting codes. We will use quadratic residue codes for this purpose. Huffman and Pless [**7**] contains more background (they assume that R is a finite field, but the coding-theoretic ideas are the same).

Throughout this section, let R denote a commutative finite chain ring with maximal ideal $J(R) = \mathrm{rad}\,R$ and residue field $\mathbb{F}_q = R/J(R)$. Our goal is to prove the following two theorems.

THEOREM 4.1. *Let n be a prime of the form $n \equiv 1 \pmod 4$, and let $(q,n) = 1$. If q is a quadratic residue modulo n, then there exists a pure $[[n,1,d]]_R$ stabilizer code with distance d satisfying $d \geq \sqrt{n}$.*

THEOREM 4.2. *Let n be a prime of the form $n \equiv 3 \pmod 4$, and let $(q,n) = 1$. If q is a quadratic residue modulo n, then there exists a pure $[[n,1,d]]_R$ stabilizer code with distance d satisfying $d^2 - d + 1 \geq n$.*

We can construct these quantum error-correcting codes using additive classical error-correcting codes. These classical codes are of interesting in their own right due to their number-theoretic properties.

Since we will go back and forth between polynomials over R and polynomials over the residue field $\mathbb{F}_q = R/J(R)$, we will use the bar notation to denote the natural projection $R[x] \to \mathbb{F}_q[x]$ given by

$$\overline{f(x)} = f(x) + J(R)[x].$$

This notation is a bit more compact.

Quadratic Residue Codes. We will construct additive classical codes C that have a lot of structure. Apart from being additive subgroups of R^n, they are also linear cyclic codes. This means that C is an R-module that contains with each codeword $(c_0, c_1, \ldots, c_{n-1})$ the cyclically shifted codeword $(c_1, c_2, \ldots, c_{n-1}, c_0)$. A linear cyclic code corresponds to an ideal in the ring $R[x]/\langle x^n - 1\rangle$ by identifying $(c_0, c_1, \ldots, c_{n-1})$ with $c_0 + c_1 x + \cdots + c_{n-1}x^{n-1} + \langle x^n - 1\rangle$ in $R[x]/\langle x^n - 1\rangle$. We will go back and forth between these two representations, as is customary in coding theory.

If R is a finite chain ring, then the truncated polynomial ring $R[x]/\langle x^n - 1\rangle$ is a principal ideal ring, see [**4**, Corollary 3.7]. Therefore, C corresponds to a principal ideal in $R[x]/\langle x^n - 1\rangle$. A polynomial $g(x)$ such that $g(x) + \langle x^n - 1\rangle$ generates the principal ideal in $R[x]/\langle x^n - 1\rangle$ corresponding to C is called the *generator polynomial* of C. If the code C is a proper subset of R^n, then the generator polynomial $g(x)$ of C must be a divisor of $x^n - 1$. We will now construct generator polynomials from quadratic residues and non-residues modulo n.

Let n be an odd prime that is coprime to q. We denote by

$$Q = \left\{r^2 \pmod{n} \,|r \in \mathbb{Z}, 1 \leq r \leq (n-1)/2\right\}$$

the set of quadratic residues modulo n and by

$$N = \{1, \ldots, n-1\} \setminus Q$$

the set of quadratic non-residues modulo n.

We choose a fixed primitive n-th root of unity α from the algebraic closure of $\mathbb{F}_q$. We define the polynomials

$$f_Q(x) = \prod_{r\in Q} (x - \alpha^r) \quad \text{and} \quad f_N(x) = \prod_{r\in N} (x - \alpha^r)$$

in $\mathbb{F}_q[x]$. Then the polynomial $x^n - 1$ factors as follows

$$x^n - 1 = (x-1)\, f_Q(x)\, f_N(x) \in \mathbb{F}_q[x].$$

We will now lift these polynomials to the polynomial ring $R[x]$. By Hensel's lemma [**4**, Theorem 2.4], there exist monic polynomials $(x - a)$, $q_Q(x)$, and $q_N(x)$ in $R[x]$ that are pairwise coprime and satisfy

$$\begin{aligned} \overline{(x-a)} &= (x-1), \\ \overline{q_Q(x)} &= f_Q(x), \\ \overline{q_N(x)} &= f_N(x), \end{aligned}$$

and $x^n - 1 = (x-a)\, q_Q(x)\, q_N(x) \in R[x]$.

We claim that we must have $a = 1$ in the previous equation for $x^n - 1$. Indeed, substituting 1 into the equation, we obtain $(1-a)\, q_Q(1)\, q_N(1) = 0$; since

$\overline{q_Q(1)} = f_Q(1) \neq 0$ and $\overline{q_N(1)} = f_N(1) \neq 0$, it follows that $q_Q(1)$ and $q_N(1)$ are both invertible elements of R, whence $a = 1$. Therefore, we have $x^n - 1 = (x-1)\, q_Q(x)\, q_N(x) \in R[x]$, as claimed.

We say that a codeword $x = (x_1, x_2, \cdots, x_n) \in R^n$ is *even-like* if and only if $\sum_{i=1}^{n} x_i = 0$. A codeword that is not even-like is called *odd-like*. We say that a code is *even-like* if it has only even-like codewords and that it is *odd-like* if it is not even-like.

The *quadratic residue codes* C_Q, C_N and their respective even-like subcodes C'_Q, C'_N are cyclic codes generated by

$$\begin{aligned} C_Q &= \langle q_Q(x) \rangle, & C'_Q &= \langle (x-1) q_Q(x) \rangle, \\ C_N &= \langle q_N(x) \rangle, & C'_N &= \langle (x-1) q_N(x) \rangle. \end{aligned}$$

The quadratic residue code C_Q and C_N have parameters $[n, (n+1)/2, d]_R$, and their respective even-like subcodes C'_Q and C'_N have parameters $[n, (n-1)/2, d']_R$, with $d' \geq d$.

We will obtain lower bounds for the minimum distance of these codes later in this section. First we need to derive some basic results about cyclic codes over finite chain rings.

Cyclic Complement. Denote by R_n the ring $R[x]/\langle x^n - 1 \rangle$. The *cyclic complement* C^C of a cyclic code C is a cyclic code satisfying $C^C + C = R_n$ and $C^C \cap C = \{0\}$.

THEOREM 4.3. *Let C be a proper cyclic code of length n over R with a monic generator polynomial $g(x)$ dividing $x^n - 1$ and a generating idempotent $e(x)$. Let C^C be the cyclic complement of C. Then C^C has a generator polynomial $\widehat{g}(x) = (x^n - 1)/g(x)$ and a generating idempotent $1 - e(x)$.*

PROOF. Since C is a proper subcode of R_n, the polynomial $g(x)$ must divide $x^n - 1$. It follows that $\hat{g}(x) = (x^n - 1)/g(x)$ is a divisor of $x^n - 1$, so $\langle \widehat{g}(x) \rangle$ is a proper cyclic subcode of R_n. Since $\gcd(n, q) = 1$, the polynomial $x^n - 1$ is separable in $\mathbb{F}_q[x]$, so $\overline{g(x)}$ and $\overline{\hat{g}(x)} = \overline{(x^n - 1)/g(x)}$ are coprime polynomials in $\mathbb{F}_q[x]$. Hensel lifting yields

$$\langle g(x) \rangle + \langle \widehat{g}(x) \rangle = R_n.$$

Suppose that $k(x)$ is a arbitrary polynomial such that $k(x) + \langle x^n - 1 \rangle \in \langle g(x) \rangle \cap \langle \widehat{g}(x) \rangle$. Each root α in a splitting field of $x^n - 1$ over $\mathbb{F}_q$ can be lifted to unique root β of $x^n - 1$ over $R[x]$. Since $\overline{k}(\alpha) = 0$, it follows that $k(\beta) = 0$. Therefore, $k(x)$ is a multiple of $x^n - 1$, which implies that

$$k(x) + \langle x^n - 1 \rangle = 0$$

in R_n. Since $k(x)$ was arbitrary, we can conclude that $\langle g(x) \rangle \cap \langle \widehat{g}(x) \rangle = \{0\}$.

Let $1 = e_1(x) + e_2(x)$, where $e_1(x) \in \langle g(x) \rangle$ and $e_2(x) \in \langle \widehat{g}(x) \rangle$. Then there exist $a(x), b(x) \in R_n$ such that $e_1(x) = a(x)\, g(x)$ and $e_2(x) = b(x)\, \widehat{g}(x)$. Then $e_1(x)^2 = e_1(x)(1 - e_2(x)) = e_1(x) - e_1(x)\, e_2(x) = e_1(x) - a(x)\, g(x)\, b(x)\, \widehat{g}(x) = e_1(x)$, so $e_1(x)$ is an idempotent of C, thus $e_1(x) = e(x)$. Similarly, $e_2(x)^2 = (1 - e(x))^2 = 1 - e(x)$ is the idempotent of C^C. □

Even- and Odd-like Codewords. In the next lemma, we collect some elementary facts about even- and odd-like codewords. Recall that the *repetition code* of length n over a finite chain ring R is given by

$$\{(c, c, \ldots, c) \mid c \in R\} \subset R^n.$$

By slight abuse of notation, we will refer to elements $a(x) + \langle x^n - 1\rangle$ of R_n as $a(x)$. This abbreviation will make the formulation of the following lemma a bit more readable.

LEMMA 4.4. *Let* $h(x) = \frac{1}{n}\left(1 + x + x^2 + \cdots + x^{n-1}\right)$, $a(x) = \sum_{i=0}^{n-1} a_i x^i \in R_n$, *and* C *a proper cyclic subcode of* R_n *with generating polynomial* $g(x)$. *Then*

(1) $h(x)$ *is the generating idempotent of the repetition code of length* n *over* R

(2) $a(x)$ *is even-like if and only if* $a(1) = 0$ *if and only if* $a(x)h(x) = 0$

(3) $a(x)$ *is odd-like if and only if* $a(1) \neq 0$ *if and only if* $a(x)h(x) = \alpha h(x)$, $\alpha \neq 0$

PROOF. Expanding $(h(x))^2$, we find that

$$\left(\frac{1}{n}\left(1 + x + \cdots + x^{n-1}\right)\right)^2 = \frac{1}{n}\left(1 + x + \cdots + x^{n-1}\right),$$

so $h(x)$ is an idempotent of R_n. Additionally, the codewords of the repetition code are all of the form $f(x) = a\left(1 + x + x^2 + \cdots + x^{n-1}\right)$ for $a \in R$, so $f(x)h(x) = \frac{a}{n}\left(1 + x + x^2 + \cdots + x^{n-1}\right)^2 = a\left(1 + x + x^2 + \cdots + x^{n-1}\right) = f(x)$, so $h(x)$ is the generating idempotent for the repetition code of length n over R. If $a(x) = \sum_{i=0}^{n-1} a_i x^i$ is in R_n, then

$$a(x)h(x) = \left(\sum_{i=0}^{n-1} a_i\right)\frac{1}{n}\left(1 + x + x^2 + \cdots + x^{n-1}\right),$$

so if $a(x)$ is even-like, $\sum_{i=0}^{n-1} a_i = 0$, so $a(x)h(x) = 0$; additionally, $a(x)$ is even-like precisely when $\sum_{i=0}^{n-1} a_i = 0$. This is the same as saying that $a(1) = 0$. If $a(x)$ is odd-like, $\sum_{i=0}^{n-1} a_i \neq 0$, then $a(x)h(x) = \alpha h(x)$, for some $\alpha \in R$, $\alpha \neq 0$, which is also the same as saying $a(1) \neq 0$. Conversely, if $a(1) \neq 0$, then $a(x)$ is odd-like by definition. □

The set of all even-like codewords in R_n forms a code. The next lemma shows that it is a cyclic code and collects some useful facts about this code. Recall that a classical code C over R is called an $[n, k]$ code if and only if it is a subset of R^n and has $|R|^{n-1}$ elements.

LEMMA 4.5. *Let* $\mathcal{E}_n$ *denote the collection of even-like codewords in* R_n. *Then:*

(1) $\mathcal{E}_n$ *is an* $[n, n-1]$ *cyclic subcode of* R_n

(2) $\mathcal{E}_n^\perp$ *is the repetition code with generating idempotent*

$$h(x) = \frac{1}{n}\left(1 + x + x^2 + \cdots + x^{n-1}\right).$$

(3) $\mathcal{E}_n$ *has generating idempotent* $1 - h(x)$

PROOF. Let $u, v \in \mathcal{E}_n$ and $a, b \in R$. Since $a\sum_{i=1}^n u_i = 0$ and $b\sum_{i=1}^n v_i = 0$, we have $\sum_{i=1}^n (au_i + bv_i) = 0$, so $(au + bv) \in \mathcal{E}_n$, and therefore $\mathcal{E}_n$ is a subcode of R_n, and must therefore be cyclic. Since R_n can be partitioned into $|R|$ equally sized partitions based on the parity of the codewords, $\mathcal{E}_n$ is an $[n, n-1]$ subcode of R_n, giving (1).

Since $\mathcal{E}_n$ is an $[n, n-1]$ cyclic code, $\mathcal{E}_n^\perp$ must be an $[n, 1]$ cyclic code, so $\mathcal{E}_n^\perp$ is the repetition code. By Lemma 4.4, the repetition code has generating idempotent $h(x) = \frac{1}{n}\left(1 + x + x^2 + \cdots + x^{n-1}\right)$.

Finally by [**15**, Theorem 2], $\mathcal{E}_n$ has generating idempotent $1 - h(x)\mu_{-1} = 1 - h(x)$. □

LEMMA 4.6. *Let C be a cyclic code over R with generating idempotent $i(x)$ and let C_e be the subcode of all even-like codewords in C. If $C \neq C_e$, then $i(x) - h(x)$ is the generating idempotent of C_e.*

PROOF. Since C_e is the even-like subcode of C, $C_e = C \cap \mathcal{E}_n$. By [**15**, Theorem 1], the generating idempotent of C_e is $i(x)(1 - h(x)) = i(x) - i(x)h(x)$. Since $i(x)$ is the generating idempotent of C, but not the generating idempotent of C_e it is necessarily odd-like, so by Lemma 4.4 $i(x) - i(x)h(x) = i(x) - \alpha h(x)$, where $\alpha = \sum_{k=0}^{n-1} i_k$ is a nonzero element of R. Let $b(x)$ be an odd-like codeword in C. Then $b(x)i(x) = b(x)$. Evaluating this equation at $x = 1$ gives $\sum_{k=0}^{n-1} b_k = \sum_{k=0}^{n-1}\left(b_k \sum_{j=0}^{n-1} i_j\right) = \alpha \sum_{k=0}^{n-1} b_k$. Since $b(x)$ is an odd-like codeword, $\sum_{k=0}^{n-1} b_k \neq 0$, so $\alpha = 1$, giving $i(x) - h(x)$ as the generating idempotent of C_e. □

Duadic Codes. Define the function $\mu_a : \mathbb{Z}_n \to \mathbb{Z}_n$, where a and n are coprime, by $\mu_a(i) = ia \pmod{n}$. This function is known as a multiplier. The multiplier can also act on polynomials by $\mu_a : R_n \to R_n,\ f(x) \mapsto f(x^a)$. Since the multiplier acts in the same way on coordinate and polynomial representations of a codeword, the notation should not cause confusion. Traditionally, the multiplier is written by a right action, so $f(x)\mu_a = f(x^a)$.

THEOREM 4.7. *Let $f(x), g(x) \in R_n$, $e(x)$ be an idempotent of R_n, and a be an integer coprime to n. Then:*

(1) *if $b \equiv a \pmod{n}$, then $\mu_b = \mu_a$*
(2) *μ_a is an automorphism of R_n*
(3) *$e(x)\mu_a$ is an idempotent of R_n.*

PROOF. All of the results follow from straightforward calculations. □

THEOREM 4.8. *Let C be a cyclic code of length n over R with generating idempotent $e(x)$, and let a be an integer coprime to n. Then $C\mu_a = \langle e(x)\mu_a\rangle$ and $e(x)\mu_a$ is the generating idempotent of the cyclic code $C\mu_a$.*

PROOF. Using Theorem 4.7,

$$\begin{aligned} C\mu_a &= \{(e(x)f(x))\mu_a | f(x) \in R_n\} \\ &= \{e(x)\mu_a f(x)\mu_a | f(x) \in R_n\} \\ &= \{e(x)\mu_a h(x) | h(x) \in R_n\} \\ &= \langle e(x)\mu_a\rangle \end{aligned}$$

as μ_a is an automorphism of R_n by Theorem 4.7. Hence $C\mu_a$ is cyclic and has generating idempotent $e(x)\mu_a$ by Theorem 4.7. □

Let $e_1(x)$ and $e_2(x)$ be two even-like idempotents in R_n with $C_1 = \langle e_1(x)\rangle$ and $C_2 = \langle e_2(x)\rangle$. The codes C_1 and C_2 form a pair of *even-like duadic codes* if

(1) the idempotents satisfy

$$e_1(x) + e_2(x) = 1 - h(x) \tag{4.1}$$

(2) there is a integer a coprime to n such that

$$C_1\mu_a = C_2 \text{ and } C_2\mu_a = C_1. \tag{4.2}$$

We say that μ_a gives a *splitting* for the even-like codes C_1 and C_2 if and only if the multiplier μ_a satisfies (4.2).

By Theorem 4.8, we have that $e_1(x)\mu_a = e_2(x)$ and $e_2(x)\mu_a = e_1(x)$ if and only if $C_1\mu_a = C_2$ and $C_2\mu_a = C_1$, so we can replace equation (4.2) by

$$e_1(x)\mu_a = e_2(x) \text{ and } e_2(x)\mu_a = e_1(x). \tag{4.3}$$

Associated to the pair of even-like duadic codes is the pair of *odd-like duadic codes*

$$D_1 = \langle 1 - e_2(x)\rangle \text{ and } D_2 = \langle 1 - e_1(x)\rangle. \tag{4.4}$$

THEOREM 4.9. *Let $C_1 = \langle e_1(x)\rangle$ and $C_2 = \langle e_2(x)\rangle$ be a pair of even-like duadic codes of length n over R. Suppose that μ_a gives the splitting for C_1 and C_2. Let D_1 and D_2 be the associated odd-like duadic codes. Then:*

(1) $e_1(x)e_2(x) = 0$
(2) $C_1 \cap C_2 = \{0\}$ *and* $C_1 + C_2 = \mathcal{E}_n$
(3) C_1 *and* C_2 *each have dimension* $(n-1)/2$
(4) D_1 *is the cyclic complement of* C_2 *and* D_2 *is the cyclic complement of* C_1
(5) D_1 *and* D_2 *each have dimension* $(n+1)/2$
(6) C_i *is the even-like subcode of* D_i, *for* $i = 1, 2$
(7) $D_1\mu_a = D_2$ *and* $D_2\mu_a = D_1$
(8) $D_1 \cap D_2 = \langle h(x)\rangle$ *and* $D_1 + D_2 = R_n$
(9) $D_i = C_i + \langle h(x)\rangle = \langle h(x) + e_i(x)\rangle$, *for* $i = 1, 2$

PROOF. (1) Multiplying equation (4.1) by $e_1(x)$ gives $e_1(x)e_2(x) = 0$ by Lemma 4.4 so 1) holds.

(2) By [**15**, Theorem 1], $C_1 \cap C_2$ and $C_1 + C_2$ have generating idempotents $e_1(x)e_2(x) = 0$ and $e_1(x) + e_2(x) - e_1(x)e_2(x) = e_1(x) + e_2(x) = 1 - h(x)$ respectively, so 2) holds by Lemma 4.5.

(3) By equation (4.2), C_1 and C_2 are equivalent (that is, are the same up to permutation of coordinates) and hence have the same dimension. By 2) and by Lemma 4.5 this dimension is $(n-1)/2$, giving 3).

(4) The cyclic complement of C_i has generating idempotent $1 - e_i(x)$ by Theorem 4.3; thus 4) is immediate from the definition of D_i.

(5) Part 5) follows from the definition of cyclic complement and parts 3) and 4).

(6) As D_1 is odd-like with generating idempotent $1 - e_2(x)$ by Lemma 4.6, the generating idempotent of the even-like subcode of D_1 is $1 - e_2(x) - h(x) = e_1(x)$. Thus C_1 is the even-like subcode of D_1; analogously, C_2 is the even-like subcode of D_2 yielding 6).

(7) The generating idempotent of $D_1\mu_a$ is $(1 - e_2(x))\mu_a = 1 - e_2(x)\mu_a = 1 - e_1(x)$ by Theorem 4.8 and equation (4.3). Thus $D_1\mu_a = D_2$; analogously $D_2\mu_a = D_1$, producing 7).

(8) By [**15**, Theorem 1], $D_1 \cap D_2$ and $D_1 + D_2$ have generating idempotents $(1 - e_1(x))(1 - e_2(x)) = 1 - e_1(x) - e_2(x) = h(x)$ and $(1 - e_1(x)) + (1 - e_2(x)) - (1 - e_1(x))(1 - e_2(x)) = 1$ respectively, as $e_1(x)e_2(x) = 0$. Thus 8) holds as the generating idempotent of R_n is 1.

(9) Finally by 3), 5), and 6), C_i is a subspace of D_i of codimension 1, as $h(x) \in D_i \setminus C_i$, $D_i = C_i + \langle h(x)\rangle$. Also, $D_i = \langle h(x) + e_i(x)\rangle$ by equations (4.1) and (4.4), which proves 9). □

THEOREM 4.10. [**2**, Proposition 4.3] *Let $D_i, i \in \{1,2\}$ be the odd-like duadic codes over R, where $D_i = \langle g_i(x)\rangle$ and $(x-1)\,g_1(x)\,g_2(x) = x^n - 1$, and let C_i be the even-like duadic codes over R, where $C_i = \langle (x-1)\,g_i(x)\rangle$. Then*

(1) *if the splitting is given by μ_{-1}, then $D_1^{\perp} = C_1$ and $D_2^{\perp} = C_2$*
(2) *if the splitting is not given by μ_{-1}, then $D_1^{\perp} = C_2$ and $D_2^{\perp} = C_1$*

Square Root Bound. We will now give a lower bound on the minimum distance of a pair of odd-like duadic codes. For an additive classical code C, the minimum distance coincides with the *minimum weight*, that is, the smallest Hamming weight of a nonzero codeword in C. The *even-like minimum weight* of C is the smallest Hamming weight of an even-like nonzero codeword in C. Similarly, the *odd-like minimum weight* of C is the smallest Hamming weight of an odd-like nonzero codeword in C.

THEOREM 4.11. *(Square Root Bound) Let D_1 and D_2 be a pair of odd-like duadic codes of length n over R. Let d_0 be their (common) minimum odd-like weight. Then the following holds:*

(1) *$d_0^2 \geq n$,*
(2) *if the splitting defining the duadic codes is given by μ_{-1}, then $d_0^2 - d_0 + 1 \geq n$.*

PROOF. Suppose that the splitting defining the duadic codes is given by μ_a. Let $c(x) \in D_1$ be an odd-like codeword of Hamming weight d_0. Then $c'(x) = c(x)\,\mu_a \in D_2$ is also odd like and $c(x)\,c'(x) \in D_1 \cap D_2$ as D_1 and D_2 are ideals in R_n. But $D_1 \cap D_2 = \langle h(x)\rangle$ by Theorem 4.9 (8). By Lemma 4.4, $c(x)c'(x)$ is odd-like and in particular nonzero. Therefore $c(x)c'(x)$ is a nonzero multiple of $h(x)$, and so $\mathrm{wt}(c(x)c'(x)) = n$. The number of terms in the product $c(x)c'(x)$ is at most d_0^2, so 1) follows.

If $\mu_a = \mu_{-1}$, then the number of terms in $c(x)\,c'(x)$ is at most $d_0^2 - d_0 + 1$ because d_0 terms contribute only to the coefficient of x^0 in $c(x)c'(x)$, whereas in a general splitting perhaps just one term does, so 2) follows. □

Quantum Codes. We will now prove the theorems concerning the quantum codes that we had announced at the beginning of this section. We construct quantum codes from the following classical codes: the quadratic residue codes C_Q, C_N and their even-like subcodes C_Q' and C_N'.

THEOREM 4.12. *Let n be a prime of the form $n \equiv 3 \pmod 4$, and let $(q,n) = 1$. If q is a quadratic residue modulo n, then there exists a pure $[[n,1,d]]_R$ stabilizer code with distance d satisfying $d^2 - d + 1 \geq n$.*

PROOF. The code C_Q has parameters $[n,(n+1)/2,d]_R$ and since $n \equiv 3 \pmod 4$, by [**7**, Lemma 6.2.4] we know that -1 is not a square modulo n, so μ_{-1} gives the splitting for C_Q and C_N. Therefore by Theorem 4.10 we know that $C_Q^{\perp} = C_Q' = \langle (x-1)q_Q(x)\rangle$, so C_Q contains its dual code. By Theorem 4.11 we know that the minimum distance d is bounded by $d^2 - d + 1 \geq n$. Furthermore, $\mathrm{wt}\left(C_Q \setminus C_Q^{\perp}\right) = \mathrm{wt}\left(C_Q\right) = d$, since the minimum weight of C_Q is its minimum odd-like weight. We can therefore construct a $[[n,(n+1)-n,d]]_R$ stabilizer code by Theorem 3.3. □

THEOREM 4.13. *Let n be a prime of the form $n \equiv 1 \pmod 4$, and let $(q,n) = 1$. If q is a quadratic residue modulo n, then there exists a pure $[[n,1,d]]_R$ stabilizer code with distance d satisfying $d \geq \sqrt{n}$.*

PROOF. The code C_Q has parameters $[n,(n+1)/2,d]_R$ and since $n \equiv 1 \pmod 4$, by [**7**, Lemma 6.2.4] we know that -1 is a square modulo n, so μ_{-1} does not give a splitting for C_Q and C_N. Therefore by Theorem 4.10 $C_Q^{\perp} = C_N'$, that is $C_Q^{\perp} \leq C_N$. By Theorem 4.11 we know that the minimum distance d is bounded by $d \geq \sqrt{n}$. Moreover, $\mathrm{wt}\left(C_Q \setminus C_N^{\perp}\right) = \mathrm{wt}\left(C_N \setminus C_Q^{\perp}\right) = \mathrm{wt}\left(C_Q\right) = \mathrm{wt}\left(C_N\right) = d$ since the minimum weight of C_Q and C_N is their (common) minimum odd-like weight. Therefore we obtain a pure $[[n,(n+1)/2+(n+1)/2-n,d]]_R$ stabilizer code by Theorem 3.3. □

5. Conclusions

We discussed the construction of quantum error-correcting codes over finite Frobenius rings. We also developed the theory of duadic codes over finite chain rings. In particular, we constructed quadratic residue codes over finite chain rings that yield $[[n,1,d]]$ quantum error-correcting codes that have minimum distance $d \geq \sqrt{n}$. These codes allow us to encode one logical quantum digit into n quantum digits.

We can use these codes as a basis for fault-tolerant quantum computing. Here all quantum digits are encoded using a quantum error-correcting code. Suppose that we want to perform an operation U on a single quantum digit in the idealized error-free model, so that a state $|x\rangle$ is mapped by A to $|y\rangle = A|x\rangle$. If the quantum error-correcting code encodes the state $|x\rangle \in \mathbb{C}^q$ as $|x'\rangle \in \mathbb{C}^{q^n}$ and the state $|y\rangle \in \mathbb{C}^q$ as $|y'\rangle \in \mathbb{C}^{q^n}$, then we need to find an operation A' in $U(q^n)$ such that $|y'\rangle = A'|x'\rangle$. Ideally, we would like A' to decompose into a tensor product

$$A' = A_0 \otimes A_0 \otimes \cdots \otimes A_0,$$

of operators $A_0 \in U(q)$, so that when an error occurs in a single quantum digit, it will not spread to other quantum digits when we apply A'. Operations with these desirable properties are called transversal. The codes given here have been developed with a view toward fault-tolerant quantum computing. For instance, the methods given in [**9**] allow us to construct transversal Fourier operations, transversal phase gates, and transversal sum gates for the quantum error-correcting codes given in this paper.

Acknowledgments. We thank both referees for carefully reading our manuscript and making many valuable suggestions that helped to improve the accuracy and presentation of our paper.

References

[1] Frank W. Anderson and Kent R. Fuller, *Rings and categories of modules*, 2nd ed., Graduate Texts in Mathematics, vol. 13, Springer-Verlag, New York, 1992. MR1245487

[2] A. Batoul, K. Guenda, and T. Gulliver, *On isodual cyclic codes over finite fields and finite chain rings: Monomial equivalence*, arXiv 1303.1870v1, 2013.

[3] A. R. Calderbank, E. M. Rains, P. W. Shor, and N. J. A. Sloane, *Quantum error correction and orthogonal geometry*, Phys. Rev. Lett. **78** (1997), no. 3, 405–408, DOI 10.1103/PhysRevLett.78.405. MR1427248

[4] Hai Quang Dinh and Sergio R. López-Permouth, *Cyclic and negacyclic codes over finite chain rings*, IEEE Trans. Inform. Theory **50** (2004), no. 8, 1728–1744, DOI 10.1109/TIT.2004.831789. MR2096841

[5] Y. Hirano, *On admissible rings*, Indag. Math. (N.S.) **8** (1997), no. 1, 55–59, DOI 10.1016/S0019-3577(97)83350-2. MR1617802

[6] Thomas Honold, *Characterization of finite Frobenius rings*, Arch. Math. (Basel) **76** (2001), no. 6, 406–415, DOI 10.1007/PL00000451. MR1831096

[7] W. Cary Huffman and Vera Pless, *Fundamentals of error-correcting codes*, Cambridge University Press, Cambridge, 2003. MR1996953

[8] T. Y. Lam, *Lectures on modules and rings*, Graduate Texts in Mathematics, vol. 189, Springer-Verlag, New York, 1999. MR1653294

[9] S. Lee and A. Klappenecker, *Generalized fault-tolerant quantum computation over nice rings*, 2016 IEEE International Symposium on Information Theory (ISIT), 2016, pp. 2279–2283.

[10] S. Nadella, *Stabilizer codes over Frobenius rings*, Master's thesis, Texas A&M University, May 2012.

[11] S. Nadella and A. Klappenecker, *Stabilizer codes over Frobenius rings*, 2012 IEEE International Symposium on Information Theory Proceedings (ISIT), July 2012, pp. 165–169.

[12] A. Sarma and A. Klappenecker, *Quantum codes over rings and quadratic algebras*, Proc. of 52nd Annual Allerton Conference on Communication, Control, and Computing, October 1-3, 2014,Allerton Retreat Center, Monticello, IL, USA, 2014, pp. 1376–1381. See also http://faculty.cs.tamu.edu/klappi/papers/quad_alg.pdf

[13] D. W. Sharpe and P. Vámos, *Injective modules*, Cambridge University Press, London-New York, 1972. Cambridge Tracts in Mathematics and Mathematical Physics, No. 62. MR0360706

[14] Peter W. Shor, *Polynomial-time algorithms for prime factorization and discrete logarithms on a quantum computer*, SIAM J. Comput. **26** (1997), no. 5, 1484–1509, DOI 10.1137/S0097539795293172. MR1471990

[15] Bijan Taeri, *Quadratic residue codes over* $\mathbb{Z}_9$, J. Korean Math. Soc. **46** (2009), no. 1, 13–30, DOI 10.4134/JKMS.2009.46.1.013. MR2475449

[16] Jay A. Wood, *Duality for modules over finite rings and applications to coding theory*, Amer. J. Math. **121** (1999), no. 3, 555–575. MR1738408

Department of Computer Science and Engineering, Texas A&M University, College Station, TX 77843-3112, USA
Email address: klappi@cse.tamu.edu

CEA-Leti, Minatec Campus, 38054, Grenoble, France
Email address: Sangjun.LEE@cea.fr

Department of Computer Science and Engineering, Texas A&M University, College Station, TX 77843-3112, USA
Email address: nemeca@tamu.edu

Contemporary Mathematics
Volume **747**, 2020
https://doi.org/10.1090/conm/747/15047

A short history of frames and quantum designs

Bernhard G. Bodmann and John I. Haas

Abstract. In this survey, we relate frame theory and quantum information theory, focusing on quantum 2-designs. These are arrangements of weighted subspaces which are in a specific sense optimal for quantum state tomography. After a brief introduction, we discuss the role of POVMs in quantum theory, developing the importance of quantum 2-designs. In the final section, we collect many if not most known examples of quantum-2 designs to date.

1. Introduction

This survey is concerned with the role of frames and designs in quantum information theory. Frames are spanning families in Hilbert spaces that permit stable expansions of vectors. In contrast to orthonormal bases, frames can incorporate linear dependencies. The flexibility in their design can be used to realize goals that would be unattainable with minimal systems, Riesz bases. There is an intriguing connection between objects studied in frame theory and in quantum information theory. A Parseval frame is a family of vectors $\{f_j\}_{j\in J}$ in a Hilbert space $\mathcal{H}$, indexed by an at most countable set J, such that the rank-one Hermitians obtained from the frame vectors resolve the identity,

$$\sum_{j\in J} f_j \otimes f_j^* = I\,.$$

Here, f^* is the linear functional $x \mapsto \langle x, f_j\rangle, x \in \mathcal{H}$. When we abbreviate the rank-one Hermitians appearing in this sum as $A_j = f_j \otimes f_j^*$, then the family $\{A_j\}_{j\in J}$ happens to be a special case of a positive operator-valued measure (POVM), a family of positive (semidefinite) operators that sum to the identity. As explained in the next section, POVMs describe the statistics for outcomes of quantum measurements, and the question of selecting an optimal one to estimate states based on the observed relative frequencies for outcomes is closely related to frame theory. For general quantum measurements, each A_j does not have to be a rank-one operator. Indeed, choosing each A_j of a POVM to be a multiple of an orthogonal projection has also been the subject of frame theory under the name of fusion frames. In the remainder of this paper, we study POVMs with more structure motivated by goals in quantum state tomography.

Following Zauner's seminal work [**37**], we define a *weighted quantum design* to be a sequence of orthogonal projections $\{P_j\}_{j=1}^n$ in a Hilbert space $\mathcal{H}$, accompanied by weights $\{w_j\}_{j=1}^n$ in $\mathbb{R}^+$. The sequence is t-coherent for $t \in \mathbb{N}$ if for each unitary

2010 *Mathematics Subject Classification.* Primary 42C15, 81P16; Secondary 05B40.

U on $\mathcal{H}$,

$$\sum_{j=1}^{n} w_j P_j^{\otimes t} = \sum_{j=1}^{n} w_j (UP_jU^*)^{\otimes t}\,.$$

A weighted quantum design that is s-coherent for all integers $1 \le s \le t$ is called a quantum t-design. Hence, a quantum t-design is also a 1-design and satisfies, by averaging over U with respect to the normalized Haar measure, $\sum_{j=1}^n w_j P_j = \sum_{j=1}^n w_j \operatorname{tr}[P_j/d] I$, where d is the dimension of $\mathcal{H}$. If the weights are scaled by an overall factor so that $\sum_{j=1}^n w_j \operatorname{tr}[P_j/d] = 1$, then such a quantum t-design is a POVM with a special structure.

2. From quantum state tomography to quantum designs

Quantum theory evolved in the early 20th century because results of physical measurements could not be explained with models of classical mechanics [**13**]. Today, the predictive power of quantum theory is unrivaled in its accuracy in physical experiments [**14**]. In the last 30 years, quantum theory has received much attention because of the computational advantages afforded by the intrinsic parallelism in quantum time evolution [**15**].

The achievements of quantum theory come with conceptual and philosophical issues about its interpretation. One of the main challenges is that the predictions are probabilistic. In order to compare classical and quantum theories, a statistical framework is necessary. Moreover, because the outcomes of experiments are formulated in language based on boolean logic, any viable quantum theory needs to include the means to describe classical systems.

We briefly review the axioms of quantum theory, here with the simplifying assumption that the underlying Hilbert space $\mathcal{H}$ is finite dimensional: States are trace-normalized non-negative operators on this Hilbert space. Observables describe the information gained from observing the outcome of an experiment. The minimal view of quantum theory is that it predicts probabilities for observing outcomes. A classical theory is embedded in the quantum theory by associating with (mutually exclusive) outcomes of an experiment mutually orthogonal subspaces of the Hilbert space.

A measurement is performed by a *quantum instrument* [**8**]. In order to keep track of the classical information gained from the measurement, we use the tensor product Hilbert space $\mathcal{H} \otimes \ell^2(\Omega)$, where $\mathcal{H}$ is used to define the quantum state and $\ell^2(\Omega)$ the outcomes of the experiment recorded by the measurement apparatus.

Associating with each outcome $\omega \in \Omega$ the diagonal projection operator D_ω on $\ell^2(\Omega)$ whose range consists of the one-dimensional subspace of functions supported on the singleton set $\{\omega\}$, a quantum instrument is modeled by a trace-preserving map $\mathcal{E}$ of the form

$$\mathcal{E}(W) = \sum_{\omega\in\Omega} \sum_{j\in J_\omega} M_{j,\omega} W M_{j,\omega}^* \otimes D_\omega$$

where W is a state operator on the Hilbert space $\mathcal{H}$, D_ω is a diagonal projection on $\ell^2(\Omega)$ while $\sum_{j\in J_\omega} M_{j,\omega} W M_{j,\omega}^*$ describes the non-normalized quantum state in $\mathcal{H}$ after the measurement with outcome ω.

More informally, the outcome of a measurement is recorded, while the quantum state undergoing the measurement is mapped by conjugating it with Kraus operators belonging to each outcome, thus creating a special form of a quantum channel. A general quantum channel on a finite-dimensional Hilbert space has the form $\mathcal{E} : W \mapsto \sum_{i=1}^n M_i W M_i^*$ where the sequence of Kraus operators $\{M_i\}_{i=1}^n$ satisfies $\sum_{i=1}^n M_i^* M_i = I$.

If one is only interested in the statistics of the outcomes, then these are given by the probabilities $p_\omega = \sum_{j \in J_\omega} \mathrm{tr}[M_{j,\omega}^* M_{j,\omega} W]$ obtained after performing a partial trace over the Hilbert space describing the quantum state. If the probabilities for the outcomes determine each state uniquely, then the measurement is called informationally complete. From now on, we assume that each outcome is associated with one Kraus operator and abbreviate $A_\omega = M_\omega^* M_\omega$. The collection $\{A_\omega : \omega \in \Omega\}$ is then a family of positive semi-definite operators summing to the identity, also known as a positive-operator valued measure.

DEFINITION 2.1. *A positive operator-valued measure $\{A_j\}_{j=1}^n$ is called informationally complete if the map $\mathcal{A} : W \mapsto (\mathrm{tr}[A_j W])_{j=1}^n$ is injective on the set of operators $\{W : W \geq 0, \mathrm{tr}\, W = 1\}$.*

Since $\mathcal{A}$ is a linear map, informational completeness is equivalent to the set $\{A_j\}_{j=1}^n$ having as its real span the space of all Hermitian matrices.

THEOREM 2.1. *A positive operator-valued measure $\{A_j\}_{j=1}^n$ is informationally complete if and only if the real span of $\{A_j\}_{j=1}^n$ consists of all Hermitian operators.*

PROOF. Assume the POVM is informationally complete and a Hermitian H gives $\mathcal{A}(H) = 0$. By summing entries of $\mathcal{A}(H)$, we get $\sum_{j=1}^n \mathrm{tr}[A_j H] = \mathrm{tr}[H] = 0$, so H has vanishing trace. We need to exclude $H \neq 0$. Splitting $H = H_+ - H_-$, where $H_+, H_- \geq 0$ yields $\mathrm{tr}[H_+] = \mathrm{tr}[H_-]$. If each of these traces is non-zero, then we can assume $\mathrm{tr}[H_+] = \mathrm{tr}[H_-] = 1$ without loss of generality, and thus H_+ and H_- are states that give $\mathcal{A}(H_+) = \mathcal{A}(H_-)$, so by completeness of the POVM, $H_+ = H_-$ and thus $H = 0$. Since the kernel is trivial the range of the adjoint map $\mathcal{A}^*$ is the orthogonal complement, consisting of all Hermitians. However, the adjoint map is simply $\mathcal{A}^*(x) = \sum_{j=1}^n x_j A_j$ for $x \in \mathbb{R}^n$, so the POVM spans the space of Hermitians.

Conversely, if the POVM is spanning, and W_1, W_2 are states with $W_1 \neq W_2$, then $H = W_1 - W_2 \neq 0$ and $\mathcal{A}(H) \neq 0$, so $\mathcal{A}(W_1) \neq \mathcal{A}(W_2)$. We conclude the POVM is informationally complete. □

The dimension of the real vector space of Hermitians on a d-dimensional complex Hilbert space is d^2. Because of the spanning property, this dimension is the minimum size required for informational completeness.

COROLLARY 2.2. *If a POVM $\{A_j\}_{j=1}^n$ on a d-dimensional Hilbert space is informationally complete, then $n \geq d^2$.*

In finite dimensions, the spanning property is equivalent to $\{A_j\}_{j=1}^n$ being a frame for the inner product space of Hermitian matrices, equipped with the Hilbert-Schmidt inner product. The reconstruction of the state then proceeds by constructing the pseudo-inverse H of the Gram matrix $G = (G_{j,l})_{j,l=1}^n$ with entries $G_{j,l} = \mathrm{tr}[A_j A_l]$. Let $\{B_j\}_{j=1}^n$ be the operator-valued canonical dual of the POVM, so if $H = (H_{j,l})_{j,l=1}^n$ has entries $H_{j,l} = \mathrm{tr}[B_j B_l]$, then $GHG = G$ and $HGH = H$.

This implies that for each operator on $\mathcal{H}$, in particular each positive, trace-normalized W,

$$W = \sum_{j=1}^{n} B_j \operatorname{tr}[WA_j] .$$

Thus, W is expressed linearly in terms of the probabilities of observing outcomes. Next, we consider the error induced by measuring relative frequencies $\hat{p}_j$ for outcomes instead of probabilities $p_j = \operatorname{tr}[WA_j]$. We quantify the error by the averaged squared Euclidean distance between W and the recovered state. In order to eliminate any bias, we randomize the measurement by conjugating W with a unitary, selected uniformly at random.

DEFINITION 2.2. *Let W be a state. The mean-squared error associated with a POVM $\{A_j\}_{j=1}^n$ and the randomized input state $W^U \equiv UWU^*$ with the unitary U chosen uniformly at random is*

$$\mathbb{E}[\|W^U - \widehat{W}^U\|^2] = \mathbb{E}[\sum_{j,l} (p_j^U - \hat{p}_j^U)(p_l^U - \hat{p}_l^U) \operatorname{tr}[B_j B_l]] .$$

The expression for the mean-squared error can be simplified substantially. We follow Scott's analysis [**31**] and assume, for simplicity, that only one outcome has been observed. Consequently, $p_j^U = \operatorname{tr}[UWU^*A_j]$ and $\{\hat{p}_j^U\}_{j=1}^n$ is a collection of $\{0,1\}$-valued random variables with $\mathbb{E}[\hat{p}_j^U] = p_j^U$ and $\sum_{j=1}^n \hat{p}_j^U = 1$.

PROPOSITION 2.3 (Scott [**31**]). *With W, $\{A_j\}_{j=1}^n$ and U as in the above definition, and a $\{0,1\}$-valued random probability vector $\hat{p}$ as described,*

$$\mathbb{E}[\|W^U - \widehat{W}^U\|^2] = \sum_j \operatorname{tr}[A_j/d] \operatorname{tr}[B_j^2] - \operatorname{tr}[W^2] .$$

PROOF. From the properties of $\hat{p}$, $\mathbb{E}[\hat{p}_j^U \hat{p}_l^U] = p_j^U \delta_{j,l}$. Consequently,

$$\mathbb{E}[\|W - \widehat{W}\|^2] = \mathbb{E}[\sum_{j,l} (p_j^U \delta_{j,l} - p_j^U p_l^U) \operatorname{tr}[B_j B_l] = \sum_j \mathbb{E}[p_j^U] \operatorname{tr}[B_j^2] - \operatorname{tr}[W^2] .$$

Finally, averaging with respect to U, we get $\mathbb{E}[p_j^U] = \mathbb{E}[\operatorname{tr}[UWU^*A_j]] = \operatorname{tr}[A_j]/d$. □

Hence, the optimal measurement for the mean-squared error minimizes the first term, while the second term remains state dependent. Scott formulated a lower bound for the first term [**31**] and characterizes cases of equality as $\{A_j\}$ being related to weighted complex projective 2-designs.

DEFINITION 2.4. *A sequence of rank-one projection operators $\{\pi_j\}_{j=1}^n$ with weights $\{w_j\}_{j=1}^n$ is a weighted projective 2-design if*

$$\sum_{j=1}^{n} w_j \pi_j \otimes \pi_j = \frac{2}{d(d+1)} \Pi_{\text{sym}}$$

where Π_{sym} is the projection onto the symmetric subspace of $\mathcal{H} \otimes \mathcal{H}$, meaning

$$\Pi_{\text{sym}} = \frac{1}{2} \sum_{j,k=1}^{d} (E_{j,j} \otimes E_{k,k} + E_{j,k} \otimes E_{k,j})$$

in terms of matrix units $E_{j,k} = e_j \otimes e_k^*$.

THEOREM 2.5 (Scott [**31**]). *Given a POVM* $\{A_j\}_{j=1}^n$ *on a* d*-dimensional Hilbert space* $\mathcal{H}$ *and the operator-valued canonical dual* $\{B_j\}_{j=1}^n$*, then*

$$\sum_j \operatorname{tr}[A_j/d]\operatorname{tr}[B_j^2] \geq \frac{1}{d} + (d^2-1)(d+1)/d\,.$$

Moreover, equality holds in this inequality if and only if $\{A_j\}_{j=1}^n$ *is a rank-one POVM and* $w_j = \operatorname{tr}[A_j]/d$, $\pi_j = A_j/\operatorname{tr}[A_j]$ *forms a weighted projective 2-design.*

PROOF. In order to find a lower bound for the first term, Scott notices that it is the trace of an operator-valued frame operator belonging to the frame $\mathcal{Q}$ with vectors $Q_j = \sqrt{\operatorname{tr}[A_j/d]}B_j$. The eigenvalues of the frame operator of $\mathcal{Q}$ are controlled by its operator-valued canonical dual $\mathcal{P}$ with vectors $P_j = \sqrt{d/\operatorname{tr}[A_j]}A_j$. To begin with, we note

$$\sum_{j=1}^n P_j \operatorname{tr}[IP_j] = d\sum_{j=1}^n A_j = dI$$

so one of the eigenvalues of $\mathcal{P}$ equals d, hence $\mathcal{Q}$ has an eigenvalue $1/d$.

We also observe $\operatorname{tr}[P_j^2] = d\operatorname{tr}[A_j^2]/\operatorname{tr}[A_j] \leq d\operatorname{tr}[A_j]$ because $0 \leq A_j \leq I$, so $\operatorname{tr}[A_j^2] \leq (\operatorname{tr}[A_j])^2$. Hence, $\sum_{j=1}^n \operatorname{tr}[P_j^2] \leq d\sum_{j=1}^n \operatorname{tr}[A_j] = d^2$.

Since the eigenvalues of $\mathcal{Q}$ are inverses of those of $\mathcal{P}$, we see that we want to optimize $\sigma(\lambda) \equiv \sum_{j=1}^{d^2} \lambda_j^{-1}$ subject to $\sum_{j=1}^{d^2} \lambda_j \leq d^2$ and $\lambda_1 = d$. By the convexity of the function $x \mapsto 1/x$, the minimum of σ cannot be smaller than for the choice $\lambda_1 = d$ and $\lambda_j = (d^2-d)/(d^2-1) = d/(d+1)$, so

$$\sigma(\lambda) = \frac{1}{d} + (d^2-1)(d+1)/d$$

Equality is achieved if and only if each A_j is of rank one and $\mathcal{P}$ has two eigenvalues (ignoring multiplicity), $\lambda_1 = d$ and $\lambda_j = d/(d+1)$. Consequently, each $\pi_j = A_j/\operatorname{tr}[A_j]$ is a rank-one projection and for each Hermitian H with vanishing trace,

$$\sum_{j=1}^n A_j \operatorname{tr}[HA_j]/\operatorname{tr}[A_j] = \frac{1}{d+1}H\,.$$

Finally, we use the characterization of weighted projective 2-designs: Assuming each A_j is rank one, let $w_j = \operatorname{tr}[A_j]/d$ and $\pi_j = A_j/\operatorname{tr}[A_j]$, then

$$\sum_{j,l=1}^n w_j w_l |\operatorname{tr}[\pi_j\pi_l]|^2 = \frac{1}{d^4}\sum_{j,l=1}^n |\operatorname{tr}[P_jP_l]|^2$$

The sum is identified as the squared Hilbert-Schmidt norm of the frame operator of $\mathcal{P}$, so

$$\sum_{j,l=1}^n w_j w_l |\operatorname{tr}[\pi_j\pi_l]|^2 = \frac{1}{d^4}(d^2 + (d^2-1)d^2/(d+1)^2) = \frac{2}{d(d+1)}\,.$$

This last identity is equivalent to $\sum_{j=1}^n w_j\pi_j \otimes \pi_j = \frac{2}{d(d+1)}\Pi_{\mathrm{sym}}$ [**31**][Theorem 6], hence equality holds in the bound if and only if the weights and rank-one orthogonal projections form a weighted complex projective 2-design. $\square$

We conclude that $\{A_j\}_{j=1}^n$ meets the bound for the first term in the mean-squared error if and only if the weighted projections associated with the POVM form a weighted complex projective 2-design.

COROLLARY 2.3. *Let $\{A_j\}_{j=1}^n$ be a POVM, then the mean-squared error for linear reconstruction from observing one outcome is bounded below by*

$$\mathbb{E}[\|W^U - \widehat{W}^U\|^2] \geq \frac{1}{d} + (d^2-1)(d+1)/d - \mathrm{tr}[W^2]$$

and equality holds if and only if each A_j in the POVM is rank one and $w_j = \mathrm{tr}[A_j/d]$, $\pi_j = A_j/|TR[A_j]$ form a weighted complex projective 2-design.

To summarize, if weighted complex projective 2-designs exist, then they are optimal for quantum state tomography with linear reconstruction using the operator-valued canonical dual of the associated POVM.

3. Known quantum 2-designs

We conclude by collecting known examples of quantum designs, highlighting connections between the design properties and geometric properties. We briefly sketch history and construction principles, occasionally demonstrating concrete examples and highlighting open problems along the way.

The presentation is organized as follows. In Sections 3.1 and 3.2, we examine quantum 2-designs comprised of rank one projectors, encoding the elements of each design as the columns of unit norm tight frames,

$$F = [f_1\, f_2\, \cdots\, f_n],$$

from which the POVM's projection arises via

$$P_j := f_j \otimes f_j^*, j \in \{1, ., 2, \dots, n\}.$$

In the first case, Section 3.1, we examine equally weighted complex projective 2-designs, a special class of POVMs whose quantum 2-design property is characterized by minimal *coherence*,

$$\mu(F) = \max_{j \neq l} |\langle f_j, f_l \rangle|,$$

in particular achieving the lower bound given by Levenstein [**19, 22, 26**] (see Theorem 3.1 below). In Section 3.2, we examine quantum designs manifesting as weighted complex 2-designs, for which minimal coherence - as verified in some cases [**4**] - is a byproduct, but not a necessary condition for the the quantum 2-design property. In these cases, the 2-design property is characterized according the weighted framework of Roy and Scott [**30**] (see Theorem 3.2). Finally, in Section 3.3, we point out families of quantum 2-designs manifesting whose values are projections of rank greater than one.

3.1. Rank one complex projective 2-designs and Levenstein's bound.

The quantum 2-design property of the three families in this section are characterized by Levenstein's 2nd degree coherence bound [**19, 22, 26**].

THEOREM 3.1. *[Levenstein; [**26**], see also [**19, 22**]] Given a unit norm frame $\Phi = \{\phi_j\}_{j=1}^n \subset \mathbb{C}^d$, then the corresponding rank one operators Φ form an equally* *weighted complex projective* 2*-design (which is at most a* 3*-design) if and only if the following lower bound on frame coherence is met with equality:*

$$\mu(\Phi) \geq \sqrt{\frac{2n - d(d+1)}{(n-d)(d+1)}}.$$

3.1.1. *SIC-POVMs and Zauner's conjecture.* Among all quantum designs, *Symmetric Informational Complex Positive Operator-valued Measures (SIC-POVMs)* seem to be the ones of prime interest to academians for several reasons. Foremost, a given SIC-POVM, $\{P_j\}^{d^2}$ for $\mathbb{C}^d$, is minimal in the sense that d^2 is just enough projections to span the space of matrices, which corresponds to the cheapest possible model for optimal measurement apparatuses. Moreover, a SIC-POVM Φ admits an appealing geometry; in particular, they correspond to maximal equiangular tight frames - meaning they are the frames of largest admissible cardinality, ie $n = d^2$, such that the absolute inner products between distinct elements is constantly equal to Welch's lower coherence bound [**34**]:

$$\mu(\Phi) \geq \sqrt{\frac{n-d}{d(n-1)}}.$$

In fact, maximal ETFs are precisely where Levenstein's bound and Welch's bound agree[**19**]. Setting $n = d^2$, both bounds evaluate as follows:

$$\sqrt{\frac{2d^2 - d(d+1)}{(d^2-d)(d+1)}} = \sqrt{\frac{d^2-d}{d(d^2-1)}} = \frac{1}{d+1}$$

SIC-POVMs have enjoyed a rich history dating back to the dissertation of Zauner [**36**] (see also [**37**] for an English translation), wherein he conjectured the existence of a SIC-POVM in $\mathbb{C}^d$ for every $d \in \mathbb{N}$, a conjecture which has captivated researchers throughout the last two decades. To date, analytic/symbolic verifications of the predicted fiducial vectors - each satisfying the structure predicted by Zauner - have been determined for $d = 1, \ldots, 21, 24, 28, 30, 31, 35,$
$37, 39, 43$ and 48 [**2, 7, 16, 25, 29, 33, 36**]; moreover, due to recent advances, numerical approximations of fiducial vectors seemingly in agreement with Zauner's conjecture have been verified for $d \leq 150$ [**17, 18, 29, 32**]. An existential or constructive proof that Zauner's conjecture holds in infinitely many dimensions remains an outstanding open problem.

EXAMPLE 3.1 (A SIC-POVM for C^2). *With ω a primitive third root of unity, the columns of the* 2×4 *unit norm frame,*

$$F = \begin{bmatrix} 1 & \frac{1}{3} & \frac{1}{3} & \frac{1}{3} \\ 0 & \frac{4}{9} & \frac{4}{9}\omega & \frac{4}{9}\omega^2 \end{bmatrix},$$

correspond to a maximal equiangular tight frame, whence generating a SIC-POVM for $\mathbb{C}^2$*,*

$$\mathcal{P} = \left\{f_j \otimes f_j^*\right\}_{j=1}^{4}.$$

This SIC-POVM manifests as a tetrahedron on the Bloch sphere.

3.1.2. *Maximal sets of mutually unbiased bases.* A pair of *mutually unbiased bases (MUBs)*, B and B', is a pair of orthonormal bases for $\mathcal{H}$ with the property that $|\langle x, y\rangle|^2 = \frac{1}{d}$ for every $x \in B$ and $y \in B'$. It is known **[12]** that if $\mathcal{B}$ is a family of pairwise mutually unbiased bases for $\mathbb{C}^d$, then $|\mathcal{B}| \leq d+1$. In the case that a maximal family of mutually unbiased bases $B_1, B_2, \ldots, B_{d+1}$ exists, then their union, written as a block matrix,

$$F = [B_1\, B_2\, \ldots\, B_{d+1}],$$

consists of $n = d(d+1)$ unit vectors, which together achieve the orthoplex bound **[4, 9, 27]**,

$$\mu(\Phi) \geq \frac{1}{\sqrt{d}},$$

a lower bound on frame coherence which holds whenever $n > d^2$. Indeed, it is known **[19]** if a unit norm frame, Φ, achieves the orthoplex bound, then $|\Phi| \leq d(d+1)$. Consequently, maximal families of MUBs are also maximal unit norm frames that achieve the orthoplex bound, and the converse, that maximal unit norm frames that achieves the orthoplex bound necessarily form families of MUBs, is also true **[24]**. Interestingly, as with the Welch bound, it is only when a maximal family of MUBs exist that Levenstein's bound agrees with the orthoplex bound. Setting $n = d(d+1)$, we see

$$\sqrt{\frac{2d(d+1) - d(d+1)}{(d(d+1) - d)(d+1)}} = \frac{1}{\sqrt{d}}.$$

Unlike SIC-POVMs, maximal sets of MUBs are a class 2-designs currently enjoying the certainty of existence in an infinitude of dimensions. Thanks to the work of citeCameronSeidel1973, we know that than maximal sets of MUBs exist whenever d is a prime number, and the authors of **[35]** later extended the result to all prime powers. Maximal MUBs of prime power order enjoy several straightforward combinatorial constructions, for example, using mutually orthogonal squares**[28]** or relative differences **[6]**. Nevertheless, upon consideration of the first composite dimension, $d = 6$, one encounters a well-known open problem:are there any composite dimensions, d, for which $d+1$ MUBs exist? Numerical evidence suggests that no more than three pairwise MUBs coexist in $\mathbb{C}^6$ **[23]**, but the answer remains unknown.

EXAMPLE 3.2 (Three MUBs in $\mathbb{C}^2$). *The columns of matrices,*

$$B_1 = \begin{bmatrix} 1 & 0 \\ 0 & 1 \end{bmatrix}, \quad B_2 = \frac{1}{\sqrt{2}} \begin{bmatrix} 1 & 1 \\ 1 & -1 \end{bmatrix} \text{ and } B_3 = \frac{1}{\sqrt{2}} \begin{bmatrix} i & i \\ i & -i \end{bmatrix},$$

correspond to a set three MUBs in $\mathbb{C}^2$, whence a quantum 2-design of six rank one projections. The six points of this example manifests as the vertices of an octahedron on the Bloch sphere.

3.1.3. *Sporadic Levenstein 2-designs.* Besides SIC-POVMs and maximal sets of MUBs, the authors of **[19]** distilled, via integrality analysis, that there are sparsely many feasible pairs, (d, n), where non-equiangular, non-orthoplectic Levenstein-equality frames exist. Indeed, to date, there are exactly four known nonequiangular, nonorthoplec unit norm frames which achieve Levenstein's 2nd degree bound, thereby forming quantum 2-designs in locations where the optimal coherence is characterized neither by the Welch bound nor orthoplex bounds. We list these four

d	n	$1/\alpha$	notes	location
4	40	3	Witting polytope [**11**]	Hoggar [**20**]
5	45	4	Shephard–Todd no. 33	Hoggar [**20**]
6	126	4	Shephard–Todd no. 34	Hoggar [**20**]
7	112	5	?	?
8	120	6	?	?
8	288	5	?	?
9	225	6	?	?
10	220	7	?	?
10	550	6	?	?
11	176	9	?	?
11	231	8	?	?
⋮	⋮	⋮	⋮	⋮
28	1624	19	?	?
28	4060	1755	Rudvalis group [**10**]	Hoggar [**20**]

TABLE 1. Admissible Levenstein equality pairs in low dimensions

examples in Table 1, along with several other feasible pairs – up to $d = 28$ – where Levenstein-equality frames might exist based on sufficiency of certain integrality relations. In the table, α denotes the squared Levenstein bound for the given pair, (d, n), and question marks indicate that no construction of feasible the 2-design corresponding to the given row is currently known.

3.1.4. *Rank one Levenstein-type quantum designs with higher design parameters.* Before moving to the next section, where we consider weighted rank one POVMs, we note that Levenstein provided an infinite family of lower coherence bounds by applying linear programming techniques to so called special orthogonal polynomials of increasing degrees. Up until now, we have only considered his second degree bound (see Theorem 3.1), as, unfortunately, it seems that examples that achieve these higher degree bounds are extremely rare, see for example [**21**]. Nevertheless, any example that achieves one of these higher degree bounds will form a complex projective t-design with a correspondingly higher parameter, t, thereby forming a quantum 2-design. Presently, we are aware of only one example, a family of six orthonormal bases for $\mathbb{C}^2$, whose union forms a projective 5-design, thus yielding a quantum 2-design of 12 projectorss [**5**]. This example manifests as the vertices of a regular icosahedron when embedded into the Bloch sphere.

3.2. Weighted rank one complex projective 2-designs. Next, we consider those quantum 2-designs which manifest, more generally, as weighted complex projective 2-designs. SIC-POVMs and maximal sets of MUBs, for example, are equally weighted complex projective 2-designs, with each $w_j = \frac{1}{n}$. The following useful theorem characterizes weighted designs in terms of their underlying geometry.

THEOREM 3.2. *[Roy/Scott; [**30**]] A $d \times n$ unit norm tight frame $F = [f_1\, f_2\, \dots\, f_n]$, along with normalized weights, $\{w_j\}_{j=1}^n$, $\sum_{j=1}^n w_j = 1$, forms a complex weighted*

2-design with rank-one projections $\pi_j = f_j \otimes f_j^*$, if and only if

$$\sum_{j=1}^{n}\sum_{k=1}^{n} w_j w_k \left|\langle f_j, f_k\rangle\right|^{2t} = \binom{d+t-1}{t}^{-1}.$$

3.2.1. *Weighted orthonormal bases.* In demonstration of Theorem 3.2, given that $d = p^s$ or $d+1 = p^s$ for some prime p and $s \in \mathbb{N}$ the authors [**30**] use nonlinear functions on nonabelian groups to construct m families of weighted orthonormal bases, where $m > d + p + 1$. Moreover, they show that whenever $d + 1 = p$ is a prime power, then a weighted complex projective 2-design comprised of $d + 2$ MUBs exist [**30**]. For example, while there is no known construction of seven equally weighted MUBs in $\mathbb{C}^6$, their weighted construction [**30**] forms quantum 2-design of 8 weighted MUBS – or 48 rank one orthogonal projections – since $d = 6$ is indeed a prime power plus one.

3.2.2. *Weighted unit norm tight frames.* Using the idea of weighting from the previous approach, the authors of [**4**] relax the requirement that a weighted design manifest as a set of bases, instead considering sets of unit norm tight frames, noting that such a union corresponds to automatically forms a projective 1-design. By exploiting DFT matrices and combinatorial objects called difference sets, the authors produce families of equiangular tight frames and biangular tight frames, which when appropriately weighted and adjoined with canonical orthonormal bases, form quantum 2-designs. More precisely, they produced two infinite families. The first family consists of $n = d^2 + 1$ rank one projections for $\mathbb{C}^d$ whenever d is a prime power plus one, and the second yields quantum 2-designs of $n = d^2 + d - 1$ rank one projectors whenever d is a prime power.

EXAMPLE 3.3. *We may think of the standard orthonormal basis and any properly scaled selection of two rows from a 3×3 DFT matrix as unit norm frames for $\mathbb{C}^2$. The union of such frames,*

$$F_1 = \begin{bmatrix} 1 & 0 \\ 0 & 1 \end{bmatrix} \text{ and } F_2 = \begin{bmatrix} 1 & 1 & 1 \\ 1 & \omega & \omega^2 \end{bmatrix},$$

along with weights corresponding to the matrix columns, $w_1 = w_2 = 1/6$ and $w_3 = w_4 = w_5 = 1/12$, form a weighted complex projective 2-design.

A striking by-product of these examples is that they happen to form minimally coherent frames, in particular achieving the orthoplex bound. In general, there is no reason to expect a weighted 2-design to have this property.

3.3. Quantum 2-designs with high rank projections.

3.3.1. *Zauner's conjecture revisited.* In [**1**], Appleby shows the existence of quantum 2-designs of arbitrary but equally ranked values, which behave like SIC-POVMs. By taking a random simplex and shrinking it appropriately to "fit" within the so-called Bloch space [**1**], he shows that for every d, there is some sub-dimension l with $1 \le l \le d-1$, and a quantum 2-design consisting of $n = d^2$ rank l orthogonal projections. By "behave like SIC-POVMs", we mean these objects provide the desired quantum 2-design properties and they are symmetric in the sense that their set of pairwise Hilbert Schmidt inner products is a singleton, ie they are equiangular.

3.3.2. *Maximal sets of mutually unbiased subspaces.* In his thesis, Zauner mentions this example of quantum 2-designs [**36**]. Given an ambient Hilbert space with dimension of even prime power, $d = 2^t$, $t > 1$, one may construct a family of $n = 2^{t+1} - 2$ orthogonal projections with rank $l = 2^{t-1}$ (ie, 'half-dimensional' subspaces) which constitute quantum 2-designs. In [**3**], the authors provide a concrete construction of this example by exploiting maximal families of MUBs and maximal sets of Johnson codes with optimal distance relations.

References

[1] D.M. Appleby. Symmetric informationally complete measurements of arbitrary rank. *Opt. Spectroscop.*, 103(3):416–428, 2007.

[2] Marcus Appleby, Tuan-Yow Chien, Steven Flammia, and Shayne Waldron, *Constructing exact symmetric informationally complete measurements from numerical solutions*, J. Phys. A **51** (2018), no. 16, 165302, 40, DOI 10.1088/1751-8121/aab4cd. MR3786021

[3] Bernhard G. Bodmann and John I. Haas, *Maximal orthoplectic fusion frames from mutually unbiased bases and block designs*, Proc. Amer. Math. Soc. **146** (2018), no. 6, 2601–2616, DOI 10.1090/proc/13956. MR3778161

[4] Bernhard G. Bodmann and John Haas, *Achieving the orthoplex bound and constructing weighted complex projective 2-designs with Singer sets*, Linear Algebra Appl. **511** (2016), 54–71, DOI 10.1016/j.laa.2016.09.005. MR3557826

[5] P. G. Casazza and J. I. Haas. On the rigidity of geometric and spectral properties of Grassmannian frames. *ArXiv e-prints*, May 2016.

[6] Peter G. Casazza, Amineh Farzannia, John I. Haas, and Tin T. Tran, *Toward the classification of biangular harmonic frames*, Appl. Comput. Harmon. Anal. **46** (2019), no. 3, 544–568, DOI 10.1016/j.acha.2017.06.004. MR3926958

[7] Tuan-Yow Chien. *Equiangular lines, projective symmetries and nice error frames.* PhD thesis, University of Auckland, 2015.

[8] Eric Chitambar, Debbie Leung, Laura Mančinska, Maris Ozols, and Andreas Winter, *Everything you always wanted to know about LOCC (but were afraid to ask)*, Comm. Math. Phys. **328** (2014), no. 1, 303–326, DOI 10.1007/s00220-014-1953-9. MR3196987

[9] John H. Conway, Ronald H. Hardin, and Neil J. A. Sloane, *Packing lines, planes, etc.: packings in Grassmannian spaces*, Experiment. Math. **5** (1996), no. 2, 139–159. MR1418961

[10] J. H. Conway and D. B. Wales, *Construction of the Rudvalis group of order* 145, 926, 144, 000, J. Algebra **27** (1973), 538–548, DOI 10.1016/0021-8693(73)90063-X. MR335620

[11] H. S. M. Coxeter, *Regular complex polytopes*, 2nd ed., Cambridge University Press, Cambridge, 1991. MR1119304

[12] P. Delsarte, J. M. Goethals, and J. J. Seidel. Bounds for systems of lines, and Jacobi polynomials. *Philips Research Reports*, 30:91, 1975.

[13] A. Einstein. Über einen die erzeugung und verwandlung des lichtes betreffenden heuristischen gesichtspunkt. *Annalen der Physik*, 322:132–148, 1905.

[14] I. Levine et al. Measurement of the electromagnetic coupling at large momentum transfer. *Phys. Rev. Lett.*, 78:424–427, 1997.

[15] Richard P. Feynman, *Simulating physics with computers*, Internat. J. Theoret. Phys. **21** (1981/82), no. 6-7, 467–488, DOI 10.1007/BF02650179. Physics of computation, Part II (Dedham, Mass., 1981). MR658311

[16] Steve Flammia. Exact SIC fiducial vectors. Available online: http://www.physics.usyd.edu.au/ sflammia/SIC/.

[17] C. A. Fuchs, M. C. Hoang, and B. C. Stacey. The SIC Question: History and State of Play. *ArXiv e-prints*, March 2017.

[18] C. A. Fuchs and B. C. Stacey. QBism: Quantum Theory as a Hero's Handbook. *ArXiv e-prints*, December 2016.

[19] John Haas, Nathaniel Hammen, and Dustin Mixon. The Levenstein bound for packings in projective spaces. *SPIE Conference Proceedings*, 2017.

[20] S. G. Hoggar, *t-designs in projective spaces*, European J. Combin. **3** (1982), no. 3, 233–254, DOI 10.1016/S0195-6698(82)80035-8. MR679208

[21] S. G. Hoggar, *Tight 4- and 5-designs in projective spaces*, Graphs Combin. **5** (1989), no. 1, 87–94, DOI 10.1007/BF01788661. MR981234

[22] John Isaac Haas IV, Nathaniel Hammen, and Dustin Mixon. Optimal line packings from levenstein's bound. *In Preparation*, 2017.

[23] Philippe Jaming, Máté Matolcsi, and Péter Móra, *The problem of mutually unbiased bases in dimension 6*, Cryptogr. Commun. **2** (2010), no. 2, 211–220, DOI 10.1007/s12095-010-0023-1. MR2719840

[24] A. Klappenecker and M. Rötteler. Mutually unbiased bases are complex projective 2-designs. In *International Symposium on Information Theory, 2005. ISIT 2005.*, pages 1740–1744, Sept 2005.

[25] Alexander Koldobsky and Hermann König, *Aspects of the isometric theory of Banach spaces*, Handbook of the geometry of Banach spaces, Vol. I, North-Holland, Amsterdam, 2001, pp. 899–939, DOI 10.1016/S1874-5849(01)80023-1. MR1863709

[26] V. I. Levenshteĭn, *Designs as maximum codes in polynomial metric spaces*, Acta Appl. Math. **29** (1992), no. 1-2, 1–82, DOI 10.1007/BF00053379. Interactions between algebra and combinatorics. MR1192833

[27] R. A. Rankin, *The closest packing of spherical caps in n dimensions*, Proc. Glasgow Math. Assoc. **2** (1955), 139–144. MR74013

[28] Asha Rao, Diane Donovan, and Joanne L. Hall, *Mutually orthogonal Latin squares and mutually unbiased bases in dimensions of odd prime power MOLS and MUBs in odd prime power dimensions*, Cryptogr. Commun. **2** (2010), no. 2, 221–231, DOI 10.1007/s12095-010-0027-x. MR2719841

[29] Joseph M. Renes, Robin Blume-Kohout, A. J. Scott, and Carlton M. Caves, *Symmetric informationally complete quantum measurements*, J. Math. Phys. **45** (2004), no. 6, 2171–2180, DOI 10.1063/1.1737053. MR2059685

[30] Aidan Roy and A. J. Scott, *Weighted complex projective 2-designs from bases: optimal state determination by orthogonal measurements*, J. Math. Phys. **48** (2007), no. 7, 072110, 24, DOI 10.1063/1.2748617. MR2337670

[31] A. J. Scott, *Tight informationally complete quantum measurements*, J. Phys. A **39** (2006), no. 43, 13507–13530, DOI 10.1088/0305-4470/39/43/009. MR2269701

[32] A. J. Scott. SICs: Extending the list of solutions. *ArXiv e-prints*, March 2017.

[33] A. J. Scott and M. Grassl, *Symmetric informationally complete positive-operator-valued measures: a new computer study*, J. Math. Phys. **51** (2010), no. 4, 042203, 16, DOI 10.1063/1.3374022. MR2662471

[34] L. R. Welch. Lower bounds on the maximum cross correlation of signals. *IEEE Trans. on Information Theory*, 20(3):397–9, May 1974.

[35] William K. Wootters and Brian D. Fields, *Optimal state-determination by mutually unbiased measurements*, Ann. Physics **191** (1989), no. 2, 363–381, DOI 10.1016/0003-4916(89)90322-9. MR1003014

[36] Gerhard Zauner, *Quantum designs: foundations of a noncommutative design theory*, Int. J. Quantum Inf. **9** (2011), no. 1, 445–507, DOI 10.1142/S0219749911006776. MR2931102

[37] Gerhard Zauner, *Quantum designs: foundations of a noncommutative design theory*, Int. J. Quantum Inf. **9** (2011), no. 1, 445–507, DOI 10.1142/S0219749911006776. MR2931102

Selected Published Titles in This Series

747 **Paul Bruillard, Carlos Ortiz Marrero, and Julia Plavnik, Editors,** Topological Phases of Matter and Quantum Computation, 2020

744 **Pieter Moree, Anke Pohl, L'ubomír Snoha, and Tom Ward, Editors,** Dynamics: Topology and Numbers, 2020

743 **H. Garth Dales, Dmitry Khavinson, and Javad Mashreghi, Editors,** Complex Analysis and Spectral Theory, 2020

742 **Francisco-Jesús Castro-Jiménez, David Bradley Massey, Bernard Teissier, and Meral Tosun, Editors,** A Panorama of Singularities, 2020

741 **Houssam Abdul-Rahman, Robert Sims, and Amanda Young, Editors,** Analytic Trends in Mathematical Physics, 2020

740 **Alina Bucur and David Zureick-Brown, Editors,** Analytic Methods in Arithmetic Geometry, 2019

739 **Yaiza Canzani, Linan Chen, and Dmitry Jakobson, Editors,** Probabilistic Methods in Geometry, Topology and Spectral Theory, 2019

738 **Shrikrishna G. Dani, Surender K. Jain, Jugal K. Verma, and Meenakshi P. Wasadikar, Editors,** Contributions in Algebra and Algebraic Geometry, 2019

737 **Fernanda Botelho, Editor,** Recent Trends in Operator Theory and Applications, 2019

736 **Jane Hawkins, Rachel L. Rossetti, and Jim Wiseman, Editors,** Dynamical Systems and Random Processes, 2019

735 **Yanir A. Rubinstein and Bernard Shiffman, Editors,** Advances in Complex Geometry, 2019

734 **Peter Kuchment and Evgeny Semenov, Editors,** Differential Equations, Mathematical Physics, and Applications, 2019

733 **Peter Kuchment and Evgeny Semenov, Editors,** Functional Analysis and Geometry, 2019

732 **Samuele Anni, Jay Jorgenson, Lejla Smajlović, and Lynne Walling, Editors,** Automorphic Forms and Related Topics, 2019

731 **Robert G. Niemeyer, Erin P. J. Pearse, John A. Rock, and Tony Samuel, Editors,** Horizons of Fractal Geometry and Complex Dimensions, 2019

730 **Alberto Facchini, Lorna Gregory, Sonia L'Innocente, and Marcus Tressl, Editors,** Model Theory of Modules, Algebras and Categories, 2019

729 **Daniel G. Davis, Hans-Werner Henn, J. F. Jardine, Mark W. Johnson, and Charles Rezk, Editors,** Homotopy Theory: Tools and Applications, 2019

728 **Nicolás Andruskiewitsch and Dmitri Nikshych, Editors,** Tensor Categories and Hopf Algebras, 2019

727 **André Leroy, Christian Lomp, Sergio López-Permouth, and Frédérique Oggier, Editors,** Rings, Modules and Codes, 2019

726 **Eugene Plotkin, Editor,** Groups, Algebras and Identities, 2019

725 **Shijun Zheng, Marius Beceanu, Jerry Bona, Geng Chen, Tuoc Van Phan, and Avy Soffer, Editors,** Nonlinear Dispersive Waves and Fluids, 2019

724 **Lubjana Beshaj and Tony Shaska, Editors,** Algebraic Curves and Their Applications, 2019

723 **Donatella Danielli, Arshak Petrosyan, and Camelia A. Pop, Editors,** New Developments in the Analysis of Nonlocal Operators, 2019

722 **Yves Aubry, Everett W. Howe, and Christophe Ritzenthaler, Editors,** Arithmetic Geometry: Computation and Applications, 2019

For a complete list of titles in this series, visit the AMS Bookstore at **www.ams.org/bookstore/conmseries/**.